D0570745

The Collins Guide to the

RARE MAMMALS OF THE WORLD

The Collins Guide to the RARE MAMMALS OF THE WORLD

John A. Burton
with the assistance of Vivien G. Burton

Illustrated by Bruce Pearson

Foreword by Sir Peter Scott

THE STEPHEN GREENE PRESS
Lexington, Massachusetts

Acknowledgements

This guide relies extensively on the research of others. It would be impossible to list all the books and scientific papers consulted, but those most regularly used, and those most likely to be of value to the user of this guide are listed in the bibliography. We have used the libraries of the Fauna & Flora Preservation Society, the Linnaean Society of London, the Zoological Society of London and the collections and library of the Mammal Section of the British Museum (Natural History). The staff of all these institutions have given their advice freely, but particular mention should be made to Gina Douglas (Linnaean Society), John E. Hills and Daphne Hill (British Museum (Natural History)); and Gillian Wills of the FFPS helped produce a listing of threatened mammals. We would like to thank Tim Inskipp for reading the final draft of the manuscript; his encyclopaedic knowledge found many of the more obvious errors and inconsistencies. However, all errors either of omission or commission are entirely the responsibility of the authors. Jillian Luff interpreted an amazing variety of extremely complex sources in order to produce the distribution maps. Finally we would like to thank Crispin Fisher, Francesca Dow and Amanda Kent of Collins for their enthusiastic support.

First published in Great Britain by William Collins Sons & Co. Ltd
First published in the United States of America in 1988 by
The Stephen Greene Press Inc.
Distributed by Viking Penguin Inc., 40 West 23rd Street,
New York, NY 10010

CIP data available

ISBN 0-8289-0658-0

Printed in Great Britain
by William Collins Sons & Co. Ltd. Glasgow
Set in 8/9 pt Times Roman

CONTENTS

FOREWORD

by Sir Peter Scott

Over the past thirty years or so interest in wildlife and its conservation has been increasing steadily, and the publication of field guides has played an important part in this growing concern. Being able to identify what you see and to find out some background facts and figures adds enormously to people's enjoyment of animals and the natural world.

This book uses the same format as the now familiar field guides, but it draws attention to the desperate crisis situation that we have reached in the second half of the twentieth century, when one species – *Homo sapiens* – is responsible for the current catastrophic increase in the rate of species extinction. It has been estimated that we could be losing one species per hour by 1990, and that between 10% and 20% of the earth's 5–10 million species are likely to become extinct by the year 2000. Most of these species will be invertebrate animals and plants, all of them playing important parts in the balance of their ecosystems. They will also, sadly, include some of the mammals described in this book. More and more people are becoming aware of the crisis, and I believe this book will make those who read it better able to take an active part in conservation and help to reduce the tragic toll of vanishing species.

Peter Scott

INTRODUCTION

The Ecology of Extinction

Extinction of species is part of the natural process of evolution. Millions of species have already become extinct, and every species now in existence will become extinct one day – including Man. The life span of individual species varies enormously; some such as the crocodiles remain remarkably similar to ancestors which walked the earth before the dinosaurs. Mammals, on the other hand, are comparatively recent, the earliest having evolved some 200 million years ago; the fauna we are familiar with today emerged during the last two million years, in which there have been seven Ice Ages, interrupted by warmer interglacial periods. The ebbing and flowing of the icecaps, and the associated widespread climatic changes have been responsible for huge numbers of extinctions, but between one and four million years ago various types of hominoids – ancestors of man – appeared and began to accelerate the rate of extinction of other species. The various types of man, Australopithecines, Neanderthalers and latterly *Homo sapiens* were all hunters, and modern *Homo sapiens* has spread to all parts of the inhabitable world, reaching Australia over 30,000 years ago, and North America at least 25,000 years ago. Contrary to popular opinion, the 'noble savage' did not necessarily live in harmony with his environment, but rapidly managed to exterminate many of the larger mammals. Following man's arrival in Australia, over 30% of the large marsupials became extinct; his appearance in North America brought about the extinction of 70% of the megafauna there. In Africa about 40% of the larger mammals disappeared, and a similar wave of extinctions took place across Europe and Asia, and in South America. As man colonised islands such as Madagascar, the same happened.

Parallel to the mass extinctions, those species that survived often evolved into much smaller ones; a tendency for dwarfing also developed among species living on smaller oceanic islands. Dwarf hippopotami are known from Mediterranean islands and a dwarf elephant is known from Malta. The reasons for these trends are not fully understood, but the smaller forms would be more agile and adaptable, particularly in the broken terrain of many islands, and require less food.

The process of recent extinctions is well illustrated in the Americas. At the end of the Pleistocene, about 10,000 years ago, North and South America were united, after some 60 million years of isolation, and mammals could cross the new bridge. Wolves and large cats from the north invaded South America and exterminated many of the South American plant-eaters, many of which had developed into giant forms, such as *Toxodon* (a rhino-like animal) and *Megatherium* (the Giant Ground Sloth). Herbivores also spread southwards, including horses, deer and the Imperial Mammoth. Some later became extinct, such as the horses and mammoth (which survived until about AD 600); and tapirs and the Spectacled Bear, although extinct in North America, still survive in the South. Conversely, some South American mammals moved northwards, but most have not spread beyond Central America; anteaters, agoutis and spider monkeys are among them, and the armadillos are still spreading. About 11,000 years ago the most recent Ice Age drew to a close, and at around that time most populations of ground sloths, mastodons, mammoths, sabre-toothed cats and dire wolves became extinct. Deer and bison invaded from Asia, across the land bridge which connected the two continents now separated by the Bering Straits, and displaced existing herbivores, including the native North American horses; these became extinct at about this time. Within a matter of only a few thousand years 35 genera of mammals occurring in North America were reduced to thirteen. Similar changes took place in Europe; and in Africa, though on a smaller scale, there were depletions of the fauna.

In the major climatic changes which occurred several times earlier in the Pleistocene, such dramatic depletions of fauna are not apparent: one factor which many scientists believe can help explain these extinctions is man. *Homo sapiens* was slowly increasing in numbers and colonising more and more of the earth's surface.

The primitive Stone Age hunters caused a massive reduction in the mammal megafauna, helping to exterminate dozens of species up to the size of woolly mammoths and woolly rhinos. Increasingly sophisticated hunting techniques ensured the steady depletion of species, until about the 17th century, when the introduction of modern firearms and the spread of western colonialism contributed to a second major wave of extinctions. This affected many species, but in particular island forms, such as the rice rats of the Galapagos Islands (p 112). European colonists brought with them rats and mice from the Old World, which were much more adaptable and aggressive than many of the native species, and were able to displace the native rodents; in many cases they predated directly on birds, reptiles and other forms of wildlife.

The 20th century has seen a wave of extinctions quite unprecedented in the history of the world. It is commonplace to describe the extinction of the dinosaurs as sudden and catastrophic: in geological terms it was, but in fact it probably took place over several millions of years – and almost certainly longer than the entire time man has been on this planet. In this context, the series of man-made and man-induced extinctions is catastrophic. In many parts of the world, the rate of habitat destruction could well have reached a point where, even if fragments remain, irreversible damage has been done. It is doubtful if national parks, marooned as islands in seas of agricultural 'desert' and 'development', will in the long run be able to maintain their original diversity. Biological theorists have established that the number of species capable of surviving on an island is dependent on the size of the island, its distance from other islands and from mainland masses, and factors such as availability of water, altitude, etc.* This island theory applies to isolated reserves and national parks, just as much as to oceanic islands. Research has shown that in national parks – such as the Parc des Volcans in Rwanda, Africa, famous for its Mountain Gorillas – a number of species has already disappeared following the reduction in the park's size, even though suitable habitat still exists. In a recent study W. D. Newmark† analysed the non-flying mammal fauna of 29 parks, to see how size and diversity of habitat affected the total number of species. While both were found to be important, and affected by numerous factors (including altitude, vegetative diversity, and amount of cover) it was concluded that the species diversity of the vegetation alone correlated poorly with mammalian diversity.

The Causes of Extinction

The most obvious cause of extinction is direct hunting. Although in recent years relatively few species have been exterminated by hunting, many have had their range seriously reduced and their numbers diminished to such low levels that any further threats would be disastrous. The Arabian Oryx was exterminated in the wild by hunting: the killing of an oryx was considered a feat of great endurance by certain desert-dwelling tribes, and they became easy prey to hunters with modern firearms and motor vehicles; fortunately they were saved in captivity. The Addax and Scimitar Oryx are in a similar position: they are easily hunted, desert-dwelling antelope, virtually extinct in the wild, but with healthy populations in captivity; reintroduction programmes are being developed. Hunting reduced the American Bison to the point of near-extinction, partly through a demand for its hide, but also as part of campaigns to deprive the native Indian peoples of their main source of food. Finally, its prairie habitat was destroyed; and recolonisation remains impossible as long as the prairies are ploughed for wheat. The Vicuna, once sheared by the Incas for its priceless wool, was hunted to extinction over a large part of its range during the early part of this century, simply because it was easier to remove the wool from dead animals. Fortunately small herds survived, because their habitat was remote and inhospitable; it has been able to rebuild its populations under protection, and shearing has started again. Tigers, hunted for sport by the Maharajas and European colonials, were reduced to a highly fragmented range; and although they are now increasing in some parks and reserves they have nowhere else to go.

The destruction of habitat is usually cited as the most important threat to endangered wildlife. Reserves and national parks are created to counterbalance this threat; but while they may help to preserve a few species in the short term, in the long term, and on their present scale, they may not save significant numbers. This is mainly because the majority of the larger reserves and parks are in marginal or 'waste' lands; for the richest and most productive lands, with good supplies of fresh water, are usually the first to be farmed. In Britain, for instance, the rich lowland south is almost entirely used for agriculture; even apparently 'natural' woodlands and heaths have generally been created and managed at some stage in the past. The largest nature reserves in the British Isles are nearly all in mountainous uplands of little value for farming. How much of lowland forest survives in Europe? Is there any reserve in the whole of the North American prairie that could support a natural climax ecosystem including bison, Pronghorn, wolves, prairie dogs, Black-footed Ferrets, and Kit Foxes? Reserves also tend to be fragmented and, in common with oceanic islands, support smaller numbers of species than larger continuous reserves would. It is significant, too, that very few reserves have been established on the basis of sound biological surveys. There are many reserves with boundaries which cut across the migration routes of the very species they are trying to protect, or where the endangered species may live close by, but not actually in the reserve. In such cases the mammals may have to cross farmland or remain unprotected outside the reserves, or may be dependent on water supplies that are only seasonally available. Where the park or reserve boundary is an actual fence (as is the case in many places in southern Africa) the consequences can be disastrous, with animals starving or dying of thirst.

Recent research has focused on the minimum critical size for reserves and national parks. Its discouraging conclusion is that an increasing number of species are doomed to extinction, in a process that probably cannot

* Information on a large number of mammal faunas of islands is contained in a recent paper by T. E. Lawlor (*Biol. Journal Linn. Soc.* 1986, 28: 99–125).

† ('Species-Area Relationship and its Determinants for Mammals in western North American National Parks', *Biol. Journal Linn. Soc.* 1986, 28: 83–98).

be halted. We may save a few of the larger and more spectacular species in zoos, but substantial numbers of bats, insectivores and rodents are doomed, simply because most people are not interested in them.

In addition to direct persecution and destruction of habitat, there are a number of other factors which either result in extinction or endanger species. Oil pollution particularly affects marine species, such as seals, sea lions, whales and dolphins. Virtually no attention has been paid to the effects of nuclear radiation pollution on wildlife, such as that which resulted from the disasters at Windscale (Sellafield), Three Mile Island and Chernobyl. Disturbance can be a significant threat, sometimes even caused by those directly interested in threatened species; large agglomerates of roosting bats are particularly affected. Propellors from speedboats maim and kill manatees; otters drown in fish traps; and huge numbers of animals are trapped or poisoned 'by accident', when mistaken for other species such as pests or fur-bearers.

The widespread introduction of machine guns, and the other implements of modern warfare, has had a significant impact in many areas, but the greatest threat to all wildlife comes from the very real danger of nuclear confrontation, followed by a nuclear 'winter'. Although there is still some debate about the extent of the aftereffects of a nuclear war, it seems generally agreed among scientists that the clouds of dust released into the atmosphere would produce wintry conditions which could last for months, even years, before anything like a normal climate returned. Even an extension of our ordinary winter for a few weeks would lead to the widespread extermination of hibernating species, which would probably not have the fat reserves necessary to survive. Furthermore, chain reactions would start to disrupt stable ecosystems once less hardly plants failed to fruit, or died, and crop failures occurred. It is probably best left to the writers of science-fiction to predict the actual results; but many biologists believe that a nuclear 'winter' could be as disastrous as the actual confrontation itself. There is little doubt, however, that of all mammals, the ever-adaptable house mice and rats would survive and proliferate in their relatively sheltered habitats; and perhaps the Age of Man will be followed by the Age of the Rodent.

A greater understanding of the world's threatened wildlife may perhaps lead to the realisation that the only real threat is man, in his ever-increasing numbers, and his ever-increasing stockpile of nuclear weapons.

The Purpose of this Guide

Interest in and concern about the world's threatened wildlife is increasing; so it is somewhat surprising that there exists no comprehensive listing of the world's threatened mammals. The discovery of rare species is a never-ending process, but in this guide we have attempted to include all the mammal species which might be considered threatened. We have given brief descriptions of as many rare mammals as possible, illustrated a wide selection, and indicated distribution with the help of maps. Some species, such as the Eurasian Badger (p 140) cannot be considered internationally threatened: it has survived centuries of persecution over most of its range – if in reduced numbers. But widespread public interest and belief in their precarious status, has prompted us to include such species. We have aimed to provide in an accessible form a reference to the majority of the world's rarest and most threatened mammals.

How Many Rare and Threatened Mammals?

One of the most disturbing facts to us is the ignorance – even among generally well-informed people – of which species are likely to be threatened. It is to the task of recording them in a single book that we have tried to address ourselves. The tropical and subtropical forests are among the world's most critically threatened habitats: the mammal species found in them, therefore, are also likely to be threatened to a greater or lesser extent. The largest groups of mammals, insectivores (about 350 species), bats (over 950 species), and rodents (over 1,500 species) all reach their greatest diversity in tropical areas, and new species are described every year. Many of them have very restricted ranges, and a large number have never been seen since they were first discovered – often over half a century ago – in habitats that are being destroyed or altered. Far too little attention has been paid to these species, and we have tried to identify as many as possible which *may* be at risk. Only a fraction of these are currently listed in the International Lists of Endangered Species (published by IUCN), and although some will subsequently be shown to be more abundant and widespread than they seem, others may already be extinct. We hope this guide will stimulate searches for these little-known species, and studies of the status and biology of all those that are rare and threatened. Nearly all the previously published literature is biased towards the large and spectacular mammals, and often ignores the smaller mammals such as bats, insectivores and rodents. This may be because funding is often more readily available for the study of certain species than for others. Primates, perhaps because of their close relationship with man, attract more sponsorship for research than almost any other group of mammals. Bats, shrews, and rodents, lacking the glamour, attract little. In an increasingly competitive professional field, many biologists may feel that their career will not make significant strides with the study of a little-known mouse from the upper reaches of the Amazon – were the funds even available.

Whatever the reasons for these discrepancies, the fact remains that approximately one in four of the world's mammals is threatened – or so little-studied that we lack even the basic information to gauge whether its survival can be assured.

It is our intention and hope that readers will be inspired to fill in some of the gaps in our knowledge of the world's threatened mammals. Many of the ** species (*see below*) are known from restricted ranges, and on further investigation, some may prove to be identical to a closely related species; but much more information on their precise distribution and ecology is needed. Many of the species mentioned occur in areas that are increasingly visited by tourists, particularly in South-East Asia. It lies well within the capabilities of most well-organised student expeditions to collect basic information on distribution, and that could add significantly to our knowledge of the majority of the species described in this book.

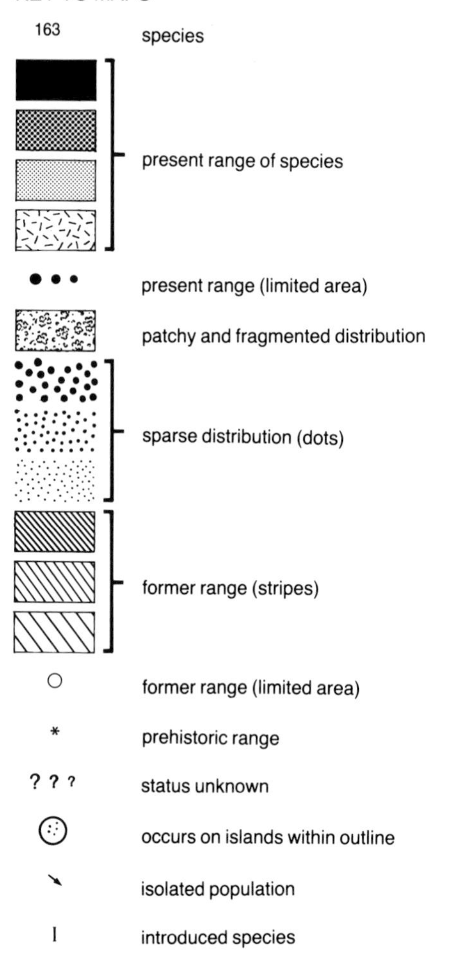

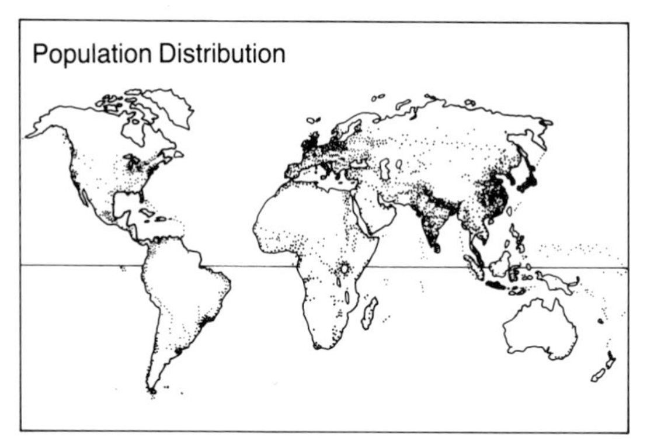

In interpreting distribution maps, it is important to bear in mind that in areas of high human density, wildlife often tends to be more sparse, or at least confined to protected areas.

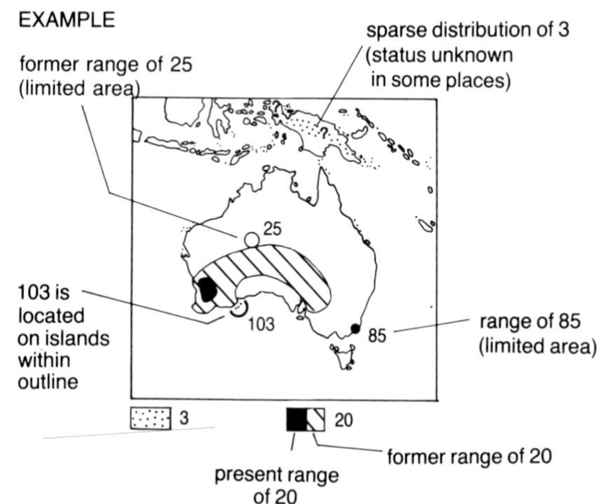

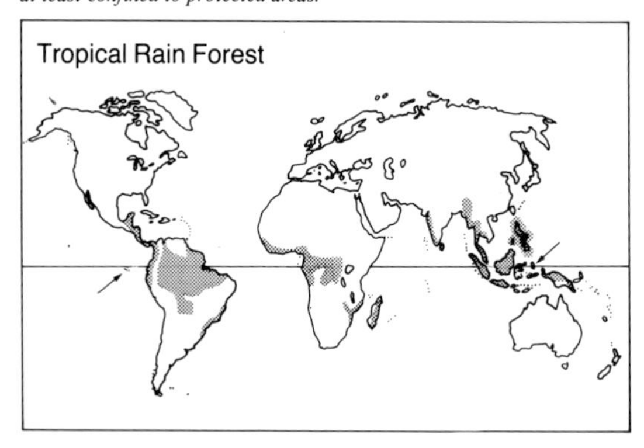

It is in these habitats that the greatest diversity of species occurs; they are also among the most threatened of habitats in many parts of the world, notably in southern and SE Asia.

HOW TO USE THIS GUIDE

The Maps

The distribution maps are intended to show overall maximum range: the species may be expected to occur *within* the area indicated, where there is suitable habitat. In most cases the precise limits of a species' range are not known, and this is often indicated by ? marks. Where information is available, we have tried to give indications of changes in status, range or relative abundance. Because of the scale at which the maps are reproduced, the range should, in all cases, be regarded as a general indication only; for detailed information on past and present distribution, the relevant specialist literature listed in the bibliography should be consulted.

The Texts

The texts give a brief description of the animal, together with notes on interesting aspects of its biology and ecology, a summary of its status both in the wild and in captivity, and notes on its protection. Where available, measurements are given as follows:

HB	Head and body length – from tip of snout to base of tail
T	Tail
FA	Forearm – given for bats, since this is one of the most commonly-used measurements for distinguishing between them
SH	Shoulder height – given for many larger mammals, such as ungulates and some carnivores, as it gives a good impression of size
TL	Head, body and tail length – used for cetaceans, as the head and tail are incorporated into the body shape
in./ins	inch/inches
ft	foot/feet
oz	ounce(s)
lb	pound(s)
ml./mls	mile(s)
sq.ml./mls	square mile(s)

Because of the extreme rarity of many of the species described, measurements have often not been taken, are incomplete, or cannot be traced in any publications. In these instances, reference is usually made to a similar-sized species.

Sources of information It would not be possible to gather the information needed to compile such a book as this at first-hand; wherever possible, the most up-to-date and comprehensive works of reference were used, but inevitably, in using secondary sources of information, the errors of earlier authors will occasionally be perpetrated. Several hundred books and papers were consulted, and the most important of these are listed in the bibliography. There is no reliable single source of information on the world's threatened mammals; the most recent attempt at listing, 1986 *IUCN Red List of Threatened Animals*, is dependent on received information and is therefore incomplete, except in respect of the better-known groups, such as primates. The main sources of information used in compiling this guide were *Walker's Mammals of the World*, by Nowak & Paradiso, and the numerous Red Data Books and regional guides to mammals. All publications consulted were published before mid-1986, although later information was inserted in a few exceptional cases. The IUCN maintains a data base on threatened mammals, which is used for compiling publications such as the *Mammal Red Data Book, part* 1, but access to this data base at the time of writing was not easily available. For one country, Australia, the main problem lay in the amount of information available, often extremely detailed, and much of it extremely recent. The reader is referred to the bibliography for further references.

Abbreviations and Acronyms In addition to those mentioned above, the following are used:

CITES	Convention on International Trade in Endangered Species of Wild Fauna and Flora
FFPS	Fauna and Flora Preservation Society
IUCN	International Union for Conservation of Nature and Natural Resources
NP	National park
RDB	Red Data Book
SSC	Species Survival Commission of IUCN (formerly Survival Service Commission)
sp.	Species (singular)
sp.nov.	New Species (as yet undescribed)
spp.	Species (plural)
ssp.	Subspecies (singular)
sspp.	Subspecies (plural)
WWF	World Wildlife Fund

The Illustrations

Living animals have been drawn wherever possible, either in the field or in a zoo. However, because of the rarity of most of the species involved, the illustration has often had to be based on a combination of photographs, museum specimens, descriptions, and observations of closely-related species. The mammals shown are approximately to scale within each plate, but the text should be consulted for precise sizes.

The Categories

In the past, various methods have been used to categorise degrees of threat to rare and threatened species, but none is entirely satisfactory. In 1984 at the General Assembly of the International Union for Conservation of Nature and Natural Resources (IUCN), its Species Survival Commission (SSC) convened a special meeting on the topic, at which one of us (JAB) presented an historical background written with Sir Peter Scott. While researching this presentation and, later, this book, it became apparent that to use any of the existing systems of measuring rarity or danger was likely to prove either difficult or confusing. The most widely known system is probably the one used in the IUCN Red Data Books. The IUCN defines the categories of threatened mammals as follows:

Ext. (Extinct): Taxa that are possibly extinct, or near extinction.
E (Endangered): Taxa in danger of extinction, and whose survival is unlikely if the causal factors continue operating.
V (Vulnerable): Taxa believed likely to move into the 'endangered' category in the near future, if the causal factors continue operating.
R (Rare): Taxa with small world populations that are not at present 'endangered' or 'vulnerable' but are at risk.
I (Indeterminate): Taxa known to be 'endangered', 'vulnerable' or 'rare', but where there is not enough information to say which of the three categories is appropriate.
K (Insufficiently known): Taxa that are suspected, but not definitely known, to belong to any of the above categories, because of lack of information.

The main difficulty in applying these definitions is that they overlap; what one scientist considers 'vulnerable' may well be considered 'endangered', 'rare' or 'indeterminate' by others. Furthermore, confusion often arises in the USA and elsewhere because the general term 'threatened' is used as an equivalent to 'vulnerable', and the IUCN categories, often with slightly differing definitions, have been used in national RDBs. We have therefore developed a 'star' rating system. The decision as to which category a species belongs often appears rather arbitrary, but in fact is no more subjective than most of the other systems currently in use. In most cases it will give at a glance a more accurate impression of the status of the species concerned.

In the Checklist (Appendix I), the most recent published listings of IUCN are given for direct comparison. Here, the terms endangered, threatened, vulnerable, or rare are used in a broad sense, and are not intended to reflect any defined category approximate to the IUCN definitions.

***** Endangered and likely to become extinct – or may already be extinct. Only survives in much reduced populations and is very vulnerable to any new threats. Eg: Javan Rhino, Blue Whale, Lion-tailed Macaque, Rodrigues Fruit Bat.

**** Endangered and likely to become extinct in the wild, but survives as a captive population or under highly controlled conditions, such as zoos or isolated reserves, or as a reintroduction. Eg: Arabian Oryx, European Bison, Cretan Goat, Bontebok.

*** Some subspecies (interpreted broadly to include discrete populations) endangered, and/or declining over most of range. Eg: Gorilla, Maned Wolf.

** Occurs in small isolated populations, is naturally rare and therefore very vulnerable to any new threats. Or only known from very few observations. ** Species could be stable, but many, particularly those occurring in parts of the tropics where forest is being destroyed, could be highly endangered. Eg: many tropical bats, shrews and rodents.

* Species considered by some authorities to be of conservation concern, but which are probably not threatened with imminent extinction.

Also in this last * category, species for which there is insufficient information to be placed in any of the other categories. Examples: several species of dolphins and small whales; some Australian small mammals. Likewise species of general public concern, which may be locally threatened, but otherwise do not fit into higher ratings. Eg: Eurasian Badger.

Taxonomy and Nomenclature

In guides to the more common species of animals, the scientific names are often superfluous, since the reader will generally know precisely which species is being described from the vernacular. In this guide, however, the scientific names are far more important than the vernacular, since standard vernaculars do not exist for many of the little-known species.

We have not followed any single taxonomic work rigidly, but *Walker's Mammals of the World*, 4th Edition, by Nowak & Paradiso, has been the most generally used work of reference, together with *Mammals Species of the World, a Taxonomic and Geographic Reference*, edited by Honacki *et al*, and *A World List of Mammalian Species*, by Corbet & Hill. These, together with the many other taxonomic works consulted, are listed in the bibliography. But since this is not intended to be a taxonomic reference of any sort, we have used the various classifications to suit the needs of conservation as we consider appropriate. In general, this means that we have tended to 'split' rather than 'lump' species. If in doubt, we have tended to conserve species which other authors may have lumped, particularly if it helps pinpoint populations. Where we have not followed one of the works named above, we have tried to define the species clearly. The checklist (Appendix I) is in the systematic order used by Corbet & Hill.

Systematic Order

Although the text and illustrations follow approximately the sequence used by Nowak & Paradiso, we have frequently deviated from this general structure, because of the constraints of keeping the text and illustrations together, and the problems of wide variations in the size of species.

The Scope and Limitation of this Book

We have tried to cover all groups of mammals, but because of the enormous variation in the quality and quantity of information available, it is impossible to approach the problem consistently. Many complete books have been written on gorillas, tigers and other popular species, whereas some insectivores, rodents and bats are only described in a single brief scientific paper, written perhaps a century ago.

This book is probably the first attempt to review the conservation status of all the world's mammals.

New Data

The very nature of the subject of this guide means that new data will alter our knowledge of many species even before it is published; Appendix II lists species which came to our attention too late for consideration, but which may justify inclusion. We welcome any new data concerning either text or illustrations: both for species included in this edition, and for species which should be considered in future editions. Data can be sent to John Burton c/o the Publishers, William Collins & Sons, 8 Grafton Street, London W1X 3LA, UK, or c/o The Fauna and Flora Preservation Society, 8–12 Camden High Street, London NW1 0JH, UK.

The Fauna and Flora Preservation Society There are many organisations concerned with national and international wildlife conservation. The FFPS was the first of the international societies to be concerned with rare and threatened species. Since its foundation in 1903, it has published a journal (since 1950 entitled *Oryx*) which is one of the most important single sources for information on threatened mammals.

IUCN The International Union for Conservation of Nature and Natural Resources is an international coordinating body for both government agencies and nongovernmental bodies involved with wildlife conservation. Two of its operations are particularly relevant to mammal conservation: its Species Survival Commission (SSC), which through a network of specialist groups and consultants develops action plans for the conservation of a variety of mammal species; the Conservation Monitoring Centre (CMC) houses a computerised data base producing IUCN's Red Data Books. The SSC can be contacted through its Executive Officer, IUCN, Gland, Switzerland; the CMC at 219c Huntingdon Road, Cambridge, CB3 0DL, UK.

World Wildlife Fund This is the largest of the international non-governmental wildlife conservation bodies, raising several million pounds a year for conservation projects all over the world. The headquarters are in Gland, Switzerland, and the UK head office is at 11–13 Ockford Road, Godalming, Surrey, GU7 1QU, UK.

Zoological Society of London The Society publishes a wide range of scientific literature concerning mammals, and organises symposia and scientific meetings. It also owns an extensive library and one of the largest collections of living mammals at its zoological gardens in London, Regent's Park, London NW1 4RY, UK and Whipsnade, The Downs, Dunstable, Bedfordshire, UK.

Mammal Society This is based c/o Linnaean Society, Burlington House, London, W1V 0LQ, UK. Although primarily concerned with British mammals, its publications and meetings often deal with exotic species. Many other countries now have their own national mammal societies often producing excellent regional guides, and atlases of distribution maps.

CITES The Convention on International Trade in Endangered Species of Wild Fauna and Flora (CITES) was concluded in Washington in 1973, and by 1986 had over 80 states belonging to it. CITES seeks to regulate the trade in threatened wildlife by co-operation between exporting and importing countries. The species subject to regulation are listed in three appendices:

Appendix I includes 'all species threatened with extinction which are or may be threatened by trade. Trade in specimens of these species must be subject to particularly strict regulation in order not to further endanger their survival and must only be authorised in exceptional circumstances.'

Appendix II includes 'all species which although not necessarily now threatened with extinction may become so unless trade in specimens of such species is subject to strict regulation ...'. Controls on species similar in appearance to threatened species are also listed here.

Appendix III includes species given strict national protection, for which party states seek international cooperation in enforcement; this appendix is little used.

The Convention is implemented by means of licences issued in member states, and no species listed in CITES can be traded without some sort of documentation. The species listed in CITES Appendices I and II are indicated in Appendix 1 of this book.

***** BROAD-STRIPED MARSUPIAL MOUSE**
Murexia rothschildi **(1)**
HB ca.4.1ins; T ca.5.7ins.
Greyish-brown above, light brown below, with a broad, dark stripe down the back. Only known from a few specimens collected at an altitude of 3250–6550ft, in SE Papua New Guinea. Will probably prove to be more widespread and abundant.

*** NARROW-STRIPED MARSUPIAL MOUSE**
Phascolosorex dorsalis **(2)**
HB 4.6–8.9ins; T 4.6–7.5ins.
Greyish-brown above, more reddish below, with a thin black stripe down the back, and brown feet. Found in montane forests, from 3950–10,150ft, in Irian Jaya, New Guinea. Thought to be relatively rare, but little known of status or habits.

*** RED-BELLIED MARSUPIAL MOUSE**
Phascolosorex doriae **(3)**
HB 5.3–6.6ins; T 4.3–6.3ins.
Very similar to (2), with which it may be conspecific, but smaller and more orange-brown in colour. Found in western New Guinea, where it occurs from 2950–6250ft. Very little known of status, but considered rare.

**** CINNAMON ANTECHINUS**
Antechinus leo **(4)**
HB 3.7–6.1ins; T 3.0–5.5ins; WT 1–3.5oz.
The largest antechinus, very similar to the widespread Yellow-footed Antechinus *A. flavipes* , but distinguished by cinnamon colouring. Found in dense vine-forests with an annual rainfall of about 47.5ins. First identified in 1979, it has a very limited range and specialised habitat. Found in forests between the Iron Range and McIlwraith Range in Cape York Peninsula, Australia, where it is locally common. A protected species, occuring in national parks and reserves, but not being maintained in captivity. A remarkable feature observed in at least 6 of the better-known species of antechinus is sudden and total mortality of males after mating.

******* DIBBLER or FRECKLED MARSUPIAL MOUSE**
Antechinus apicalis **(5)**
HB 5.5–5.7ins; T 3.7–4.5ins; WT 1.4–3.5oz.
Greyish and speckled, with distinctive white eye-ring. Found in dense *Banksia* heathlands feeding on small insects and nectar. In early 19th century widely distributed over Western Australia. Now only known from a small area around Cheynes Beach and Jerdacuttup in SW Australia, where, until its rediscovery in 1967, it had not been seen for over 80 years. Survival threatened by human disturbance, as habitat near to human habitations and cats and foxes probably predate. It is protected, and there are reserves within its range. Although young have been born in captivity, none are currently being maintained.

*** GODMAN'S or ATHERTON ANTECHINUS**
Antechinus godmani **(6)**
HB 3.6–6.3ins; T 3.3–4.9ins; WT 1.9–4.4oz.
Similar to the widespread Yellow-footed Antechinus, but with rufous belly, greyish back, and cinnamon rump. Confined to rainforest in the vicinity of Ravenshoe, Atherton Tableland, northern Queensland. Although range is restricted, this species is abundant in some localities. The * **Swamp Antechinus** *A. minimus* **(7)**, from SE Australia and Tasmania, is thought to be threatened by grazing, and other habitat changes. The ** **Kultarr** *A. laniger* **(8)** exists only in scattered populations and is generally rare; isolated populations in Queensland and New South Wales are presumed extinct.

***** INGRAM'S PLANIGALE**
Planigale ingrami **(9)**
HB 2.2–2.6ins; T 1.7–2.4ins; WT up to 0.2oz.
The smallest living marsupials, living mainly in savannah woodland among grass tussocks and in cracks in the ground. Voracious predators, feeding on large insects, spiders and also small mammals and lizards. Subspecies *P.i.subtilissima* only known from a few specimens from the Kimberley district in Western Australia; *P.i.brunnea* has a restricted range in Queensland, and although much more widespread in Queensland and Northern Territory, *P.i.ingrami* is uncommon over most of its range.

*** NARROW-NOSED PLANIGALE**
Planigale tenuirostris **(10)**
HB 2–3ins; T 2–2.6ins; WT 0.14–0.31oz.
Formerly only known from a handful of specimens from New South Wales, Australia, where it was believed threatened; now known to be more widespread, though only sparsely distributed.

The * **Paucident Planigale** *P.gilesi* **(11)** is similarly rather sparsely populated and may have declined; it is found in inland SE Australia.

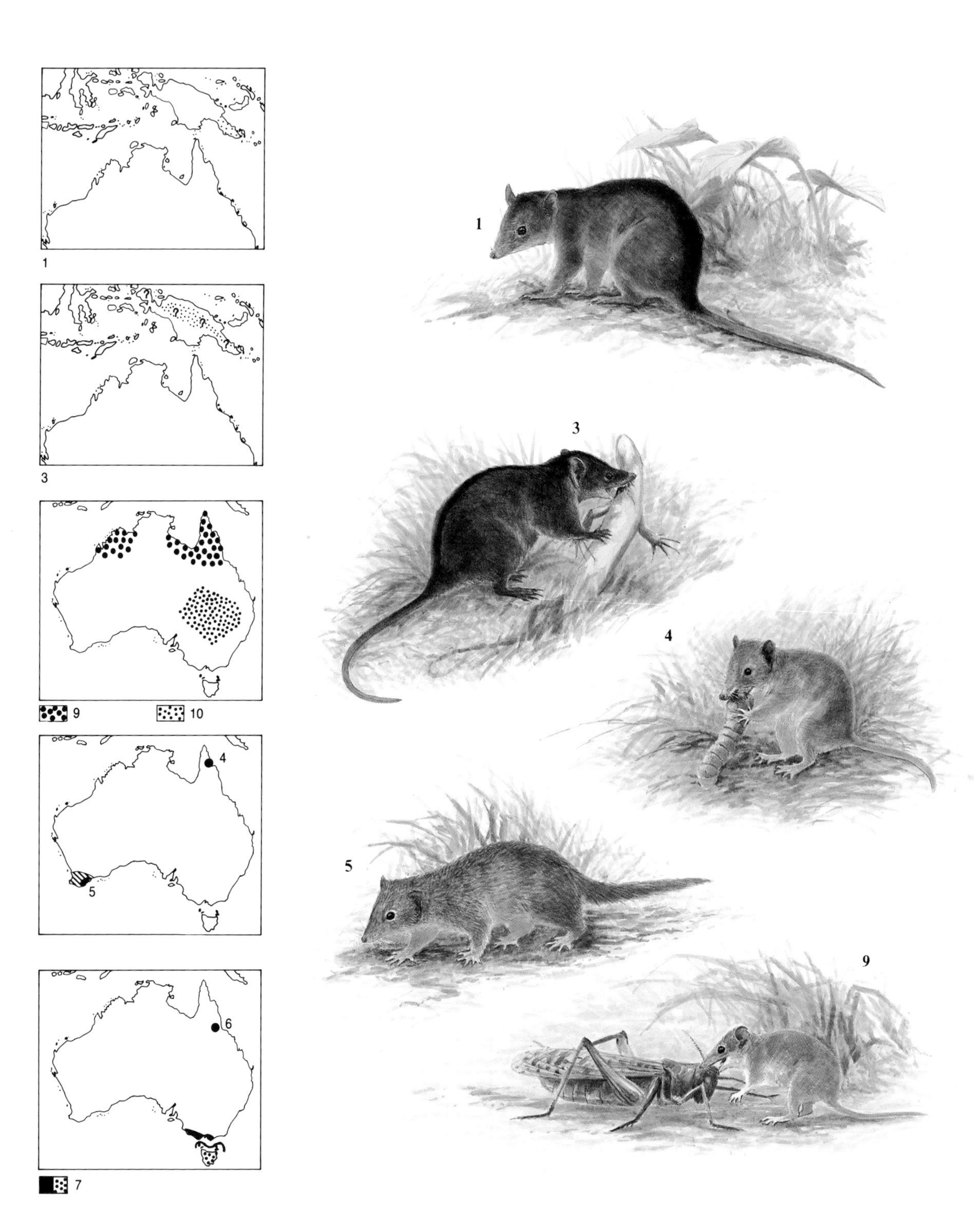
1
3
9
10
4
5
6
7
1
3
4
5
9

******* TASMANIAN WOLF or THYLACINE**
Thylacinus cynocephalus (**12**)
HB 39.4–44.5ins; T 19.7–25.6ins; SH 23.6ins.
Very dog-like, with a remarkably wide gape and very long incisors. Fur is short and coarse. Females have a backward-opening pouch, where 2–4 young are reared until about 3 months old. Have been found in open forest, grasslands and also dense forest. The Thylacine is one of the rarest mammals in the world. It was widespread, occurring over much of Australia and New Guinea; by about 1000 BC it became extinct, probably due to competition from introduced domestic dogs (Dingo and Singing Dog), only surviving in Tasmania. From 1840 bounty hunting, because of alleged attacks on sheep and poultry, drastically reduced the Tasmanian population. By 1860 they were restricted to the more inaccessible parts of the islands; by 1914 they were extremely rare and were last positively recorded in the wild in the 1930s; in recent years they have been allegedly sighted on Tasmania, and the rear view photographed on mainland Australia, but not all scientists are convinced by the data available. Fully protected from 1938, and in 1966 a reserve of 2588 sq.mls was established partly to protect it,should it survive. Thylacines have been known to live over 8 years in captivity, but have never been bred there.

***** EASTERN QUOLL or NATIVE CAT**
Dasyurus viverrinus (**13**)
HB 11–17.7ins; T 6.7–11ins; WT 1.8–4.4lb.
Rather variable in colour, ranging from greyish olive-brown to brown or black, with buff or white spots on the back and sides. Distinguished from other quolls by lack of a first toe on the hind foot. Found in a variety of habitats including forest and open country, often close to human habitations. Nocturnal animals, and although mainly terrestrial can climb well. They feed on a wide variety of small animals, including rats, mice and insects, as well as some plants, and occasionally raid poultry. A common species in SE Australia until the 1930s, even around Adelaide and Melbourne, but have since become very rare. During 1901–1905 they suffered an overall decline due to an epidemic disease, which remains a serious threat to the Tasmanian population. Although still common in Tasmania, they may be extinct on mainland Australia. Small numbers are kept in zoos, and are regularly bred.

***** KOWARI**
Dasyuroides byrnei (**14**)
HB 5.3–7.1ins; T 4.3–5.5ins; WT 2.5–4.9oz.
A small, grey carnivorous marsupial, with a brush of black hairs on the lower part of the tail. Found in the gibber deserts, where they are sparsely populated and their range has contracted.

*** TIGER CAT**
Dasyurus maculatus (**15**)
HB 15.8–29.9ins; T 13.8–22.1ins; WT 4.4–6.6lb.
Colour variable, but greyish or brownish above, with pale buff or white spots extending onto the tail. Found in moist forest, and in general habits are similar to the preceding species. Although still found on mainland Australia as well as Tasmania, they are now confined to eastern parts of their range, and are virtually extinct in Queensland and the more western parts of their range. Occasionally exhibited in zoos.

***** MULGARA**
Dasycercus cristicauda (**16**)
HB 4.9–8.7ins; T 2.8–5.1ins; WT up to 6.1oz.
Upperparts buff to reddish-brown, underside creamy or white. Fur is soft and dense, and tail covered with shiny, coarse, dark hairs forming a crest towards the tip. Found in arid, stony desert or spinifex grassland, feeding on insects, small mammals, reptiles etc. Able to survive without drinking by metabolising water from prey. Although formerly widespread in Queensland and South Australia, now confined to Western Australia. Generally rare and thought to be declining, due to habitat disturbance and introduced predators. However, during plagues of house mice they are believed to increase in numbers, feeding on mice. They have been bred occasionally in captivity.

***** WESTERN QUOLL, CHUDITCH or NATIVE CAT**
Dasyurus geoffroii (**17**)
HB 10.6–14.2ins; T 8.1–11ins; WT 1.5–4.6lb.
Rather variable, but generally greyish or yellowish-brown, with a pale face and belly, and a dark end to the tail; upperparts blotched or spotted with buff or white. Preferred habitat is open savannah, dry forest and woodland. In habits it is similar to other quolls. Originally widespread throughout much of Australia, but now only survives in Western Australia. The causes of decline are not fully understood, but an epidemic in 1901–5 was a major factor. Species also occurs in Papua New Guinea; here they do not appear to be threatened but little is known of status or distribution. Only occasionally exhibited in zoos.

**** LONG-NOSED ECHIDNA**
Zaglossus bruijni (**18**)
HB up to 31.5ins; T vestigial; WT up to 22lb.
The hair often nearly hides the spines on the back; the underside usually lacks spines. Found in humid montane forests at 6250–9850ft in New Guinea, and possibly Salawati Island. Mainly nocturnal, feeding on earthworms and other invertebrates on the forest floor. Echidnas are egg-laying mammals (monotremes) and usually lay a single egg into a pouch. Three subspecies have been described, which have sometimes been regarded as separate species. They are hunted for human food; in settled areas they have become extinct or scarce through hunting and forest destruction. Listed on CITES App.II and protected in both Irian Jaya (Indonesia) and Papua New Guinea, although the PNG legislation does not prevent traditional hunting. Rarely kept in captivity; have been known to live for over 30 years, but do not appear to have bred there.

In Australia another egg-laying mammal, the * **Duck-billed Platypus** *Ornithorhynchus anatinus* (**19**), although still common is considered vulnerable since its freshwater habitat is increasingly threatened.

18
15
14
13
16
12(ca. 1800)
17
12
13
16
15
17
18

***** NUMBAT or BANDED ANTEATER**
Myrmecobius fasciatus **(20)**
HB 8.3–10.8ins; T 6.3–8.3ins; WT 9.9–19.2oz.
Colour rather variable; in SW Australia (*M.f.fasciatus*) greyish brown with white bars, while in the eastern part of the range (*M.f.rufus*), the ground colour was rich rufous; this population is probably extinct. Found in open woodland, where the tree *Eucalyptus wandoo* predominates. Unlike most marsupials they are active by day, feeding primarily on ants and termites, and occasionally other invertebrates, eating up to 20,000 termites daily. Usually solitary, with normal litter size 2–4. Numbats have declined drastically since Europeans first settled in Australia, and are now extinct in New South Wales and South Australia. Main causes of this include land clearance for agriculture, brush fires and introduced predators such as foxes, cats and dogs. Now probably confined to Western Australia, west of Ongerup and south of York. Numbats are protected and occur in several nature reserves and state forests. Have only occasionally been kept in captivity; a breeding colony is maintained at Wanneroo, W Australia, where the first young were bred in 1985.

DUNNARTS
Sminthopsis spp.
HB 2.8–4.7ins; T 2.2–5.1ins; WT up to ca.0.7oz.
The fur is soft and dense; in some the tail is used as a fat store. Several species have restricted ranges and are little-known.

**** JULIA CREEK DUNNART**
Sminthopsis douglasi **(21)**
Discovered in 1931, but not recognised as a full species until 1979. The handful of specimens known are from the Cloncurry River area in NC Queensland, Australia. Protected under Queensland legislation.

*** WHITE-TAILED DUNNART**
Sminthopsis granulipes **(22)**
Similar to other dunnarts, but with a flattened tail, and fine granulations on the toe pads. Until recently, known from a few specimens from Western Australia between Albany Bay, Lake Grace and Jurien Bay. Remains have been found in coastal caves in SW Australia. Probably more abundant than previously believed.

*** HAIRY-FOOTED DUNNART**
Sminthopsis hirtipes **(23)**
Distinguished from other dunnarts by bristly hairs on the soles of the feet, and large ears. A little-known species from arid areas in central Australia. Probably much commoner than often described.

******* LONG-TAILED DUNNART**
Sminthopsis longicaudata **(24)**
Distinguished by a long tail with a tuft on the end. Known only from three specimens prior to 1975, but since then a few more have been found. Species occurs in the Gibson Desert Nature Reserve in Western Australia.

******* SANDHILL DUNNART**
Sminthopsis psammophila **(25)**
Greyish or reddish-buff, with white underparts; distinguished from other dunnarts by crest of hairs along last quarter of tail. Discovered in 1894 near Lake Amadeus, Northern Territory, then remaining further unknown until 1969, when it was rediscovered at Mamblyn and Bonderoo on Lake Eyre Peninsula, South Australia; since then remains have also been found at Ayers Rock, but between 1969 and early 1980s not seen again. Reasons for its rarity are unknown. Bonderoo is close to Hambidge Conservation Area, where similar habitat exists, but it has not been shown to occur there yet.

**** RED-TAILED PHASCOGALE or WAMBENGER**
Phascogale calura **(26)**
HB 3.7–4.8ins; T 4.7–5.7ins; WT 1.3–2.4oz.
Both the Wambenger, and the closely related and more widespread * **Brush-tailed Phascogale** *P.tapoatafa* **(27)** are greyish above and whitish below. The Wambenger has a black tail with an orange-brown base. Arboreal and nocturnal, both live in humid forests, feeding on insects and nectar. Nests are built in holes or forks in trees and are thought to have 3–8 young per litter. Although likely to be widespread in forests, they are now confined to small populations in SW Western Australia. The decline is probably due to longterm climatic changes, resulting in increasing dryness, but they are also hunted by feral cats. The Wambenger is protected and occurs in the Dryandra State Forest and other reserves. * *P.tapoatafa* has declined locally in Eastern Australia.

*** NINGAUIS**
Ningaui spp.
First described in 1975, they are rather similar to the dunnarts, *Sminthopsis* spp., but smaller with longer hair, broader feet, as well as differences in teeth and skull. The few specimens known have all been collected in arid grasslands and savannah. * *N. timealeyi* **(28)** is known from NW Western Australia, * *N. ridei* **(29)** is known from C Western Australia and * *N. yvonneae* **(30)** from S South Australia, C Queensland, and W Victoria. Although only recently discovered, they are probably not threatened.

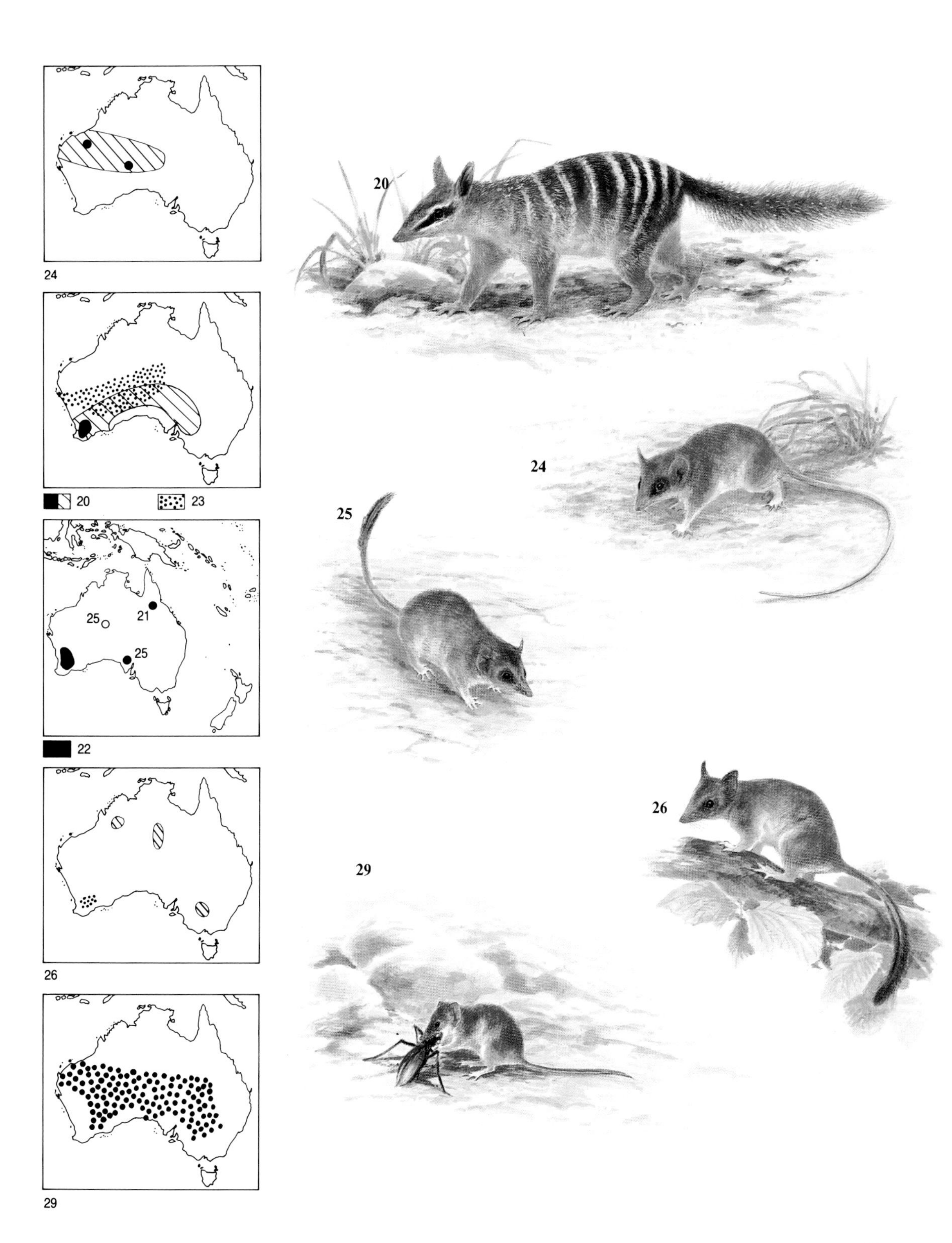
24
20
23
25
21
25
22
26
29
20
24
25
26
29

**** NEW GUINEA MOUSE BANDICOOT**
Microperoryctes murina **(31)**
HB 5.9–6.9ins; T 4.1–4.3ins.
Greyish above, paler below. One specimen had a dark dorsal stripe. Found in Moss Forest between 6250–6550ft. Only known from the Weyland Mountains and the Vogelkop Peninsula of New Guinea but probably more widespread than observations suggest.

****** WESTERN BARRED BANDICOOT**
Perameles bougainville **(32)**
HB 7.9–11.8ins; T 3–4.7ins; WT 6.6–8.7oz.
Smallest of the long-nosed bandicoots, with distinctive markings on rump, and dark upperside to the tail. Ears large (up to 1.6ins long), and thin. Formerly occurred in a wide variety of habitats, from stony deserts to sand-plains, heath and woodland. They are mainly nocturnal and insectivorous with 2–3 young per litter. Originally found across South Australia from the islands of Shark Bay to the Liverpool Plains of New South Wales. By 1925 believed extinct in South Australia; now almost certainly extinct on mainland but still fairly abundant on Bernier and Dorré Islands. Causes of decline included predation by cats and other introduced predators, and habitat changes caused by livestock and rabbits.

***** EASTERN BARRED or GUNN'S BANDICOOT**
Perameles gunnii **(33)**
HB 10.6–13.8ins; T 2.8–4.3ins; WT 15.7–31.5oz.
Distinguished from other long-nosed bandicoots by distinct dark bars across rump, and a white tail with a grey or brown patch at the base. Found in woodland and open country with thick ground cover. Once abundant and widespread but now restricted to Hamilton, Victoria, in suburban gardens and farmed areas. Their digging leads to their persecution. Still fairly abundant in Tasmania (up to 2125 per sq.ml.) where they are widespread and not in danger; they occur in several national parks and reserves. There was a rapid decline in 1970s on mainland Australia, and by mid-1980s they were critically endangered. Rarely seen in zoos, but have been maintained in a research centre. The **Desert Bandicoot** *Perameles eremiana* is probably extinct and the * **White-lipped Bandicoot** *Echymipera clara* **(34)** is found in rain forest up to 3950ft. They feed on insects, figs and *Pandanus* fruits, and are confined to C New Guinea, north of the Central Mountains, and Japen Island, Geelvink Bay. They may be threatened by hunting and agriculture.

*** RUFOUS BANDICOOT**
Echymipera rufescens **(35)**
HB 15–17.7ins; T 3.5–4.7ins; WT 1.6–1.9lb.
Generally rather spiny fur, and tail often missing or broken. Terrestrial, nocturnal and omnivorous, the Rufous Bandicoot lives in rainforest up to 2650ft. Discovered in Australia in 1932, then not seen again until the 1970s. Found in E Cape York Peninsula, north of Coen, but only a small part of its range is in reserves or parks, and elsewhere it is vulnerable to forest clearance. Although once thought rare, research since 1970 suggests this species is reasonably abundant within its restricted range. Also found in New Guinea and the Aru and Kai Islands, where it is fairly widespread and abundant, and not threatened.

******* GOLDEN BANDICOOT**
Isoodon auratus **(36)**
HB 7.5–11.6ins; T 3.3–4.7ins; WT 9.3–23.4oz.
Glossy fur, with distinct coarse rich golden hairs, marked with black; white throat and belly. Formerly found in arid sandy plains, grasslands and open woodlands of C and N Australia; now extinct throughout most of the mainland, confined to Western Australia in NW Kimberleys, on Barrow Island, Double Island and Augustus Island; only known to occur on mainland in Prince Regent Flora & Fauna Reserve. Decline probably caused by introduced predators and changes in land use.

***** SOUTHERN BROWN BANDICOOT**
Isoodon obesulus **(37)**
Slightly larger than Golden Bandicoot (36) and found in SW and SE Australia, Tasmania and in eastern NE Queensland. Has declined in many areas since the arrival of Europeans, largely due to the loss of vegetation through cattle and sheep grazing. It is still common in some parts of its range.

**** SERAM ISLAND LONG-NOSED BANDICOOT**
Rhynchomeles prattorum **(38)**
HB 12.6ins; T 5.1ins.
Dark chocolate-brown, with a white patch on chest. Tail blackish-brown, and almost bare of fur. Found in dense forest on Mount Mannsela on the island of Seram (Indonesia) at 6250ft. Only 4 specimens known; has not been recorded since its discovery in 1920.

***** COMMON or GREATER RABBIT-BANDICOOT or BILBY**
Macrotis lagotis **(39)**
HB 11.4–21.7ins; T 7.9–11.4ins; WT 1.8–4.8lb.
Males about twice the weight of females. Long, silky fur, greyish above, often tinged with fawn or bluish, and reddish along the sides; underside white, middle of tail black, white end with a crest of long hairs. Found mainly in dry habitats including woodland, savannah, shrubby grassland, and more arid habitats. They dig extensively and distribution depends on suitable soils. Nocturnal mammals, feeding on insects, small animals such as lizards, and also some plant matter. Normally solitary, with litters of 1–3 young. Once extremely abundant, the Common Bilby occurred in suitable habitat from the coast of Western Australia to the Great Dividing Range in the east. During the early 1900s species began to decline, through hunting for its pelt, competition with rabbits for burrows, and predation by foxes. Now reduced to scattered pockets in S Queensland, Western Australia and Northern Territory. Four of the six subspecies are now extinct. Although occasionally kept in captivity there are no self-sustaining populations.

******* LESSER RABBIT-BANDICOOT or BILBY**
Macrotis leucura **(40)**
Smaller than the Common Bilby (39). Formerly occurred in C Australia but has not been definitely seen for over 50 years, and is possibly extinct.

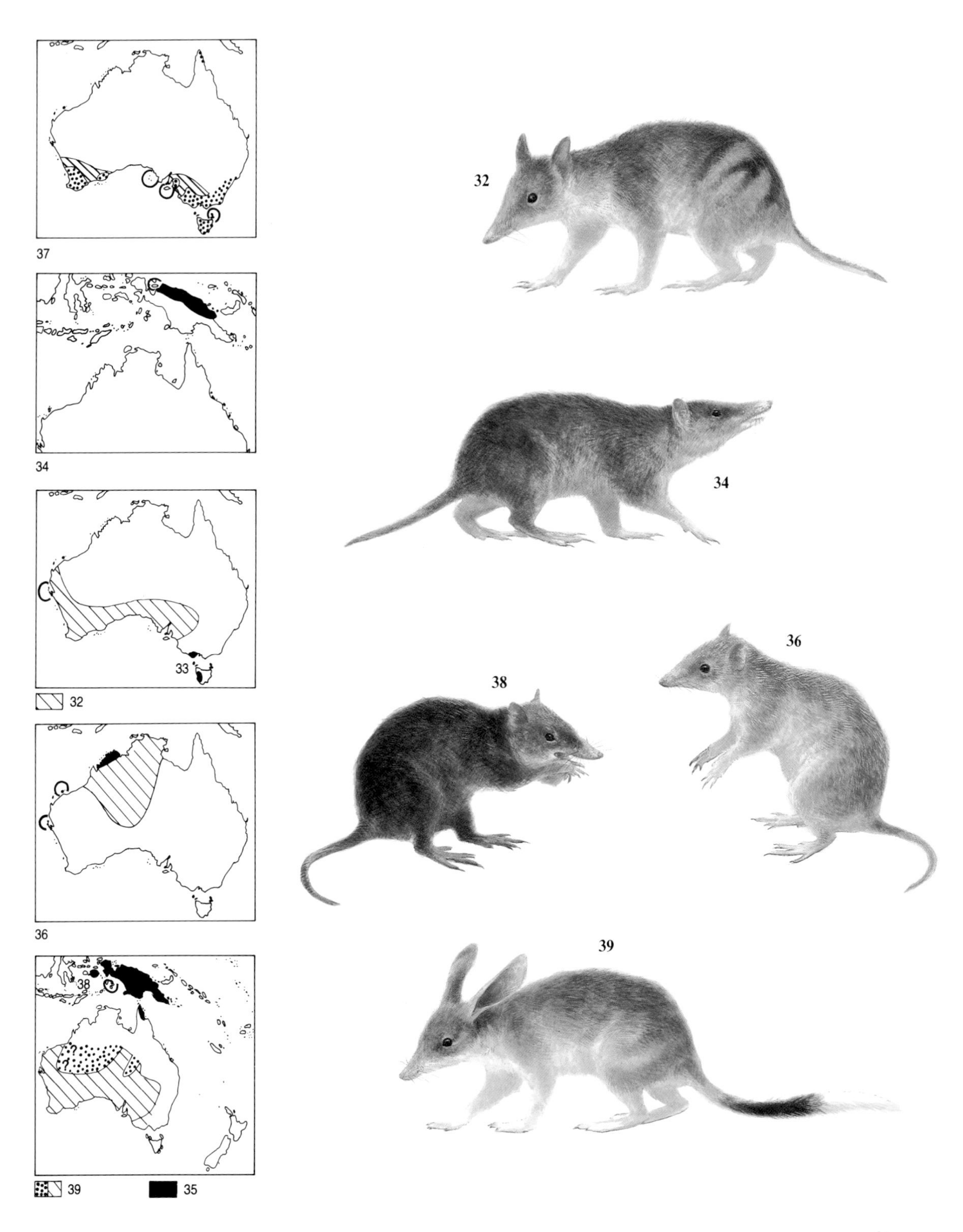
37
34
33
32
36
38
39
35
32
34
36
38
39

*** WATER OPOSSUM or YAPOK**
Chironectes minimus (**41**)
HB 10.6–15.8ins; T 12.2–16.9ins; WT 1.3–1.8lb.
Fur is short and dense, marbled grey and black on the upperparts and white on the underparts. Tail furred at base. Particularly well-adapted to an aquatic life, with webbed hind feet and water-repellent fur. Both sexes have a pouch; the female can close her pouch when it contains young (up to 5), and the male withdraws the scrotum into the pouch when swimming. A carnivorous species, feeding on fish, crustaceans and other aquatic animals. Found in streams and lakes in tropical and subtropical South America, at altitudes of up to almost 6550ft, from S Mexico to N Argentina. Scarce throughout its range; hunting for its fur (which has limited commercial value) could easily endanger it.

**** BLACK FOUR-EYED OPOSSUM**
Philander (=Metachirops) mcilhennyi (**42**)
Known only from Loreto, C Peru, and described in 1972. Probably similar in habits to the more widespread *P. opossum*.

**** PATAGONIAN OPOSSUM**
Lestodelphys halli (**43**)
HB 5.1–5.7ins; T 3.2–3.9ins.
The fur is soft and dense, dark grey above, white below with a darker face; the hands and feet more white. Little is known of this species which has only been found in three localities in the Argentine pampas – further south than any other marsupial. It is thought that it may occur in the Petrified Forest reserve in Santa Cruz Province in Argentina. Believed to feed on birds and mice.

**** MOUSE OPOSSUMS**
Marmosa spp.
About 47 species are known, occurring in Central and South America. Most are forest-dwelling but others occur in a variety of habitats; several species are known only from a single locality, or only from a few specimens and may be threatened; these include: *M. andersoni* (**44**) which is known only from Cuzco, Peru, and was first described in 1972; *M. cracens* (**45**) known only from Venezuela, and was first described in 1979. *M. handleyi* (**46**) is known only from 2 specimens from Colombia, described in 1981. *M. tatei* (**47**) is known only from a single locality in Ancash, Peru, described in 1956. Many others may prove to have restricted ranges, and there are probably new species awaiting description.

**** BUSHY-TAILED OPOSSUM**
Glironia venusta (**48**)
HB 6.3–8.1ins; T 7.7–8.9ins.
Fawn or cinnamon-brown above, paler below, with a dark facial mask; the tail is well furred, flecked white. Virtually unknown in the wild but thought to feed on insects, berries etc. Only five specimens are known – four from commercial animal dealers and all from tropical humid forests in the upper Amazon regions of Ecuador, Peru and Bolivia. Subsequent study could show it to be more widespread and abundant than present data suggest.

**** BOLIVIAN SHORT-TAILED OPOSSUM**
Monodelphis kunsi (**49**)
Known only from Beni, Bolivia, and described in 1975. The status of *M. maraxina*, confined to Marajo Island in the Amazon Delta, Brazil, is not known, and several other species of *Monodelphis* are very little known too.

**** PERUVIAN SHREW-OPPOSUM**
Lestoros inca (**50**)
HB 3.5–4.7ins; T 3.8–5.3ins; WT 0.9–1.1oz.
The long fur is dark brown all over, sometimes paler below. Has been found in moist forest in Cuzco, Peru, at altitudes of 9200–13,150ft, where it is mainly active at night feeding on insects, worms, the young of small mammals and other animals. May not be as rare as recorded occurrence suggests.

**** CHILEAN SHREW-OPOSSUM**
Rhyncholestes raphanurus (**51**)
HB 4.3–5ins; T 2.6–3.4ins; WT ca.0.7oz.
The soft fur is dark brown all over, with a blackish tail, which is shorter and thicker than the Peruvian Shrew-opossum (50). Has been found in moist forests, where it is probably terrestrial and nocturnal. Only known from two localities: Chiloe Island and Llanquihue Province, in Chile.

*** SHREW-OPOSSUMS**
Caenolestes caniventer (**52**) and *C. tatei* (**53**)
Both species are only known from single specimens, from Ecuador, and are probably only slightly aberrant forms of more widespread species of *Caenolestes*; if they are distinct, they are possibly very localised.

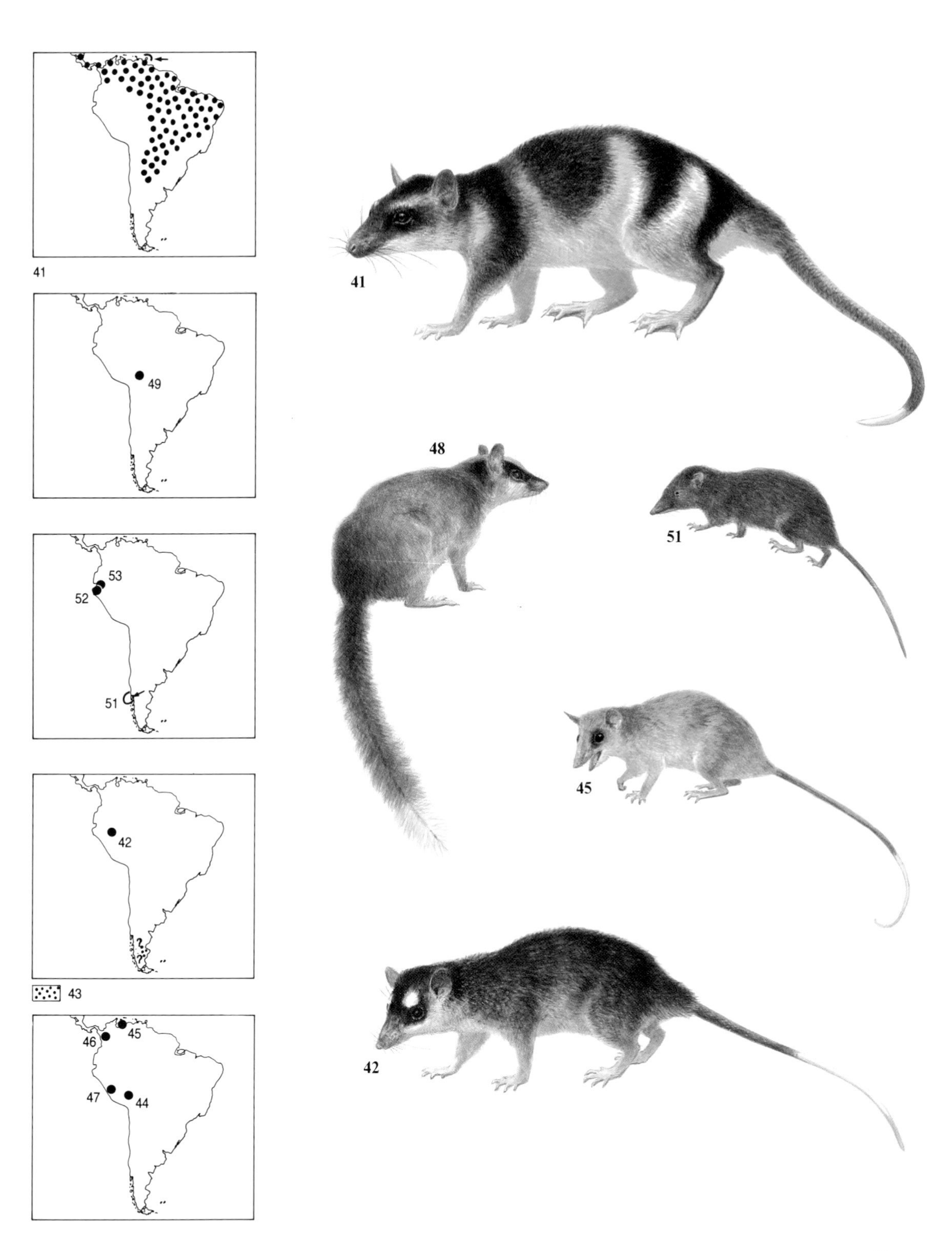
41
49
53
52
51
42
43
45
46
47
44
41
48
51
45
42

***** NORTHERN HAIRY-NOSED WOMBAT

Lasiorhinus krefftii (**54**)

HB ca.39.5ins; T ca.2ins; WT up to 79.4lb.

Fur is finer and softer than the Common Wombat, ears more pointed and nose hairy. Found in dry, open country, they dig complex burrows, connected with paths above ground. They feed almost exclusively on grasses. The single young or twins stay in the mother's pouch for about 6–7 months, and are weaned at 8–9 months. Both species of hairy-nosed wombat declined with the spread of European settlers in Australia, from direct persecution, (due to burrowing habits), and also through competition with introduced rabbits. Also susceptible to disease and drought. Have disappeared from New South Wales and SE Queensland, and only remaining population found in EC Queensland: some 60–70 burrows containing an unassessed number of animals. Part of its range (about 9.5sq.mls) is in the Epping Forest NP. Listed on CITES App.I, and fully protected in Australia.

*** SOUTHERN HAIRY-NOSED WOMBAT

Lasiorhinus latifrons (**55**)

HB 30.3–37ins; T 1–2.4ins; WT 41.9–70.6lb.

Closely related to (54) and similar in habits, it has also suffered serious declines. Although still locally abundant, populations are now fragmented and often isolated. Suffers from competition from rabbits and occasional human persecution.

* COMMON WOMBAT

Vombatus ursinus (**56**)

HB 35.5–47.3ins; T 1in; WT 48.5–86lb.

Coarse fur varies from whitish-buff to yellowish, or dark brown to blackish. Nose hairless; ears rather rounded. This species digs extensive burrows, usually with a single entrance. The single young or twins stay in the mother's pouch for about 6 months. Found in forests, often on rocky hillsides, feeding on roots, grass, fungi and other vegetable matter. Declined through human persecution, but still abundant in a few areas, notably in Tasmania. Extinct on the Bass Strait Islands, except Flinders Island. Within their range they occur in many reserves, and are protected. Occasionally kept in zoos and bred; they have lived up to 26 years.

* KOALA or NATIVE BEAR

Phascolarctos cinereus (**57**)

HB 26.8–32.3ins; T vestigial; WT 8.8–29.8lb.

The Koala is the prototype of the 'teddy bear' with soft fur, for which it was once nearly exterminated. Confined to eucalyptus forest, they feed almost exclusively on leaves and bark of the trees, together with some mistletoes and Australian box *Tristania*. The single young (occasionally twins) stays in the mother's pouch for 7 months and is then carried on her back. At the turn of the century, they were widespread and numbered millions, but forest clearance, fires, and natural diseases, together with excessive hunting, greatly reduced its range. At least 2 million skins were exported in 1924, and the species was on the verge of extinction in New South Wales, Victoria, and South Australia. The remaining Queensland population was exploited next (6,000,000 in 1927), and was soon reduced to scattered remnants. Over the last half century, with protection and reintroduction, it has recovered in some parts of its range.

*** TREE KANGAROOS

Dendrolagus spp.

HB 19.7–31.9ins; T 16.5–36.8ins; WT up to 22lb.

Fur is fairly long, and varies considerably in colour. Found mainly in montane forest at altitudes of up to 9850ft, where they are arboreal, leaping up to 30ft between trees. They live in small groups; and usually produce a single young. The **Grizzled Tree Kangaroo** *D.inustus* (**58**), is restricted to the western two-thirds of Papua New Guinea, to the north of the Central Mountains, where it is not particularly rare, but likely to be threatened by logging. The **Vogelkop Tree Kangaroo** *D.ursinus* (**59**), is restricted to the western fifth of Papua New Guinea, overlapping with the Grizzled Tree Kangaroo. Both species are listed on CITES App.II. **Bennett's Tree Kangaroo,** *D.bennettianus* (**60**), and **Lumholtz's Tree Kangaroo,** *D.lumholtzi* (**61**) are both confined to NW Queensland, Australia, where they are threatened by loss of their rainforest habitat. The range of Bennett's is known to have contracted by at least 75mls. The **Unicoloured Tree Kangaroo** *D.dorianus* (**62**) and **Ornate Tree Kangaroo** *D.goodfellowi* (**63**), although still widespread, are apparently increasingly under pressure from hunting in the Central Highlands of New Guinea, and have declined markedly in many areas; little is known of their precise status. Although tree kangaroos have lived for over 20 years in captivity and are exhibited in some of the larger zoos, they do not have self-sustaining populations. **Matschie's Tree Kangaroo** *D.matschiei* is probably only a population of *D. goodfellowi* (63).

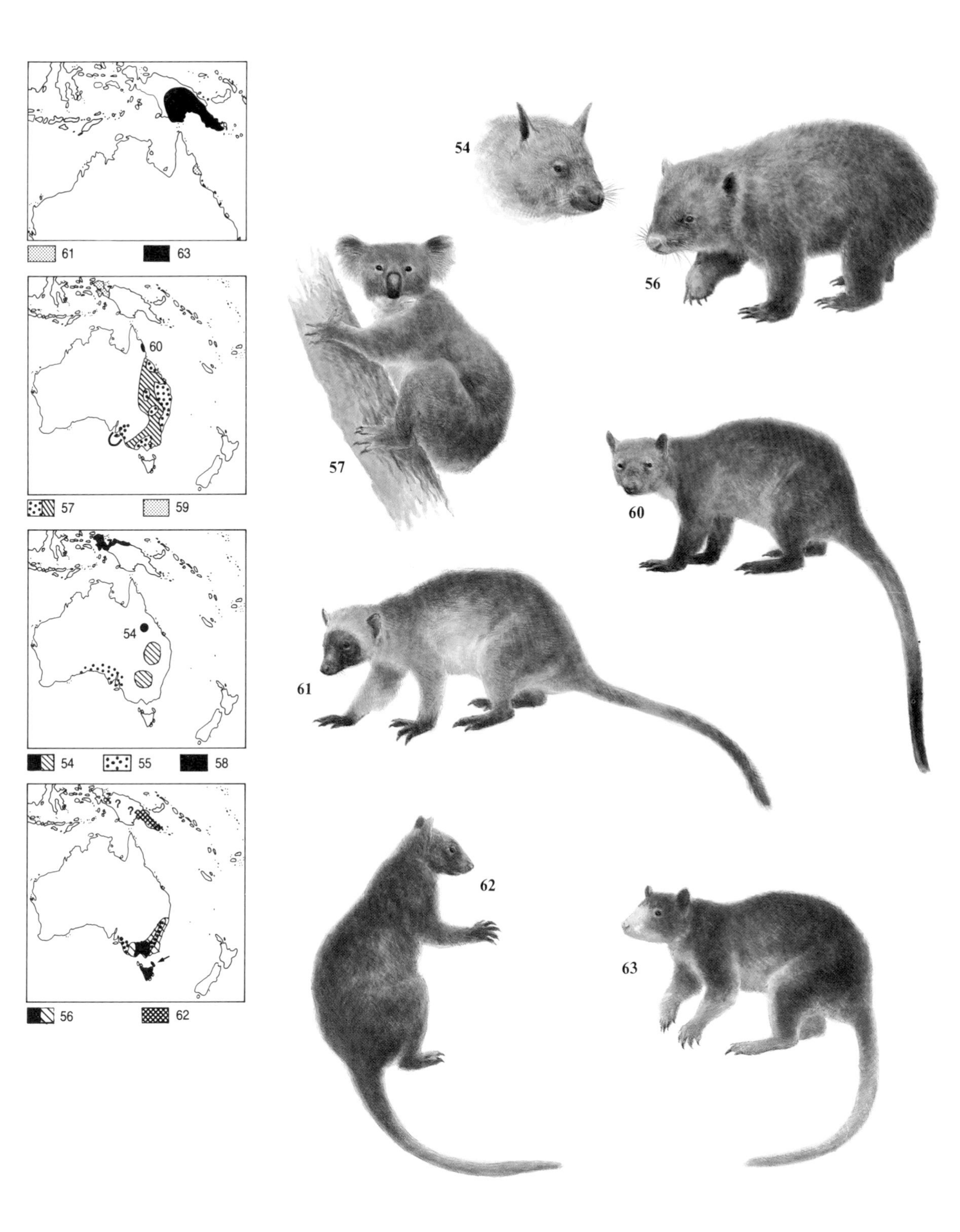
61
63
60
57
59
54
55
58
56
62
?
?
54
56
57
60
61
62
63

**** SCALY-TAILED POSSUM**
Wyulda squamicaudata (**64**)
HB ca.15.8ins; T ca.11.8ins; WT 2.8–3.9lb.
Short and soft fur. The prehensile tail is thickly furred at the base, but scaly for the rest of its length. A nocturnal species, feeding on fruits, nuts, flowers, insects and small animals. Only a few specimens recorded, from rocky, wooded and spinifex country in the Kimberleys region of Western Australia. Habitat rather inaccessible, and it may be more abundant and widespread.

***** LEMUR-LIKE RINGTAIL**
Pseudocheirus (=Hemibelideus) lemuroides (**65**)
HB 12.2–13.8ins; T 13.2–14.6ins; WT 1.8–2.9lb.
Uniform dark grey above, paler below; the tail is bushy with a naked tip. Confined to tropical rainforest between the Daintree and Herbert Rivers in E Australia. A small part of its range, on Mount Spurgeon, is within a national park, but the remainder is in state forests where logging threatens it. The closely related **Mongan** or **Herbert River Ringtail Possum** *P. herbertensis* occurs in two separate and relatively small populations in E Australia; although not presently endangered it could become so. The * **Arfak Ringtail** *P.schlegeli* (**66**) from New Guinea, is a scarce and little-known species. The * **Common Ringtail** *P.peregrinus* (**67**) has apparently declined in parts of Victoria and Western Australia, and in Tasmania its fur has been exported in large quantities – 7.5 million skins between 1923–1955.

*** TOOLAH or GREEN RINGTAIL**
Pseudocheirus archeri (**68**)
HB 13.4–15ins; T 12.2–13ins; WT 2.4–3.1lb.
One of few mammals with greenish fur, and two silvery stripes run down the back, with white patches under the ears and eyes. Although not immediately threatened, confined to a relatively small area of Queensland, Australia, in rain forests which may soon be logged; mining and other developments may also threaten. Found in Bellenden NP.

******* LEADBEATER'S POSSUM**
Gymnobelideus leadbeateri (**69**)
HB 5.9–6.7ins; T 5.7–7.1ins; WT 3.5–5.8oz.
The soft fur is greyish or brownish above, paler below, with a dark stripe down the back. The tail, which is laterally flattened, bushing out and blackish at the tip, is used for balance. Arboreal and nocturnal animals in dense wet forest at altitudes of up to 3950ft; feed mainly on insects. Although discovered in 1867, only known from 5 specimens collected before 1909, until rediscovery in 1961 in Mountain Ash Forest, near Healesville, Victoria. Although its range is fairly extensive (193sq.mls) much is threatened by logging, and bush fires. Protected by state law; habitat destruction is its main threat.

*** YELLOW-BELLIED or FLUFFY GLIDER**
Petaurus australis (**70**)
HB 10.6–11.8ins; T 16.5–18.9ins; WT 15.7–24.5oz.
Greyish-brown above, with darker stripe down centre of back; limbs darker, and underside whitish or orange-yellow. Live in colonies roosting in hollow trees in forests and woodlands in E Australia. Although widespread, becoming rare in some areas; survival depends on the maintenance of large tracts of woodland or forest.

*** HONEY POSSUM**
Tarsipes spenserae (=rostratus) (**71**)
HB 1.6–3.7ins; T 1.8–4.3ins; WT up to 48.5lb.
Generally greyish-brown with 3 dark stripes along the back, the central stripe extending the entire length of the back. Underside paler, and legs reddish with white paws. Long muzzle (1in.) used for feeding on nectar and pollen, which form the main part of its diet, together with insects and grubs. Although not yet threatened, the Honey Possum is dependent on the flowers of *Banksia, Callistemon* and *Eucalyptus*: only known from SW Western Australia, where its habitat is being destroyed with the spread of urbanisation.

***** GREY CUSCUS**
Phalanger orientalis (**72**)
HB 13.8–15.8ins; T 11–13.8ins; WT 3.3–4.8lb.
Greyish-brown above, stripe in centre of back, off-white on underside; prehensile tail naked for most of its length. Confined to a small area of tropical rainforest in Queensland, Australia, but more widespread and abundant up to an altitude of 3950ft from Timor through New Guinea to the Solomon Islands. They feed on blossoms, nectar, leaves and fruit. Their range in Australia is threatened by logging, mining and other encroachment. *** **Woodlark Cuscus** *P.lullulae* (**73**), confined to the island of Woodlark, off E New Guinea, and * **Black-spotted Cuscus** *P.rufoniger* (**74**) from mainland of New Guinea, are both only known from a handful of specimens. Little is known of the status of * **Stein's Cuscus** *P.interpositus* (**75**) from scattered localities at altitudes of 3950–4900ft.

***** BURRAMYS or MOUNTAIN PYGMY POSSUM**
Burramys parvus (**76**)
HB ca.4.3ins; T ca.5.5ins; WT ca.1.7oz.
Fur is greyish above, paler below, extending onto the lower third of the prehensile tail. Terrestrial, but agile and active; nocturnal, feeding on seeds and nuts. Discovered as a fossil from Wombeyan Caves, New South Wales, Australia, ca.15,000 years old; in 1966 it was discovered alive on Mount Hotham, Victoria. Has subsequently been found in the Bogong High Plains, Victoria, and Kosciusko NP, Victoria. Found at altitudes of 4900–5900ft, in dense shrub, snowgum trees, in the alpine and subalpine zones. They breed readily in captivity.

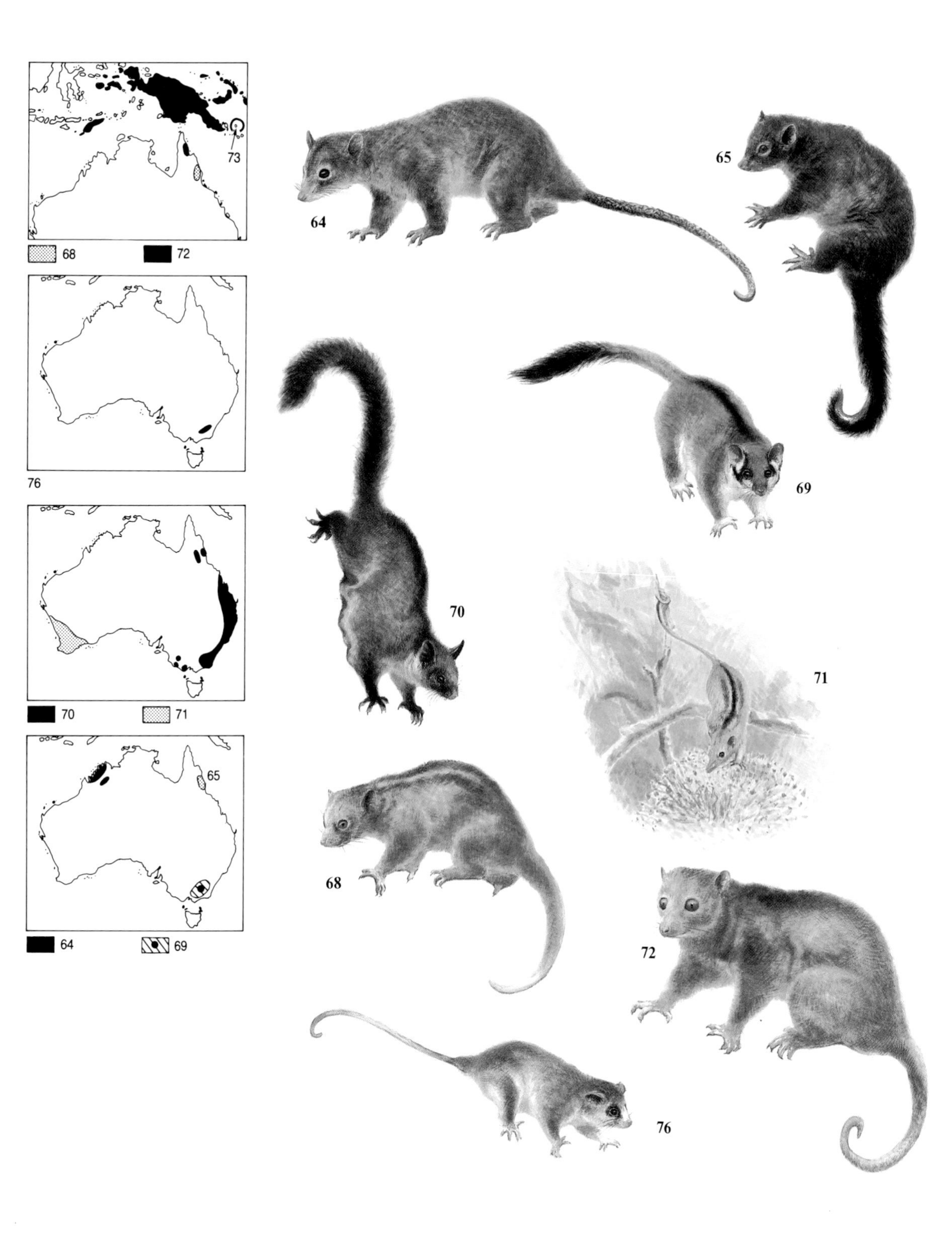
68
72
73
76
70
71
65
64
69
64
65
70
69
71
68
72
76

***** LITTLE PYGMY POSSUM**
Cercartetus lepidus (**77**)
HB 2–2.6ins; T 2.4–3ins; WT 0.2–0.3oz.
Greyish-brown above, greyish below, with a tail which is used as a fat store. One of the smallest possums. Found in mallee heathlands on mainland Australia and forests in Tasmania; also on Kangaroo Island. Although the Tasmanian populations are safe, those on the mainland could be threatened, and fossil remains indicate its range was contracting before the arrival of Europeans.

*** LONG-TAILED PYGMY POSSUM**
Cercartetus caudatus (**78**)
HB 3.9–4.3ins; T 5.1–5.9ins; WT 0.9–1.4oz.
Grey above, white below, with a distinctive dark eye patch. Until 1970 only known from 9 specimens in Australia, although probably much more widespread in New Guinea. They occur within various national parks in Australia, but little is known of their status.

******* DESERT RAT-KANGAROO**
Caloprymnus campestris (**79**)
HB 10.6–17.3ins; SH 13.8–15ins; WT 1.9–2.3lb.
Discovered in S Australia in 1843, not recorded again until 1931, when found in an area of 404 x 155mls in Lake Eyre Basin on the Queensland/South Australia boundary. Remains also found in caves in Western Australia and New South Wales. Believed to have lived in arid areas with sparse saltbush cover, but has not been sighted since 1935. May be extinct, or possibly very rare; could reappear should the right conditions develop. Listed on CITES App.I and protected in Australia.

***** WOYLIE, BETTONG or BRUSH-TAILED RAT-KANGAROO**
Bettongia penicillata (**80**)
HB 11.8–15ins; T 11.4–14.2ins; WT 2.4–3.5lb.
Although common in the Wandoo forests of SW Australia, has declined over most of the rest of its range, which once extended over much of the southern half of Australia, where in the mid-19th century they were described as abundant. By the 1920s they were extinct over most of their range and are currently restricted to a few small areas in the extreme SW tip of Western Australia, where they are found in brush woodlands. A largely nocturnal species, they build nests of dry grasses, which they carry coiled in their tail. They are adapted to surviving bush fires which regularly devastate their range. Have been re-established on Francis and Wedge Islands, South Australia.

***** EASTERN BETTONG**
Bettongia gaimardi (**81**)
HB 12.4–13ins; T 11.4–13.6ins; WT 2.6–5lb.
Larger and more greyish than the Woylie (80). Although still common in Tasmania, have not been recorded on mainland Australia this century.

***** BOODIE, TUNGOO or SHORT-NOSED RAT-KANGAROO**
Bettongia lesueur (**82**)
HB 11–15.8ins; T 8.5–11.8ins.
Formerly one of the most common and widespread native mammals in Australia, has become extinct over almost all of its mainland range since the 1930s, and now confined to Barrow, Boodie, Dorré and Bernier Islands in Shark Bay, Western Australia. The only kangaroo that lives in warrens or burrows which it digs for itself.

******* NORTHERN RAT-KANGAROO**
Bettongia tropica (**83**)
Closely related to the Woylie (80), and only distinguished by differences in the skull. Only known from 6 specimens, the last of which was collected in 1930s, this species was thought to be extinct. Rediscovered in the late 1970s in the Davies Creek NP, Queensland, Australia.

***** LONG-NOSED POTOROO**
Potorous tridactylus (**84**)
HB 13.4–16.2ins; T 6.3–10.2ins; WT 1.5–2.6lb.
Soft greyish fur is pale on the belly, tail often tipped white. Found mainly in damp grasslands, or wet forests, feeding largely on fungi, roots and other vegetation, and some insects, which they dig up in the ground debris. Its range has contracted considerably, and the population from SW Western Australia (*P.t.gilberti*) has not been seen this century. They are extinct in Clark Island, and rare on Flinders and King Islands in the Bass Straits. Still reasonably abundant in Tasmania (*P.t.apicalis*); the mainland population (*P.t.tridactylus*) is threatened by bush fires and forest clearance.The closely related **Broad-faced Potoroo** *P. platyops,* from SW Australia, became extinct in the 1870s.

**** LONG-FOOTED POTOROO**
Potorous longipes (**85**)
HB 15–16.4ins; T 12.4–12.8ins; WT 3.5–4.8lb.
First collected in 1968 and described as a distinct species in 1980, it is only known from a handful of specimens from E Victoria, Australia, where the forest adjacent to the only known sites where they occur is threatened by logging. A small captive colony has been started.

*** MUSKY RAT-KANGAROO**
Hypsiprymnodon moschatus (**86**)
HB 6–10.8ins; T 4.8–6.3ins; WT 0.8–1.5lb.
A dark brownish rat-kangaroo, with a naked tail, it is believed to be the most primitive living kangaroo, and the only one to have 5 toes on the hind foot and to give birth regularly to twins. Confined to a very small area of rain forest in Queensland (186 x 40mls), where they are rarely seen and must therefore be considered highly vulnerable.

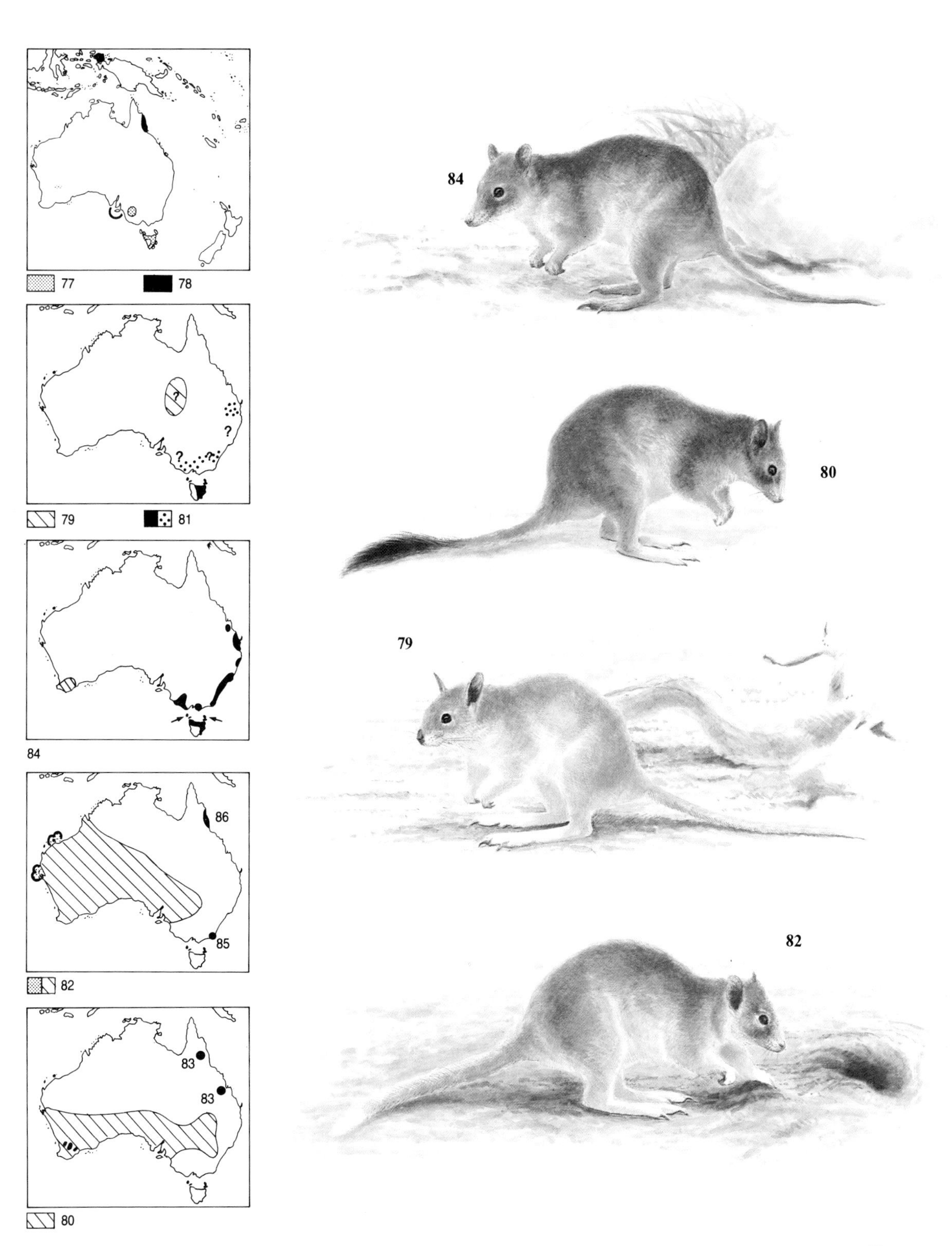

77
78
79
81
84
86
85
82
83
83
80
84
80
79
82

***** BRIDLED NAIL-TAILED WALLABY, FLASHJACK or MERRIN

Onychogalea fraenata (**87**)

HB 16.9–27.6ins; T 14.2–21.3ins; SH ca.19.7ins; WT 8.8–13.2lb.

A graceful wallaby with delicate black and white markings on the hips, cheeks and back, and a horny tip to the tail. Formerly abundant in forests, mainly brigalow *Acacia harpophylla* and Cypress pine *Callitris* spp. and in shrublands. Confined to Australia where its range once extended west of the Great Divide, from SW Queensland through New South Wales to NE Victoria. Despite its former abundance in the mid-19th century, by the 1930s it was believed extinct. Decline was caused by clearance of its habitat for sheep and cattle grazing, and also possibly predation by introduced foxes and feral dogs. It was rediscovered in 1973 in an area of about 39sq.mls near Dingo in E Queensland; part of the area is now a reserve. A captive colony has also been established. The closely related ***** **Crescent Nail-tailed Wallaby** or **Warrung** *O.lunata* (**88**) has not been definitely recorded since 1930. It was once fairly abundant, in what is now the wheat belt of Australia, and may survive in the arid parts of Western and C Australia.

*** WESTERN, RUFOUS HARE WALLABY or MALA

Lagorchestes hirsutus (**89**)

HB 12.2–15.4ins; T 9.7–12ins; WT 1.8–4.4lb.

Generally a rich sandy-buff with longer reddish-orange hair on the back and rump, making it appear rather shaggy. Hare wallabies are fast, very agile jumpers, and, like hares, squat in 'forms'. When first seen in the early part of the 19th century, they were reasonably abundant and ranged over most of the semi-arid and arid interior of Australia; for unknown reasons they began to decline in the 1940s and 1950s. Now confined to two populations in desert areas interspersed with salt pans in the Tanami Desert, Northern Territory, and to heath and spinifex habitats on Dorré and Bernier Islands, Shark Bay, Australia. In 1978 it was estimated that each population in the Tanami Desert contained 6–10 individuals. They feed mainly on grasses, and one of the causes of their decline is thought to be more intense fires in recent years. All populations are protected by law, and they occur in nature reserves. Wildlife on Dorré has been protected since 1892, and on Bernier since 1919. A small number of the mainland population is being bred in captivity.

*** SPECTACLED HARE WALLABY

Lagorchestes conspicillatus (**90**)

HB 15.8–18.5ins; T 14.6–19.3ins; WT 3.5–9.9lb.

The most brightly coloured of the hare wallabies, with a conspicuous orange eye-ring. In general habits similar to other hare wallabies. Populations are now fragmented, occurring in three main areas: Queensland, Northern Territory and Western Australia and Barrow Island (The species is extinct on Hermite Island.) However, it has declined less than any other hare wallaby.

***** CENTRAL HARE WALLABY

Lagorchestes asomatus (**91**)

Known from a single specimen from C Northern Territory, Australia, in an area that remains largely unexplored. If it survives, its status is unknown.

*** QUOKKA

Setonix brachyurus (**92**)

HB 15.8–21.3ins; T 9.7–12.2ins; WT 5.9–9.3lb.

One of the smallest wallabies; the fur is short, greyish-brown, and fairly coarse. The first marsupial to be studied in depth, and until the 1930s it was a common species in SW Australia, where it lived in swampy thickets. Now very rare on the mainland of Australia, surviving in only a few isolated colonies, but still reasonably numerous on the islands of Rottnest and Bald, with populations of over 600 on Rottnest, slightly fewer on Bald Island.

*** BANDED HARE WALLABY

Lagostrophus fasciatus (**93**)

HB 15.8–17.7ins; T 13.8–15.8ins; WT 2.9–4.6lb.

Easily distinguished from other hare wallabies by banded coloration and naked muzzle; fur very thick, long, and soft. Unlike other hare wallabies this one is sociable, often congregating under bushes during the day and emerging at night to feed on plants and fruit. Formerly occurred on mainland Australia, in South Australia, and SW Australia; the last mainland specimen was collected in 1906. Healthy populations survive on Dorré and Bernier Islands, Shark Bay, and a reintroduction programme was commenced on Dirk Hartog Island, in 1974; it had been exterminated by the introduction of sheep in the 1920s. Legally protected, its remaining range is within nature reserves. Have been bred in captivity for the Dirk Hartog Island reintroduction programme. After a fire in 1973 on Dorré Island, 4 males and 7 females were transferred to enclosures on Dirk Hartog; by 1976, 3 pairs of the colony of 17 males and 18 females were released.

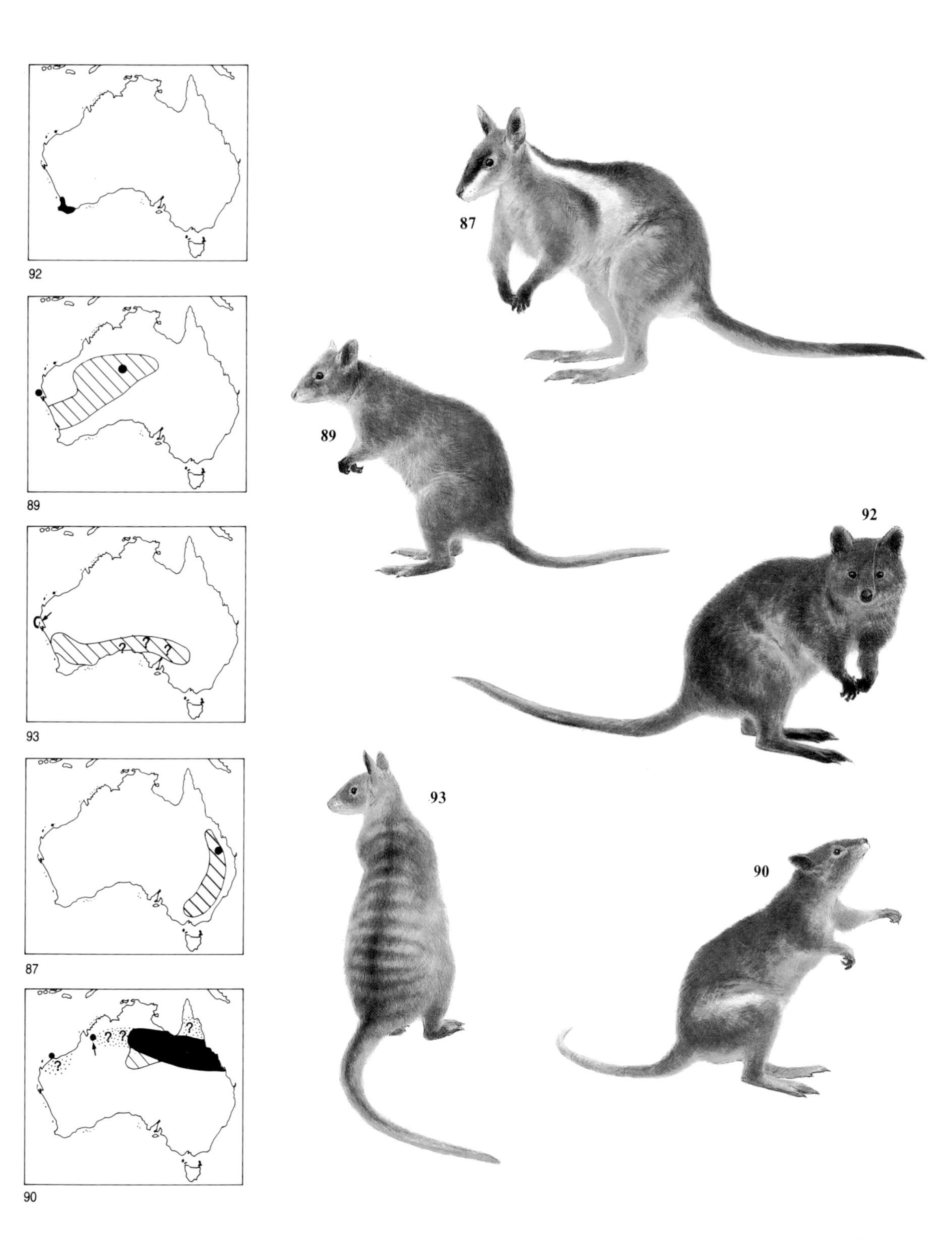
92
89
93
87
90
?
87
89
92
93
90

*** GREAT KANGAROOS**
Macropus spp.
HB up to 90.5ins; T up to 39.5ins; WT up to 132lb.
The **Eastern Grey** *M.giganteus* **(94)**, the **Western Grey** *M. fuliginosus* **(95)** and the **Red Kangaroo** *M.rufa* **(96)**, have all been exploited for their skins. They have declined in some areas and increased in others; throughout most of their range there has been no proper management plan for cropping. Also evidence that they are threatened by hunting in some areas. Most species of kangaroo and wallaby have also been exploited and locally persecuted, but remain widespread and abundant, receiving varying degrees of protection.

***** LITTLE ROCK WALLABY or NABARLEK**
Petrogale (=Peradorcas) concinna **(97)**
HB 11.4–13.8ins; T 8.7–12.2ins; WT 2.4–3.7lb.
Very distinctive, with rufous fur above, whitish below. The teeth move forward along the jaw during the animal's lifetime and are shed in succession; a unique feature among kangaroos. The three populations have very small restricted ranges.

***** RED-BELLIED or TASMANIAN PADEMELON**
Thylogale billardierii **(98)**
HB 22.1–24.8ins; T 12.6–19ins; WT 5.3–26.5lb.
Soft, dense greyish fur, tinged with olive and reddish-orange on the belly. A social species, living in small groups. Originally found in scrub areas of South Australia, Tasmania and most of the islands in the Bass Straits. In 1863 described as 'so numerous that the thousands annually destroyed make no apparent diminution of its numbers'. In the 19th century they were exterminated on the Australian mainland, and several of the islands. Still widespread and abundant in Tasmania. Until recently extensively hunted for skins: between 1923 and 1960 some 2.5 million were exported from Tasmania. The * **Red-legged Pademelon** *T.stigmatica* **(99)** and the * **Red-necked Pademelon** *T.thetis* **(100)**, found in forests in E Queensland, E New South Wales and SC New Guinea, have declined in some parts of their range.

*** BLACK FOREST-WALLABY**
Dorcopsis atrata **(101)**
HB 28.9–39.2ins; T 11.2–15.6ins; WT 6.2–14.5lb.
Blackish-brown, or blackish above, blackish-brown below. Front limbs relatively robust;unlike most wallabies do not appear specialised for jumping. Found in rainforest on Goodenough Island, off SE Papua New Guinea, feeding mostly on leaves. Eaten by natives, and due to their restricted range are considered threatened. The related * **Papuan Forest-wallaby** *Dorcopsulus (=Dorcopsis) macleayi* **(102)** is a little-known species confined to mountainous areas near Port Moresby, Papua New Guinea, at altitudes of 3950–4250ft.

***** TAMMAR or DAMA WALLABY**
Macropus eugenii **(103)**
HB 20.5–26.8ins; T 13–17.7ins; WT 8.8–22lb
One of the smallest of the *Macropus* wallabies, greyish with reddish shoulders and a slight dorsal stripe. Found in rather dry habitats such as Casuarina scrublands, they supplement their water with succulent plants and even sea water. They live in small colonies, and there are at least 11 separate populations; on Kangaroo Island, and in the Tuttanning Reserve and the Recherche Archipelago; have also been introduced onto Kawau Island, New Zealand. Extinct over most of their mainland range, as well as on Flinders and St Peter's Islands. Breed well in captivity and occur in several protected areas.

***** BLACK WALLAROO**
Macropus bernardus **(104)**
HB 23.4–28.6ins; T 21.5–25.2ins; WT 28.7–48.5lb.
Males are dark, sooty, almost black; females paler. Live in arid rocky country in Arnhem Land, Northern Territory, Australia. At one time believed extinct, since they were not seen from 1914–1969. Although numbers are low and range restricted, they do not appear to be declining at present.

*** PARMA or WHITE-THROATED WALLABY**
Macropus parma **(105)**
HB 17.5–20.9ins; T 16–21.5ins; WT 7.1–13lb.
Dark brown with white underparts, and a white upper lip; a dark dorsal stripe as far as the mid-back. Confined to rain forest and scrub. In the early days of the colonisation of Australia, species said to be common around Wollongong, New South Wales, but after clearance of its forest habitat was declared extinct in 1932. However, in 1965 an introduced colony was discovered on the island of Kawau, near Auckland, New Zealand, which had been established in the 19th century. Animals from this colony were dispersed among the world's zoos, where they now thrive. In 1967, natural populations were rediscovered on the Australian mainland in an area between the Hunter and Clarence Rivers in New South Wales. Their future could be threatened by exploitation of the forests in which they live. They are common in captivity.

***** YELLOW-FOOTED ROCK WALLABY**
Petrogale xanthopus **(106)**
HB 18.9–25.6ins; T 22.3–27.6ins; WT 13.4–16.5lb.
Strikingly marked with yellow ears, and yellow and brown rings on tail. Extensively hunted for pelts, and goats also compete for both food and shelter in rocky caves, forcing the wallabies into the heat of the day; they are also predated by foxes. While still locally abundant their range has been considerably reduced. Although generally more widespread, the *** **Brush-tailed Rock Wallaby** *P.penicillata* **(107)** appears easily eliminated by disturbance; a colony in Warrumbungle NP was destroyed within 5 years of opening a nature trail nearby. It had disappeared from Victoria by 1905 (except for 2 small colonies) and much reduced in Eastern and South Australia. The ***** **Proserpine Rock Wallaby** *P. persephone* **(108)**, a greyish wallaby related to the Yellow-footed Rock Wallaby, was discovered in 1976, from a restricted area of Queensland rainforest. Only known from two localities, and little known of status or biology. * **Godman's Rock Wallaby** *P. godmani* **(109)** occurs in open grassy forests of N Queensland, where it appears to be declining in competition with other wallabies. The * **Black-footed Rock Wallaby** *P. lateralis* **(110)** may comprise several species, some of which have already become extinct; the remaining populations are fragmented, and in SW Australia it is increasingly rare. *** **Rothschild's Rock Wallaby** *P. rothschildi* **(111)** is little-known, occurring on islands in the Dampier Archipelago, and Western Australia. A population on Lewis Island is probably extinct. The ** **Warabi** *P. burbidgei* **(112)** was discovered in 1963 and only found in a very small area in the Kimberley region, and adjacent islands in Western Australia. No evidence of decline but range is small.

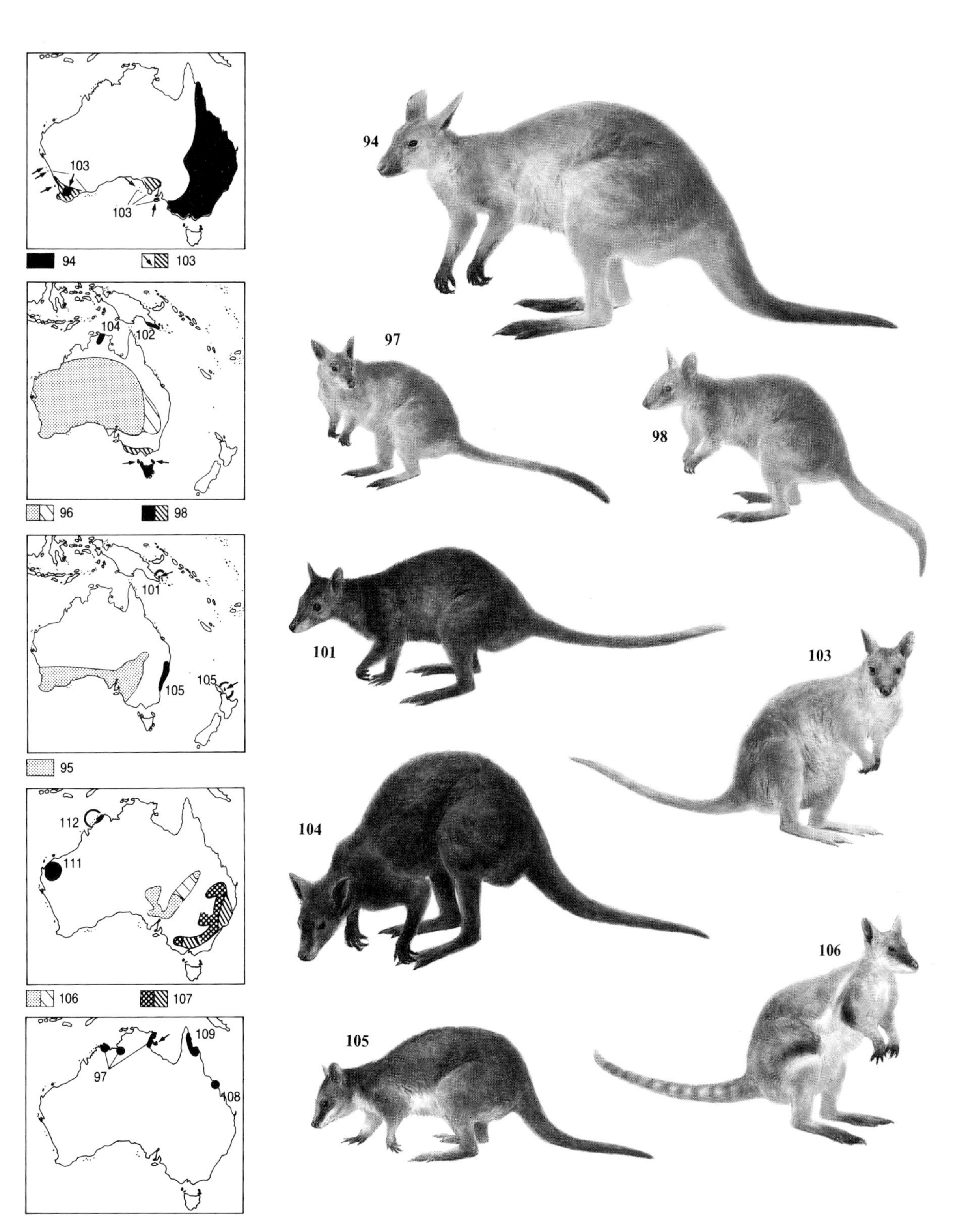
103
103
94
103
104
102
96
98
101
105
105
95
112
111
106
107
109
97
108
94
97
98
101
103
104
106
105

*** MINDANAO MOON RAT**
Podogymnura truei **(113)**
HB 5.1–5.9ins; T 1.6–2.8ins.
Known also as the **Philippines Gymnure** and the **Wood Shrew**. Fur is long and soft and they have well developed canine teeth. Found at altitudes of 5250–7550ft, on the island of Mindanao in the Philippines, where they live in forests with dense undergrowth. Believed to have habits similar to true shrews, although related to hedgehogs. Much of their forest habitat has been destroyed by logging, and also by slash-and-burn agriculture. They occur in 3 separate populations, around Mt Apo and Mt McKinley in Davao Province, and Mt Katanglad, Bukidon Province, and are protected in the Mt Apo NP.

The ** **Hainan Moon Rat** *Neohylomys hainanensis* **(114)** was not described until 1959, and little appears to be known of its status.

***** HISPANIOLAN SOLENODON**
Solenodon paradoxus **(115)**
HB 12.6ins; T 9.5ins.
Colour varies from buff and blackish to reddish, with a white spot on the nape. Found in forests and brush country, as well as around plantations. They are mainly nocturnal, hiding during the day in rock clefts, hollow trees, or burrows which they excavate for themselves. Despite complete protection, they are declining due to loss of their forest habitat, and also predation by feral cats and dogs. They still occur in both the Dominican Republic and Haiti, but are particularly threatened in the latter. Have been kept in captivity, and have lived for over 11 years (particularly long-lived for an insectivore), but there are no self-sustaining populations.

***** CUBAN SOLENODON**
Solenodon cubanus **(116)**
HB 12.2ins; T 6.7ins.
Generally dark brown, otherwise rather similar to the Hispaniolan Solenodon, but with longer, finer fur. At one time believed extinct, it has been rediscovered in the east of Cuba, but is still rare. *S.marcanoi* has been found in recent fossil deposits in Haiti and the Dominican Republic; a closely related group of insectivores, the 6 species of West Indian shrews, *Nesophontes*, are only known from skeletal remains, mostly found in owl pellets, and are believed to have become extinct in the 16th century, soon after the arrival of the Spaniards, since remains of *Nesophontes* species have been found with those of rats and mice, which were unknown before the arrival of the Spaniards. However, in the 1930s comparatively fresh remains were found, so the possibility of a small population of one or more species surviving cannot be ruled out.

*** GIANT AFRICAN WATER SHREW or GIANT OTTER SHREW**
Potamogale velox **(117)**
HB 11.4–13.8ins; T 9.7–11.4ins; WT ca.2lb.
The fur is short and dense with coarser guard hairs. In appearance it is remarkably like an otter *Lutra* sp. The powerful tail is flattened and it swims by moving it from side to side. Found near slow-flowing lowland streams, as well as fast, clear mountain streams, at altitudes of up to 5900ft. Mostly nocturnal, they hide in burrows with the entrance below water, and are very fast swimmers. They feed mainly on large invertebrates such as crabs, as well as fish and amphibians. The two young may be born any month of the year. Throughout their range they appear to be rather thinly distributed, occurring in rainforest from Nigeria to W Kenya and south to N Zambia and C Angola. The ***Dwarf Otter Shrew** *Micropotamogale ruwenzorii* **(118)**, which is less than half the size of the Giant Otter Shrew, is confined to a very small area in the Ruwenzori Mountains; although little known, with few specimens in museums, there is no particular reason to believe they are currently threatened; further information is needed.

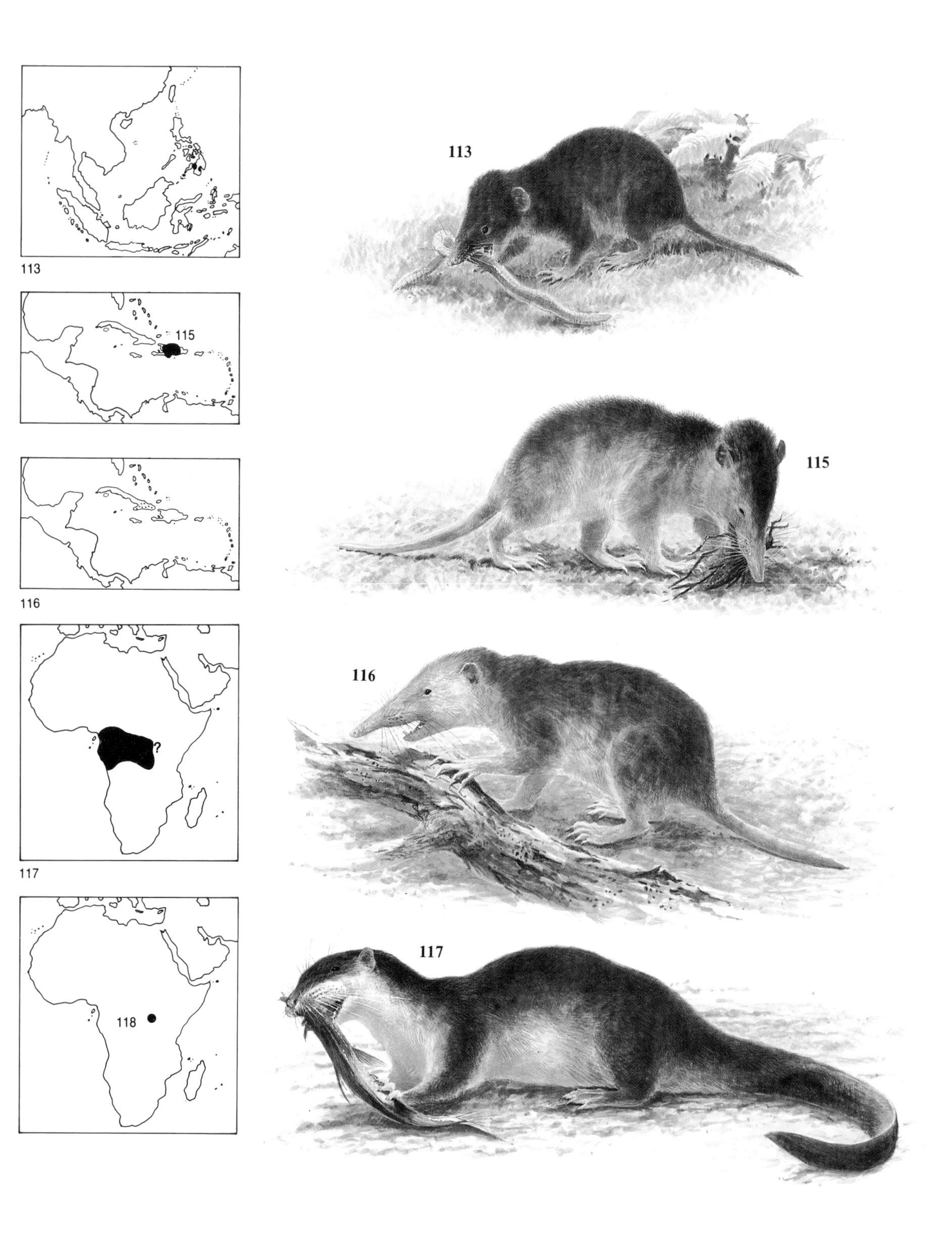
113
113
115
116
117
115
116
?
117
118
113
115
116
117

MADAGASCAN TENRECS

The Family Tenrecidae contains about 33 species, of which 29 are confined to Madagascar, where several species are threatened or even extinct. Because of their often secretive and nocturnal behaviour, little is known about many of them, and even their precise distribution is poorly documented, with very little published information about the majority of species. Most species can be bred in captivity, although those kept in zoos are generally the more widespread and abundant species. The main threats come from extensive deforestation as well as hunting and the increasing use of pesticides.

** RICE TENRECS or VOALAVONARABO

Oryzorictes spp.
HB 3.3–5.1ins; T 1.2–2ins.
Rather mole-like, dark brown or greyish-brown above, paler below. They are modified for a burrowing existence, inhabiting marshy areas, including the embankments of rice paddies, where they may be threatened by increasing use of pesticides, and drainage of their natural habitat. *O.hova* **(119)** is found in C Madagascar; *O.talpoides* **(120)** in NW Madagascar; and *O.tetradactylus* **(121)** in C Madagascar.

SHREW-LIKE TENRECS

Microgale spp.
About 18 species are known, each occurring in a relatively restricted part of Madagascar, in a particular microhabitat. Some are arboreal, others terrestrial. Because of their high degree of specialisation, several species are probably extremely vulnerable to alterations to their forest habitat. They are also apparently vulnerable to competition from the introduced Black Rat *Rattus rattus* and the Oriental House Shrew *Suncus murinus*. The following are believed to be under possible threat: ** *M.crassipes* **(122)**, known only from a single locality near Tananarive; ** *M.melanorrhachis* **(123)**, which is only known from two localities in E Madagascar; ***** *M.pusilla* **(124)**, which is only known from the Ikongo Forest and has not been recorded this century; and ** *M.longicaudata* **(125)** and ** *M.cowani* **(126)**, which are only known from the Ankafana Forest in E Madagascar.

*** WEB-FOOTED TENREC or VOALAVONDDRANO

Limnogale mergulus **(127)**
HB 4.7–7.9ins; T 4.7–6.3ins.
They are specialised for an aquatic existence, having webbed feet and a laterally compressed tail, and well developed vibrissae or whiskers. Found in streams, marshes and lakes at altitudes of 1950–6550ft, where they feed on aquatic plants, as well as crayfish and other aquatic invertebrates and fish. Their distribution and status are poorly known, but they would be badly affected by the pollution of rivers and silting.

***** *Dasogale fontoynonti* (128)

HB 3.7ins; T vestigial.
A spiny hedgehog-like tenrec, known only from two poorly preserved specimens described in 1928, from the forests of E Madagascar. It has been suggested that they may be young *Setifer setosus*, one of the most widespread and abundant species of tenrec.

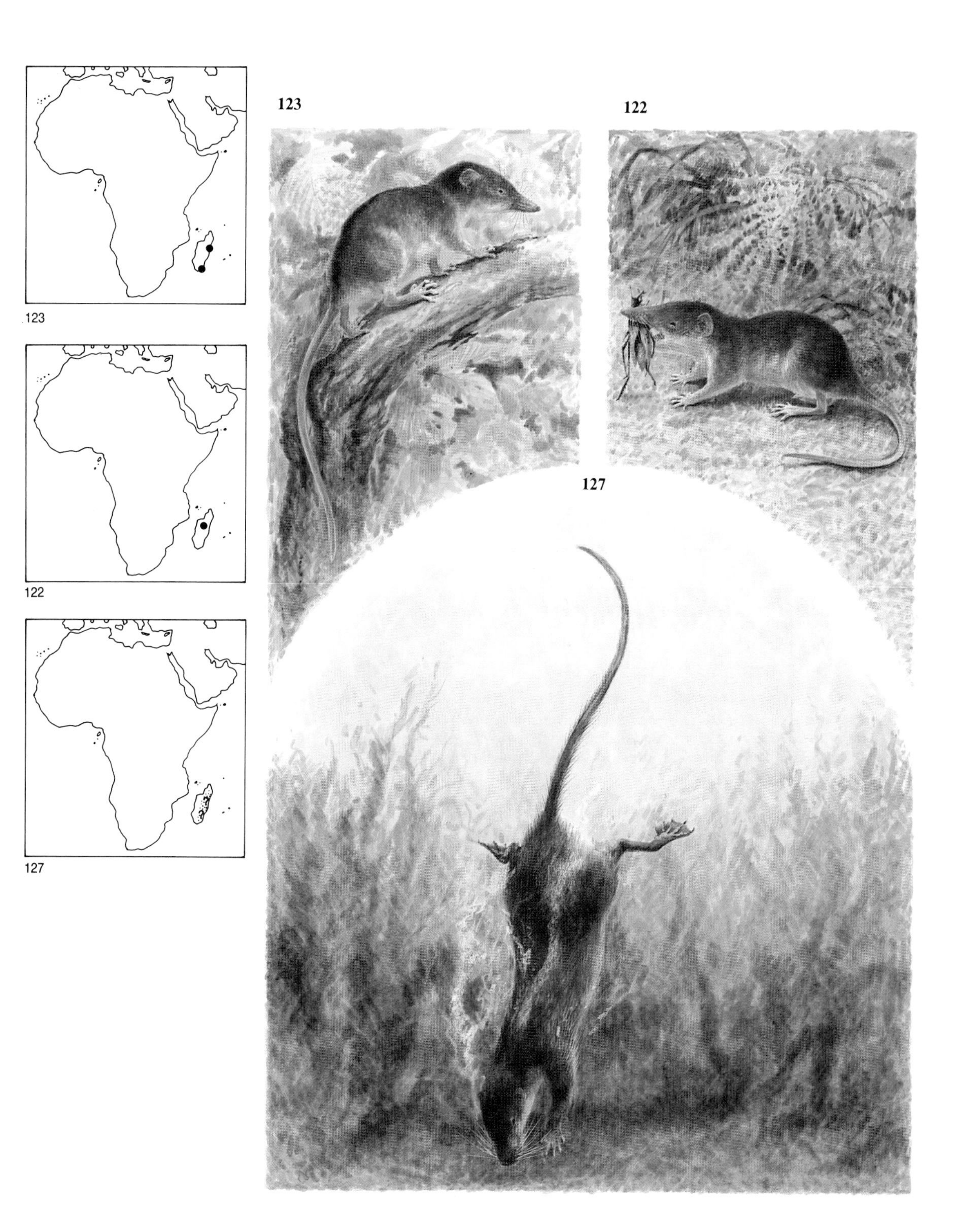
123
122
127
123
122
127

GOLDEN MOLES

Chrysochloris spp.

HB 3.5–5.5ins; T vestigial; WT up to 1.9oz.

Three species known: ** **Visagie's Cape Golden Mole** *C.visagei* **(129)** is only known from one specimen collected in Cape Province, South Africa, described in 1950; it is probably closely related to the widespread Cape Golden Mole *C.asiatica*. * *C.stuhlmanni* **(130)** is found in N and E Zaire, W Uganda, Kenya and E Tanzania. Although the range is wide, some populations are fairly isolated, but there is no evidence to suggest that they are seriously threatened, although they may disappear from areas being cultivated.

***** DE WINTON'S GOLDEN MOLE

Cryptochloris wintoni **(131)**

HB ca.3.5ins; T vestigial.

Upperparts yellowish-fawn, with a silvery sheen. Confined to a very restricted area near Port Noloth, Cape Province, South Africa, where they live in coastal dunes in burrows up to 100ft long. They feed on insects and their larvae; also legless lizards, *Typhlosaurus vermis*. Believed to desert their runs when disturbed, and because of their rarity and restricted range are considered threatened.

*** VAN ZYL'S GOLDEN MOLE

Cryptochloris zyli **(132)**

HB ca.3ins; T vestigial.

Dark grey above with a violet sheen, slightly paler below, and whitish on sides of face. Only known from a single specimen described in 1938, from near Lambert's Bay, Cape Province, South Africa.

*** GIANT GOLDEN MOLE

Chrysospalax trevelyani **(133)**

HB up to 9.1ins; T vestigial.

By far the largest of the golden moles; glossy brown above with two dull yellowish patches where the eyes should be. Found in and around forested areas where they excavate burrows up to 15.5ins deep, pushing up mounds of earth sometimes 23.5ins in diameter. They feed on giant earthworms *Microchaetus* spp. Nothing is known of their breeding habits. Confined to a series of isolated populations in South Africa, from East London, Cape Province, to Port St John, Transkei.

*** GRANT'S DESERT GOLDEN MOLE

Eremitalpa granti **(134)**

HB 3–3.5ins; T vestigial; WT ca.0.5oz.

The fur is long and silky, pale greyish-yellow with a silvery sheen above, and paler and more yellow below. They are nocturnal, living in sand dunes, foraging over distances of up to 3.7mls. In loose sand they 'swim', and their runs disappear behind them; in firmer ground they dig tunnels as deep as 17.5ins. They feed on a wide range of small animals up to the size of geckos, crickets and large beetles; also moths, spiders, legless lizards, termites and ants. Little is known of their breeding habits. *E.g.namibensis* is confined to the coastal areas of the Namib Desert in Namibia and South Africa, *E.g.granti*, from Walvis Bay, south to W Cape Province. They are thought to be strictly nocturnal, and their remains are occasionally found in the food pellets of the Barn Owls *Tyto alba*.

AFRICAN GOLDEN MOLES

Amblysomus spp.

HB 3.4–5.1ins; T vestigial.

There are about 10 species of African golden mole, some widespread and abundant, others with very resticted ranges.

** **Duthie's Golden Mole** *A.duthiae* **(135)** is very dark with a greenish sheen, and only known from a few specimens from coastal South Africa, between Knysna and Port Elizabeth; nothing is known of its habits, other than its occurrence in sandy soils.

** **Arend's Golden Mole** *A.arendsi* **(136)** is also very dark or glossy black, with a greenish or purple sheen; occurring in a very small area on the borders of Zimbabwe and Mozambique, south of the Zambezi River, in montane grasslands.

** **Gunning's Golden Mole** *A.gunningi* **(137)** is golden brown above with a rich brownish-bronze sheen. Only known from a very restricted area in E Transvaal, South Africa. From its description in 1908 until the mid-1970s, it was only known from the Woodbush Forest; in 1974 it was discovered in the Agatha Forest Reserve, 12.5mls away. Found in both montane forest and grasslands.

** **Juliana's Golden Mole** *A. julianae* **(138)** is rather small cinnamon-brown above, but rather variable. Found in sandy soils and open country. Only known from three localities in South Africa: Willows, Pretoria; Numbi Gate; and the Kruger NP.

** **Somali Golden Mole** *A. tytonis* **(139)** is only known from a single specimen described in 1968 from Giohar, Somalia.

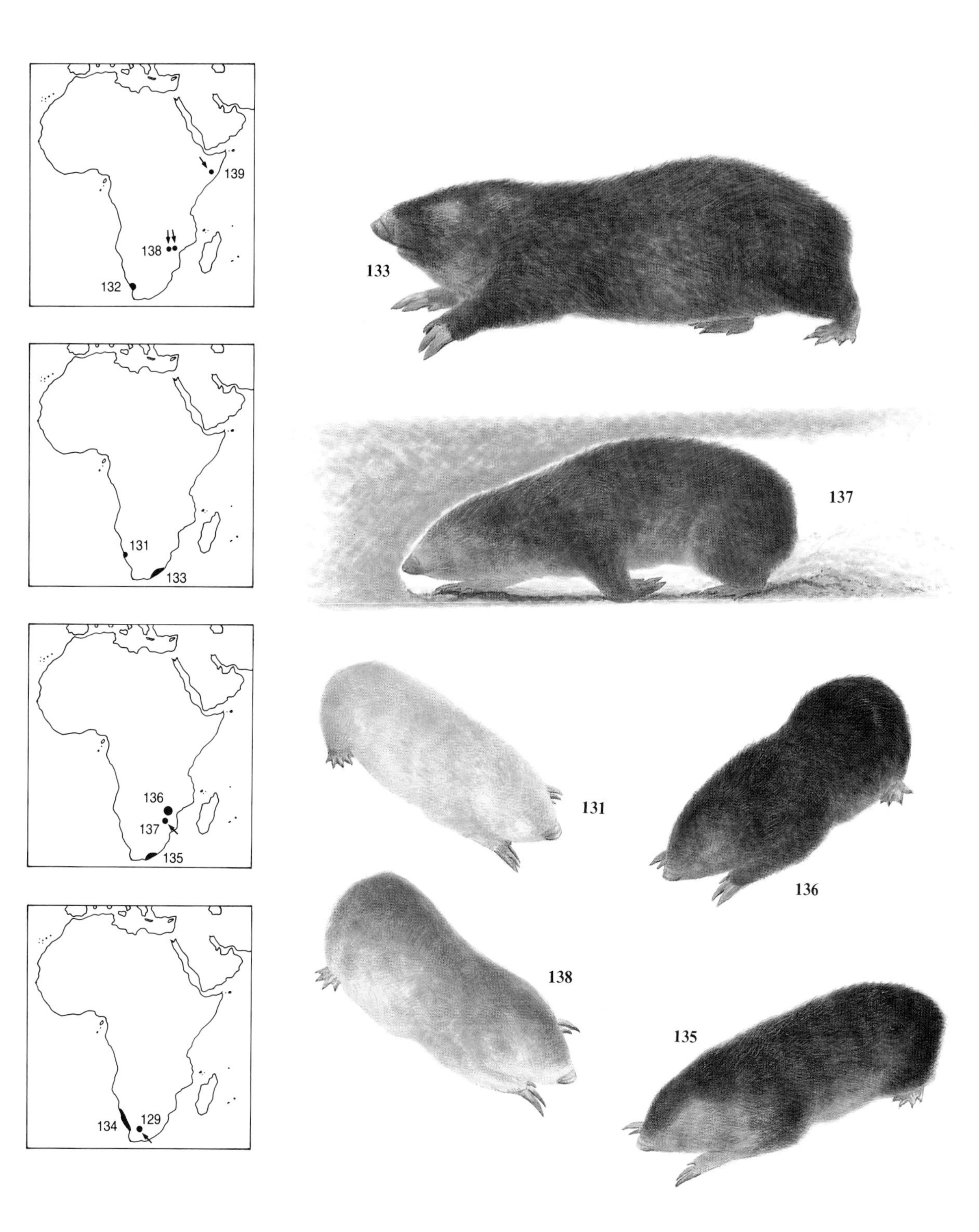
139
138
132
133
131
133
137
136
137
135
131
136
138
135
134
129

***** HOY'S PYGMY SHREW**
Microsorex hoyi (**140**)
HB ca.2ins; T 1.1–1.4ins; WT 0.08–0.13oz.
The smallest and one of the rarest North American mammals, reddish-brown above, pale below, greyer in winter. Found in woods, open bush country and bogs from Alaska and most of Canada, south to a few northern States of the USA. Very little known of precise distribution or habits. Those found in Appalachians may be a separate species, *M. winnemana*.

**** ENDER'S SMALL-EARED SHREW**
Cryptotis endersi (**141**)
Only known from Bocas del Toro, Chiriqui, Panama. Superficially rather similar to *Sorex* shrews.

***** AMERICAN WATER SHREW**
Sorex palustris (**142**)
HB 3–3.6ins; T 2.5–2.8ins; WT 0.4–0.7oz.
Has dense fringe of hairs on hind feet. Blackish above, paler below. Found along banks of streams in wooded areas; semi-aquatic. Although it covers a wide range in North America, it is much fragmented, and in some states is considered threatened.

*** MERRIAM'S SHREW**
Sorex merriami (**143**)
HB ca.2.5ins; T 1.3–1.5ins; WT 0.14–0.24oz.
Pale ash-grey above, pale buff or whitish below. A little-known shrew found in isolated populations in western USA from North Dakota to New Mexico, in sagebrush, grasslands and woodland; often found in association with Sagebrush Voles *Lemniscus curtatus*.

*** SOUTH-EASTERN or BACHMAN'S SHREW**
Sorex longirostris (**144**)
HB 2–2.7ins; T 0.8–1.7ins; WT 0.07–0.14oz.
Similar to the more common Masked Shrew *S. cinereus*, but with longer, hairier tail. Although discovered in 1837, it remains a little-known species, found in moist woods, fields and brush country in SE USA. Recently discovered to be fairly abundant in the Carolinas. Due to changes to its habitat, the Dismal Swamp population *S. l. fischeri* is being genetically swamped by the more numerous *S. l. longirostris*. There are several other North American species and numerous subspecies of shrews with restricted or isolated ranges which may be vulnerable to ecological change, particularly the introduction or spread of closely related species or predators. * **Preble's Shrew** *S. preblei* (**145**) has a rather restricted range and is considered threatened in Oregon, USA. The ** **Ashland Shrew** *S. trigonirostris* (**146**) is confined to the western slope of the Grizzly Mountains, Ashland, Oregon. The ** **Suisun Shrew** *S. sinuosus* (**147**), an almost completely black species, is confined to Grizzly Island and adjacent marshes, Solano County, California. ** **Socorro Shrew** *S. juncensis* (**148**) is only known from Socorro, Baja California, Mexico. The ** **Mt Lyell Shrew** *S. lyelli* (**149**) has a restricted range in Tuolumne and Mono Counties, California. The ** **Pribilof Shrew** *S. pribilofensis* (**150**) is only known from St Paul, one of the Pribilof islands west of Alaska, and ** **St Lawrence Shrew** *S. jacksoni* (**151**) is only known from St. Lawrence Island, Alaska. The * **Rock Shrew** *S. dispar* (**152**) is a rare North American species, found in montane areas from Maine to Carolina and Tennessee; the closely related ** **Gaspar Shrew** *S. gaspensis* (**153**) is only known from SE Quebec, New Brunswick, Nova Scotia and Cape Breton Island. ** *S. stizodon* (**154**), ** *S. sclateri* (**155**) and ** *S. milleri* (**156**) all have restricted ranges in Mexico.

*** SOUTHERN SHORT-TAILED SHREW**
Blarina carolinensis (**157**)
Closely related to the more common and widespread Short-tailed Shrew, *B. brevicauda*, it is a little-known species found mainly in woodlands in SE USA; *B. c. shermani*, known only from a single locality in SW Florida, may already be extinct.

**** BORNEAN MUSK (BLACK) RAT**
Suncus ater (**158**)
HB 3ins; T 2.2ins.
Dark blackish-brown above, slightly paler below. Only known from one specimen at 5600ft, from Mt. Kinabalu, Borneo.

**** KELAART'S LONG-CLAWED SHREW**
Feroculus feroculus (**159**)
HB 4.2–4.6ins; T 2.2–2.9ins; WT ca.1.2oz.
Uniformly slaty or blackish above, paler below, with whitish forefeet. Very few specimens have been seen since its discovery in 1850, in the central mountains of Sri Lanka, at altitudes of 6050–7050ft. * **Pearson's Long-clawed Shrew** *Solisorex pearsoni* (**160**) is also restricted to the central highlands of Sri Lanka, at altitudes of 3600–6050ft, and has rarely been seen.

**** MOUSE-SHREWS**
Myosorex spp.
M. geata (**161**) is known only from two specimens from the Uluguru Mountains, Tanzania; *M. zinki* (**162**) is confined to the higher altitudes of Mt Kilimanjaro, Kenya/Tanzania; *M. norae* (**163**) is confined to the Aberdare Mountains, Kenya; *M. polulus* (**164**) is confined to Mt.Kenya, but is abundant within its limited range; *M. preussi* (**165**) is only known from a single specimen from W Cameroon; *M. polli* (**166**) is known only from one locality in Kasai, Zaire. *M. longicaudatus* (**167**) is confined to a small area in the extreme southern tip of Cape Province, South Africa; none known to be in any immediate danger despite very restricted ranges.

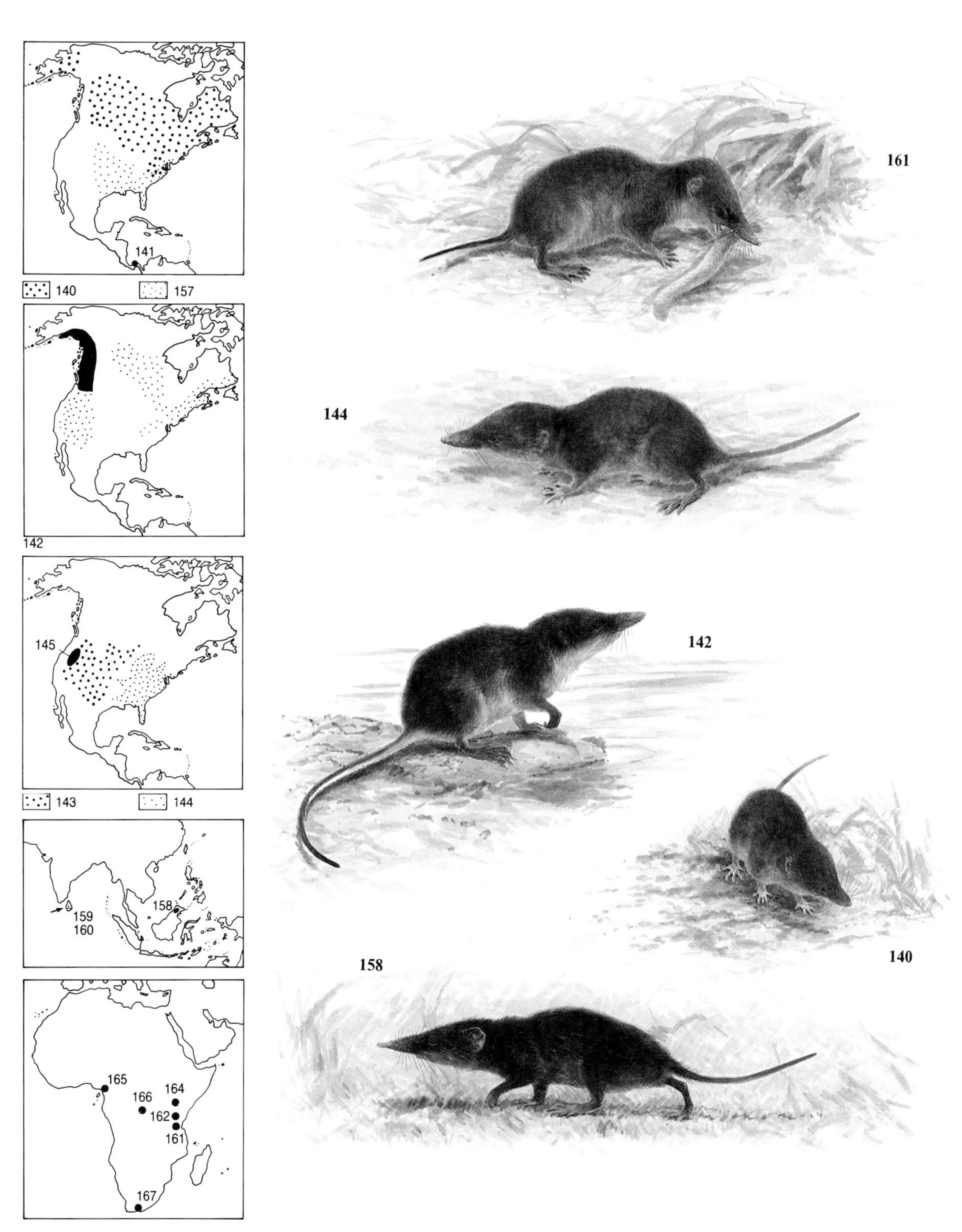

141
140
157
142
145
143
144
158
159
160
165
166
164
162
161
167
161
144
142
140
158

** WHITE-TOOTHED SHREWS

Crocidura spp.

The following all have restricted ranges and virtually nothing is known of their status in the wild. Some may prove to be more widespread, while others could merely be varieties of better known species. However, some may be endangered or even already extinct. *C. beatus* **(168)** from Mindanao, Philippines; *C. bloyeti* **(169)** from Kondoa, Irangi, in Tanzania; *C. bovei* **(170)** from Vivi in Zaire; *C. caliginea* **(171)** from Medje in Zaire; *C. cinderella* **(172)** from Gemenjulla, Gambia; *C. crenata* **(173)** from Belinga and Makokou, Gabon; *C. edwardsiana* **(174)** from Jolo, Sulu Islands, Philippines; *C. eisentrauti* **(175)** from Mt Cameroon, Cameroon; *C. elongata* **(176)** from near Lake Tondano, Sulawesi, Indonesia; *C. grandis* **(177)** from Grand Malindang Mountain, Mindanao, Philippines; *C. lea* **(178)** from Temboan, Sulawesi, Indonesia; *C. luluae* **(179)** from Kananga, Zaire; *C. macowi* **(180)** from Mt Nyiro, near Lake Turkana, Kenya; *C. mindorus* **(181)** from Mt Halcon, Mindoro, Philippines; *C. palawanensis* **(182)** from Sir J.Brooke Point, Palawan, Philippines; *C. parvacauda* **(183)** from Cotabato, Mindanao, Philippines; *C. phaeura* **(184)** from Mt Guramba, Sidamo, Ethiopia; *C. picea* **(185)** from Assumbo, Cameroon; *C. susiana* **(186)** from Dezful, SW Iran; *C. tephra* **(187)** from Torit, Sudan; *C. vulcani* **(188)** from Mt Cameroon, Cameroon; *C. zimmeri* **(189)** from Bukoma, Lualaba River, Zaire; *C. levicula* **(190)** only known from Pinedapa, Sulawesi, Indonesia. Many others have ranges that are almost as restricted; those cited are examples only.

** KANSU MOLE

Scapanulus oweni **(191)**

HB 3.9–4.1ins; T 1.4–1.5ins.

A dull silvery grey, superficially similar to the American moles *Scapanus* spp and *Scalopus* spp.. Very little is known about their habits or status, and only a few specimens are known. Apparently occurs in fir forest on the border of Kansu, Shensi, Szechuan and Tsinghai in C China, at an altitude of 8850ft. This species may prove to be more widespread and abundant when more closely studied.

*** RUSSIAN DESMAN

Desmana moschata **(192)**

HB 7.1–8.7ins; T 6.7–8.5ins.

Best described as an aquatic mole, the desman has numerous adaptations for its aquatic life: flattened tail, dense fur, webbed hind feet, with a fringe of bristles on the long, flexible snout. Inhabiting freshwater streams and lakes, it excavates a den among bushes, above water level, but with the entrance submerged, ventilation provided through roots. It feeds on insects, crustaceans, fish, amphibians and other small animals. Fairly social, having up to 5 young and with as many as 8 living in a den. Fossil remains indicate that Russian Desmans were once found over most of Europe as far west as the British Isles, but within historic times have been confined to Russia. Until the end of the last century they were fairly abundant in the Don, Volga and Ural river systems, but demand for their skins, together with extensive drainage, drastically reduced their populations. In more recent years pollution and further drainage, as well as competition from coypu and muskrats, have affected them. Now protected, and have also been introduced into the Dniepr and Ob river systems.

* GREATER JAPANESE MOLE

Talpa robusta **(193)**

HB up to 8ins; T vestigial.

A little-known species of mole. The taxonomy of this, and several closely related species is far from clear; the main differences are in dentition. It is sometimes placed in the genus *Mogera*. The precise ranges of this and its close relatives are imperfectly known, but in Siberia it is considered threatened. A closely related species *T.wogura* occurs on Japanese islands and Taiwan; their relationships need clarification.

*** PYRENEAN DESMAN

Galemys pyrenaicus **(194)**

HB 4–6.1ins; T 4.9–6.1ins; WT 1.2–2.8oz.

Similar in general appearance to the Russian Desman (192) but smaller with a proportionally longer snout, and a round tail. Pyrenean Desmans are not social and will fight, even with members of the opposite sex. They have 2–4 young and live to at least 3.5 years in the wild. Food consists of a wide variety of small aquatic animals, which are brought to land before eating. Their range includes the French and Spanish Pyrenees, and the mountains of N Spain and Portugal. They have disappeared from a large part of their former range and are still declining, due to pollution of the streams in which they survive.

** LONG-TAILED MOLE

Scaptonyx fusicaudus **(195)**

HB 2.6–3.5ins; T up to 2ins.

In appearance, resembles a cross between a mole and a shrew, with velvety fur, dark slate in colour, and large paws. Nothing is known of its habits and only a few museum specimens have been collected since its discovery in 1872. Has been found on Tsinghai, Shensi, Szechuan and Yunnan, in China, and in N Burma. Possibly more abundant than records would indicate.

* YELLOW-RUMPED ELEPHANT SHREW

Rhynchocyon chrysopygus **(196)**

HB 9.3–12.4ins; T 7.5–10.4ins; WT ca.1.2lb.

Long-legged with a long snout; rich tawny above with a golden-yellow rump. Closely related to the Giant Elephant Shrew *R. cirnei*, and often regarded as merely a well marked subspecies. Confined to two very small areas on the Kenyan coast. Feed mainly on ants and termites, as well as beetle larvae and other insects. They give birth to a single offspring, but bear young 4 or 5 times a year. Not thought to be in any immediate danger, but a very restricted range makes them vulnerable to habitat change and other factors.

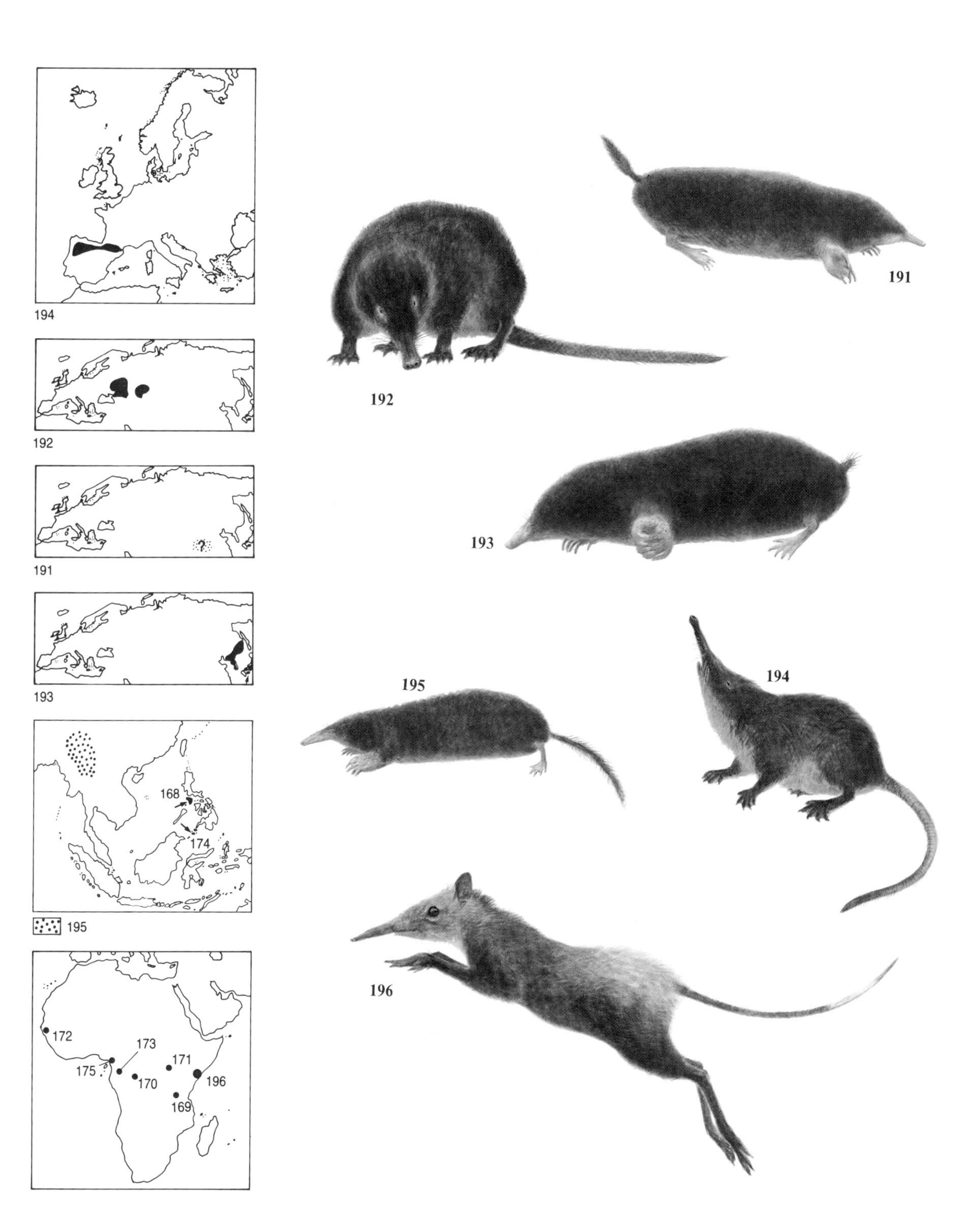
194
192
191
193
168
174
195
172
173
171
175
196
170
169
191
192
193
195
194
196

*** SAO THOME LITTLE COLLARED FRUIT BAT**
Myonycteris brachycephala (**197**)
Likely to be vulnerable because of its restricted distribution. The closely related and little-known ***Relict Little Collared Fruit Bat** *M. relicta* (**198**) has a restricted range in SE Kenya and Tanzania.

**** BONE'S FRUIT BAT**
Boneia bidens (**199**)
HB 5.4ins; T 1in.; FA 3.7ins.
This little-known fruit bat is dark brown on the back with a russet rump, paler underneath and golden brown around the neck. They are closely related to *Rousettus* bats. Bone's Fruit Bat is only known from specimens collected in the Bone Mountain range, near Gorontala and Menado in N Sulawesi.

***** INDIAN OCEAN ISLAND FLYING FOXES**
Pteropus spp.
The largest of the bats, they have a wingspan of up to 5.6ft and weigh up to 2lb. Although found on islands in the W Indian Ocean, they have not managed to colonise the African mainland. They feed on soft fruit and often roost in large groups. The species are as follows: *P. rodricensis* (**200**) found only on Rodiguez, Mascarenes, where it is seriously endangered, but there is a flourishing captive colony at Jersey Zoo. *P. niger* (**201**) found on Mauritius and extinct on Reunion, Mascarenes. *P. subniger* was found on Mauritius and Reunion, Mascarene and is now extinct. *P. rufus* (**202**) is confined to Madagascar; *P. seychellensis* (**203**) is found on the Seychelles, Aldabra, Comoros and Mafia, off NE Tanzania. *P. livingstonei* (**204**) is confined to Johanna Island in the Comoros is considered highly endangered; *P. voeltzkowi* (**205**) is found only on Pemba, off NE Tanzania. Although not all are endangered, they should probably all be considered vulnerable, due to their extremely restricted and isolated ranges.

*** PACIFIC OCEAN ISLAND FLYING FOXES**
Pteropus spp.
Similar in general appearance to the fruit bats of the Indian Ocean, several species are endangered, and because of their isolation many others are vulnerable. These include the following: *P. mariannus* (**206**) from Okinawa, Ryukyu Islands, Guam, Rota and Saipan; this species is heavily exploited for food on Guam (USA), and has decreased markedly, but lacks any protection; *P. tonganus* (**207**) occuring on several islands, including Karkar, the Solomons, New Hebrides, Tonga, Samoa, Niue, and the Cook Islands has an almost unknown status but a marked decline, associated with forest destruction and the introduction of shotguns, has been noted on Niue. *P. tokudae* (**208**) is confined to Guam (USA), where it is on the verge of extinction, having been hunted for food despite legal protection. Others with restricted ranges which would be vulnerable to hunting include *P. ornatus* (**209**) and *P. vetulus* (**210**) from New Caledonia; *P. auratus* (**211**) from the Loyalty Islands; *P. pelewensis* (**212**) and *P. pilosus* (**213**) from Palau Islands; *P. yapensis* (**214**) from Yap and Mackenzie Islands; *P. ualanus* (**215**) from the E Carolines; *P. phaeocephalus* (**216**) from Mortlock in the Carolines; *P. molossinus* (**217**) from the Carolines; *P. vanikorensis* (**218**) and *P. tuberculatus* (**219**) from Vanikoro Island; *P. pselaphon* (**220**) from Bonin Islands; *P. nitendiensis* (**221**) from Santa Cruz Islands; *P. fundatus* (**222**) from Banks' Islands; *P. samoensis* (**223**) from Fiji and Samoa; *P. howensis* (**224**) from Lord Howe Island. In the western Pacific there are several other little-known populations.

*** SE ASIAN FLYING FOXES**
Pteropus spp.
Like the Indian Ocean and Pacific Ocean, the islands of the Philippines and other parts of SE Asia also have populations of fruit bats, many of which have very restricted ranges. Little is known of their distribution, and it is likely that many of the populations on smaller islands are threatened, both by hunting and by destruction of their habitat. The following are among those that have relatively restricted ranges in the Philippines: *P. speciosus* (**225**), *P. mearnsi* (**226**), *P. pumilus* (**227**), *P. balutus* (**228**), *P. tablasi* (**229**). In Sulawesi and nearby islands there are also several little-known species of fruit bat which may be threatened.

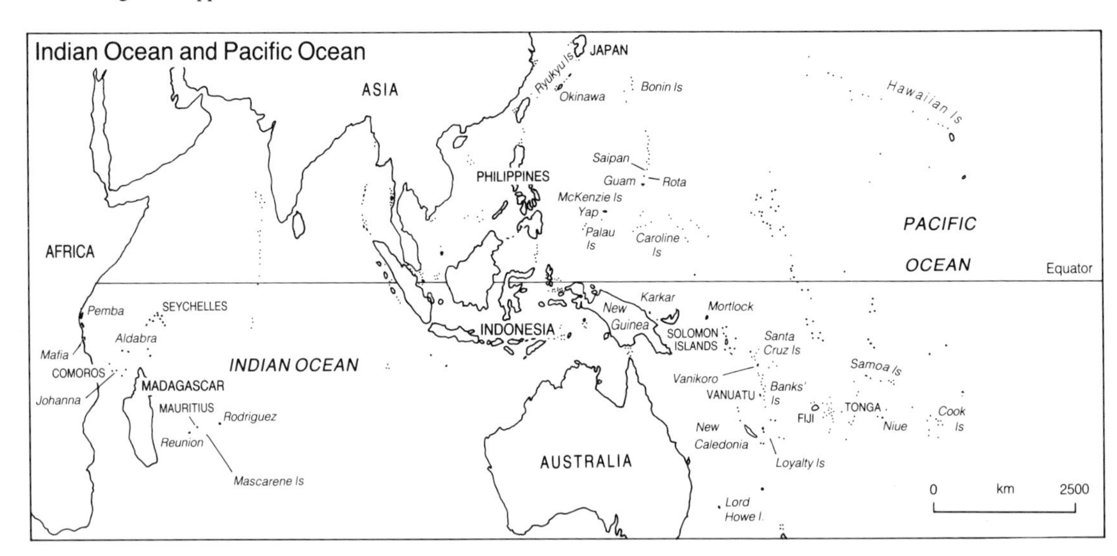

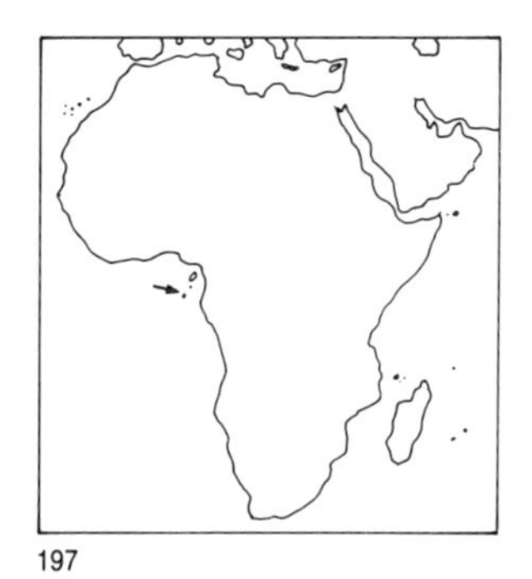
197

199

**** BARE-BACKED FRUIT BAT**
Neopteryx frosti **(230)**
HB 4.1ins; T absent; FA 4.3ins.
The fur is short, thick and reddish-brown, with a creamy and sepia muzzle. The wing membranes spread to near the middle of the back, so that it appears to have bare skin on the back. Only known from specimens collected in the 1930s at 3300ft altitude at Tamalanti in W Sulawesi.

***** FOSSIL FRUIT BAT**
Aproteles bulmerae **(231)**
The only recent specimen was found in 1975, in the Hindenburg Ranges of W Papua New Guinea, killed by a native hunter. It has subsequently been lost, and the species is only known from fossil remains. The large numbers of these, 9000–12,000 years old, found at sites of human habitation, suggest that this species was exterminated by human hunting. Fossil remains are known from the central highlands of Papua New Guinea. It is thought that some animals may survive in remoter parts of Papua New Guinea.

**** SHORT-PALATE FRUIT BAT**
Casinycteris argynnis **(232)**
HB 3.5–3.7ins; T vestigial; FA 2.2–2.5ins.
The fur is light brown with the muzzle, eyelids and ears orange-brown; there are tufts of white hair at the base of the ears. Has been found from southern Cameroon to eastern Zaire, and although only known from a handful of specimens, may be more common.

**** SHORT-NOSED FRUIT BAT**
Thoopterus nigrescens **(233)**
General colour is greyish-brown, and it is almost tail-less. Only three specimens are known, one from Morotai Island in Maluku and two from Sulawesi. There are also unconfirmed reports from the Philippines.

******* HALCON FRUIT BAT**
Haplonycteris fischeri **(234)**
T absent; FA ca.2ins.
The only specimen known is brownish with a silvery tinge in the middle of the belly. It has not been recorded since its discovery in 1937 on the slopes of Mt Halcon, Mindoro, in the Philippines.

**** LONG-HAIRED FRUIT BAT**
Otopteropus cartilagonodus **(235)**
HB 2.2–2.7ins; T absent; FA 1.8–1.9ins.
The general colour is dark brown above, paler and greyer below. The hairs of the underside have white tips. The eyes are large, and the ears are marked with reddish thickenings.The few specimens known are all from mountainous regions in Sitio Pactil, NC Luzon, where it was first described in 1969, and more recently in Isabela Province in Luzon. It may prove to be more widespread.

** *Alionycteris paucidentata* **(236)**
HB ca.2.5ins; T absent; FA 1.7–1.8ins.
Small, long-haired (presumably adapted for high altitudes), brownish-black above with thickly-haired legs and feet, and dark ears. Only known from Mt Katanglad, Bukidion, in Mindanao, Philippines.

*** FRUIT BATS**
Acerodon spp.
Externally these bats are almost indistinguishable from *Pteropus* spp., but have different dental characters. *A. celebensis* **(237)** is restricted to the islands of Sulawesi, Saleyer and Sula Mangoli; *A. humilis* **(238)** to the Talaud Islands between Sulawesi and Philippines; and *A. lucifer* **(239)** to Panay Island (Philippines). Little known of their status.

*** BARE-BACKED FRUIT BATS**
Dobsonia spp.
HB up to 1.2ins; T up to 1.2ins; WT up to 1.1lb.
Variable in colour, but mostly brownish, olive or blackish. Unlike most other fruit bats they are often found in caves and also in tree holes; they feed on tree fruits inforests. They occur on many islands throughout SE Asia, to the islands around New Guinea, and N Australia. Although large colonies are known for some species, the rapid destruction of forest cover on many islands has probably caused several populations to suffer. They are also extensively hunted for human food in much of the Pacific region. Although one of the more widespread species, *D. moluccensis* **(240)**, in Australia, the subspecies *D. m. magna*, is restricted to a small area of rainforest, and has only rarely been encountered.The **** Polillo Dog-faced Fruit Bat** *Cynopterus archipelagus* **(241)** is known only from a single specimen from the island of Polillo in the Philippines; it may prove to be a form of *C. brachyotis*.

*** SALIM ALI'S FRUIT BAT**
Latidens salimalii **(242)**
FA ca.2.7ins.
Medium-sized bat (FA ca.2.7ins), lacking a tail. The fur is brownish and greyish, and is long (up to 0.2in). Nothing is known of the species in the wild, and the only specimen ever found was in the Wavy Mountains, Madurai, in S India.

**** NORTHERN BLOSSOM BAT**
Macroglossus lagochilus (=minimus) **(243)**
A small solitary blossom-feeding bat that ranges from Sulawesi south to New Guinea, Bismarck Archipelago and N Australia. Rarely seen in Australia and precise status unknown. Occurs in coastal areas from N Western Australia, east through Northern Territory to N Queensland.

**** SULAWESI DAWN BAT**
Eonycteris rosenbergi **(244)**
A small fruit bat known only from a single specimen from N Sulawesi.

***** SEYCHELLES SHEATH-TAILED BAT**
Coleura seychellensis **(245)**
FA 2.1–2.2ins.
Probably originally occurred throughout the Seychelles, but now extremely rare, and extinct on most islands.

***TUBE-NOSED FRUIT BATS**
Nyctimene spp. **(246)**
The 10 species are found mostly on islands from Sulawesi south to N Australia, New Guinea and east to the Solomons. May be threatened by destruction of forest habitat, but some populations are able to adapt to cultivated areas, and secondary forest.

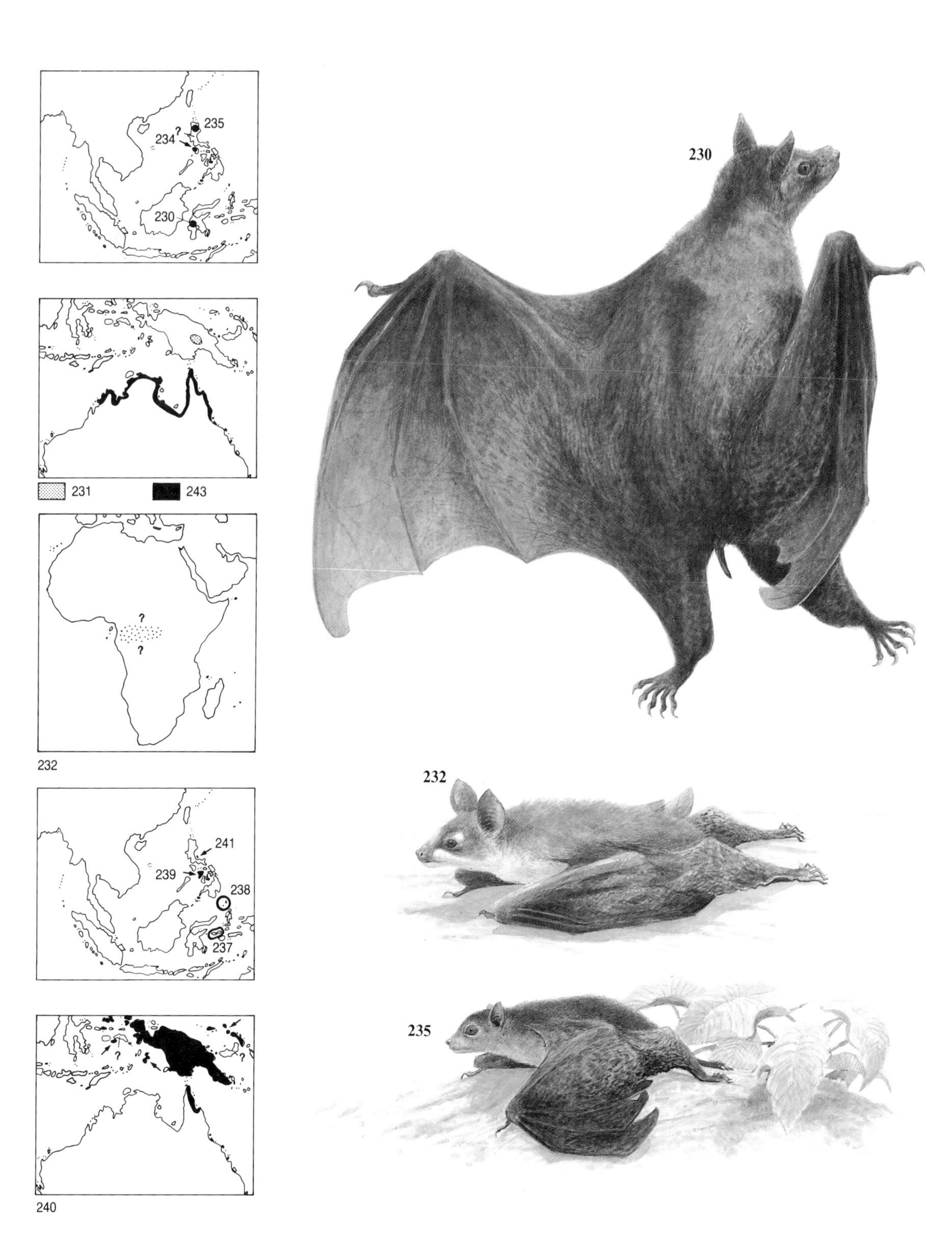
235
?
234
230
231
243
?
?
232
241
239
238
237
?
?
?
240
230
232
235

**** PACIFIC SHEATH-TAILED BAT**
Emballonura sulcata (**247**)
FA ca.1.9ins
Confined to the Mariana and Caroline Islands in the Pacific. Probably once abundant on Guam, now extremely rare, and the Rota island population *E. s. rotensis* is believed extinct.

*** AUSTRALIAN SHEATH-TAILED BATS**
Taphozous spp.
The * **Papuan Sheath-tailed Bat** *T. mixtus* (**248**) and the * **Naked-rumped Sheath-tailed Bat** *T. nudicluniatus* (**249**) both have more extensive ranges outside Australia, but in Australia are confined to Queensland where they appear to be rather rare. The * **Little Sheath-tailed Bat** *T. australis* (**250**) is rare in Australia, occuring in C and NE Queensland, but little is known of its status; it is even rarer in New Guinea. The ** **White-striped Sheath-tailed Bat** *T. kapalgensis* (**251**), a very distinctive species, brownish-orange with two white strips on the flanks, was only discovered in 1979. Although its known range is small, it may prove to be more widespread.

***** KITTI'S HOG-NOSED (BUMBLEBEE) BAT**
Craseonycteris thonglongyai (**252**)
HB 1.1–1.3ins; T absent; FA 0.9–1in; WT ca.0.07oz.
The world's smallest mammal, the size of a large bumblebee; first discovered in 1973. There are two colour phases: reddish to brownish, or greyish above. The ears are large, the face rather pig-like, and the wings relatively long and wide. They roost widely separated from each other, deep inside limestone caves, emerging to feed at dusk. Area where they occur was once forested; now cleared for agriculture. Human disturbance is the main threat (Buddhist shrines within the roost caves do not help). In 1982 the population was estimated at 160, but recently proved to be more widespread and abundant.

***** GHOST BAT (AUSTRALIAN FALSE VAMPIRE)**
Macroderma gigas (**253**)
HB 3.9–5.1ins; T absent; FA 4–4.4ins; WT 4.9–5.8oz.
With a wing span of up to 23.5ins this is the largest Microchiropteran bat. The fur on the back is white with greyish tips, and the rest of the body is whitish, although in some areas it is darker. It roosts in caves and mines and is carnivorous, preying on small insectivorous bats and mice, as well as birds, frogs and insects, but its most common food is house mice. Once widely distributed over most of Australia as far south as 34° S; huge guano deposits indicate abundance in their former haunts. However, with increasing aridity their range has contracted and they are now restricted north of 29°S. Although relatively common in a few areas threatened by human activities such as limestone quarrying; they are easily disturbed and abandon roosts. Protected by law and recorded in a few nature reserves within its range. Occasionally kept and bred in (Australian) zoos.

***** GREATER HORSESHOE BAT**
Rhinolophus ferrumequinum (**254**)
HB 2.3–2.7ins; FA 2–2.4ins; WT 0.5–1.2oz.
One of the larger species of horseshoe bat the Greater Horseshoe Bat lives in caves and mines during the winter, and in summer in roofs, barns, church towers and also caves. Females form large nursery colonies, and in winter both sexes may migrate distances of over 40mls to winter roosts. Here they congregate in large clusters in hibernacula, often used for several centuries. Found over most of Europe (except Scandinavia) through SC Asia to Japan, and south to Morocco, but within this range has undergone serious declines in many areas, particularly in NW Europe. The British population had dropped to an estimated 2200 by 1983, a 98% decline in a century. Similar declines have been noted in other parts of Europe. Extinct in the Netherlands and Poland, they are protected in all European countries, and many of the hibernacula and breeding sites are now nature reserves. Many other species of *Rhinolophus* spp. are declining due to similar factors, but most species are poorly studied. The *** **Lesser Horseshoe Bat** *R.hipposideros* (**255**) has undergone a massive decline in Germany and most other parts of Europe, * **Mehely's Horseshoe Bat** *R.mehelyi* (**256**), * **Blasius's Horseshoe Bat** *R.blasii* (**257**) and the * **Mediterranean Horseshoe Bat** *R.euryale* (**258**), although less well studied, have also been noted to be declining in most parts of their range. Blasius's Horsehoe Bat is particularly at risk in Israel, where extensive fumigation of caves to exterminate fruit bats also kills this and other species of insectivorous bats. ** *R. imaizumi* (**259**) from Iriomote Island, * *R.philippinensis* (**260**) from the Philippines (Mindoro, Luzon, Mindanao and Negros) and Kai Islands, S Sulawesi, N Borneo, Timor and NE Australia, are examples of species which have very restricted ranges or are probably declining but about which very little information is available. ** *R. marshalli* (**261**), described in 1972, is a little-known species from Chanthaburi Province, Thailand, which may prove to be more widespread, and ** *R. paradoxolophus* (**262**) is known only from Thailand and Vietnam.

**** COX'S LEAF-NOSED BAT**
Hipposideros coxi (**263**)
FA 2.1–2.2ins; WT ca.0.3oz.
Dark brownish, with nose, wings and ears very dark. Known only from SW Sarawak, Borneo.

*** RIDLEY'S LEAF-NOSED BAT**
Hipposideros ridleyi (**264**)
T 1–1.1ins; FA 1.8–1.9ins; WT 0.2–0.3oz.
Dark grey wing membranes, nose and ears. Fur dark brown. Has a very restricted range in Peninsular Malaysia where it has been found in lowland peat forests which have been heavily logged. None found between 1910 and its rediscovery in 1975, when a nursery colony was found in a culvert beneath a road. They have also been found in a very few localities in Sabah, Borneo. This species may prove to be more widespread. ** **Boonsong's Leaf-nosed Bat** *H.lekaguli* (**265**) described in 1974, but range seems more extensive than first thought.

*** CANTOR'S LEAF-NOSED BAT**
Hipposideros galeritus (inc.*cervinus*) (**266**)
HB 2–2.2ins; T 1.2–1.8ins; FA 1.9–2ins; WT 0.2–0.3oz.
Dark brownish with pink nose. Roosts in caves and although widespread in India and many other parts of SE Asia, has declined locally due to disturbance of roosts; because of its concentrations, it is particularly vulnerable. Other *Hipposideros* spp. with restricted ranges include * *H.turpis* (**267**) from the Ryukyus and Thailand; and the * **Lesser Wart-nosed Horseshoe Bat** *H.stenotis* (**268**) and the * **Greater Wart-nosed Bat** *H. semoni* (**269**) are possibly threatened in Australia.

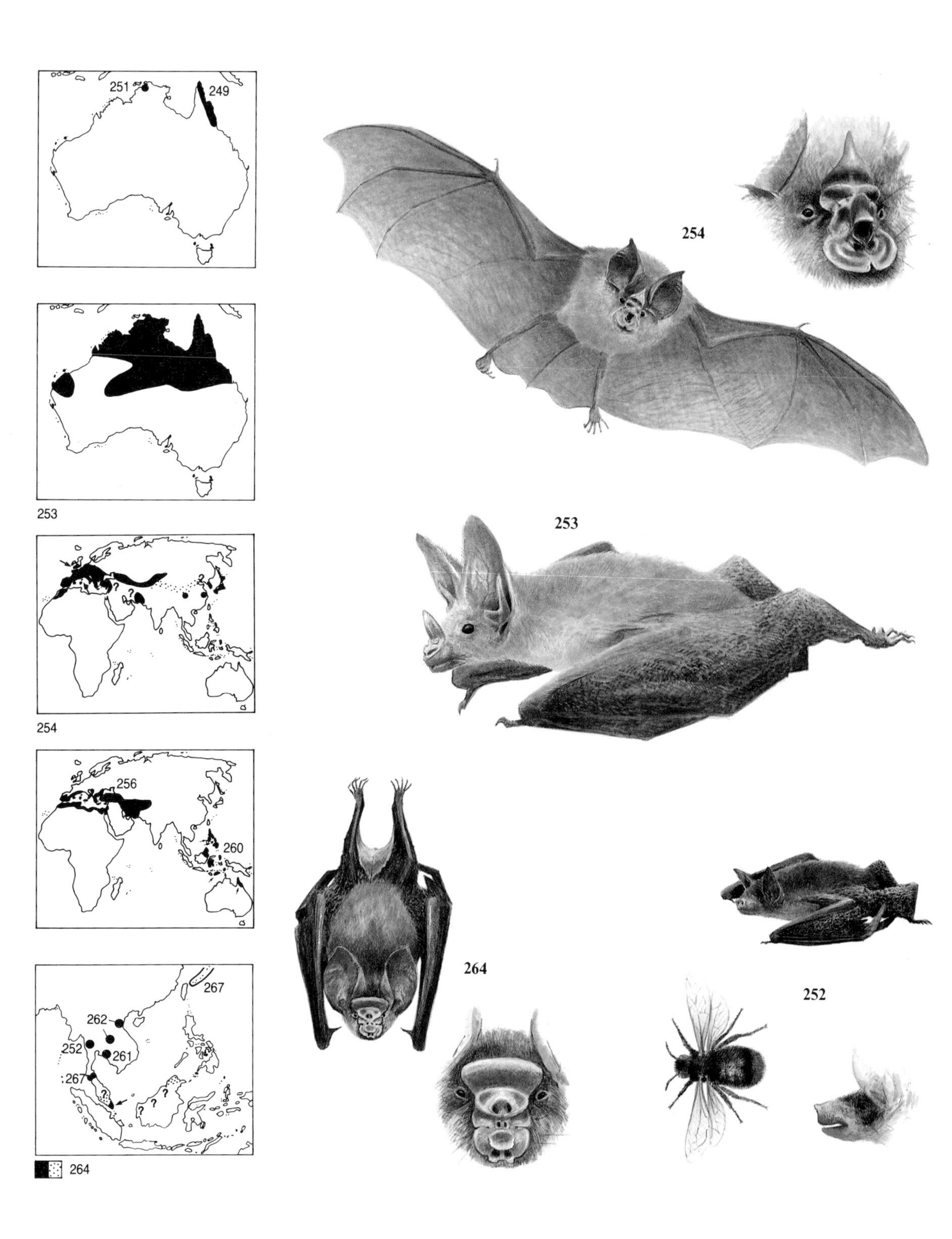
251
249
253
?
254
256
260
267
262
252
261
:267
264
254
253
264
252

**** SWORD-NOSED BAT**
Lonchorhina marinkellei (**270**)
Known from a single locality near Mitu, Durania, Colombian Amazon; it may prove to be a well-marked population of the more widespread * *L. aurita* (**271**). The latter may be declining, as their large roosting colonies in caves are susceptible to disturbance. They are found from S Mexico to Peru and N Brazil, Trinidad, and may occur on New Providence and other Caribbean Islands.

** *Platalina genovensium* (**272**)
HB 2.8ins; T 0.4in; FA 2 ins.
Generally brownish, with a diamond-shaped nose-leaf, and a long extensible tongue for feeding on nectar. Only a few specimens are known, all from Peru.

*** JAMAICAN FIG-EATING BAT**
Ariteus flavescens (**273**)
HB 2–2.6ins; FA 1.6–1.7ins; T absent; WT 0.3–0.5oz.
Light reddish above with a small patch of white on each shoulder. Fruit eaters, and known to take naseberries and rose apples. Only rarely been observed, thought to be threatened by agricultural development.

**** RED FRUIT BAT**
Stenoderma rufum (**274**)
HB 2.1–2.9ins; T absent; FA 1.8–2ins.
Upperparts reddish brown, paler below. Described from a single specimen from an unknown locality in 1820, and then discovered as a fossil on Puerto Rico in 1916; since then a few living specimens have been captured on Puerto Rico, and St. Thomas and St. John in the Virgin Islands. They were found feeding on fruit above the forest canopy.

***** FLOWER BATS**
Phyllonycteris spp.
HB 2.5–3.3ins; T 0.3–0.5ins; FA 1.7–2 ins; WT up to ca.0.7oz.
There are 4 species known; they are cave-dwelling, feeding on fruit, nectar, pollen and insects. ***** *P. major* (**275**) is only known from bones in caves in Puerto Rico, and is possibly extinct; * *P. poeyi* (**276**) occurs in large concentrations in Cuba. *** *P. aphylla* (**277**) and *** *P. obtusa* (**278**) from Jamaica and Hispaniola respectively, were thought to be extinct, but were recently rediscovered; little is known of their status. Concern has also been expressed over the * **Brown Flower Bat** *Erophylla sezekorni* (**279**), which occurs in the Greater Antilles and associated islands, where some of the very large roosts may be under threat.

**** FLOWER-FACED BAT**
Anthops ornatus (**280**)
HB 2ins; FA 1.9–2ins.
The fur is long and silky, greyish to buff. Rather similar to the leaf-nosed bats, with a much shorter tail. Virtually nothing is known about this species, which was described in the 1880s from 6 specimens collected in the Solomon Islands. Since then only a few more have been seen.

*** AFRICAN TRIDENT-NOSED BAT**
Cloeotis percivali (**281**)
HB 1.3–2.2ins; T 0.9–1.3ins; FA 1.2–1.4ins.
There are 3 pointed, trident-like, processes behind the nose-leaf. Although this species has been found in very large numbers in cave roosts, it is known from relatively few locations, and because of its concentrations, is probably vulnerable to disturbance and destruction of its roosting habitat.

***** GOLDEN (ORANGE) HORSESHOE BAT**
Rhinonicteris aurantius (**282**)
HB 1.8–2.1ins; T 0.9–1.1ins; FA 1.9–2ins; WT 0.3–0.35oz.
Fine and silky fur, bright golden or brown above, and paler below. The nose-leaf is large and scalloped. They live in caves during the wet season and have also been found in buildings. Their range is restricted to the northern Kimberley area (Western Australia), Arnhem Land in Northern Territory and Camooweal, NW Queensland. The largest known population is in Cutta Cutta Cave, near Katherine Northern Territory, but the development of the cave as a tourist attraction has caused a decline.

***** TRIPLE NOSE-LEAF BAT**
Triaenops persicus (**283**)
HB 2–2.2ins; T 1–1.3ins; FA 2–2.2ins; WT 0.3–0.5oz.
Very variable in colouring, ranging from greys to browns and reds. Mainly cave-dwelling, with a range which extends from S W Iran, Oman and Yemen, south along the coast of the Horn of Africa to Mozambique; also found sporadically inland in Zaire and around Lake Baringo and Tsavo NP, Kenya. Often congregate in large numbers, and are vulnerable to disturbance in some of the coral caves in areas where there is extensive tourist development.

**** FUNNEL-EARED BAT**
Paracoelops megalotis (**284**)
HB 1.8ins; T absent; FA 1.7ins; WT 0.2oz.
Long fur on the back, long ears (1.2ins); top of head is bright golden, contrasting with the duller brown back. The tail membrane is long (1.2ins) and since there is no tail it is supported by long heel bones. Known only from a single male collected at Vinh, Annam in Vietnam in 1945.

AFRICAN LONG-EARED BATS
Laephotis spp.
HB 1.8–2.3ins; T 1.7–1.8ins; FA 1.3–1.5ins; WT 0.2–0.3oz.
Little is known of the 4 species. Only 2 specimens of the ** **Namib Long-eared Bat** *L. namibensis* (**285**) from Gobabeb, Namibia have been found; the ** **Botswana Long-eared Bat** *L. botswanae* (**286**) is known from a small number of specimens from scattered localities in SE Zaire, NW Botswana, W Zambia and NW Zimbabwe, where they were all found in Savannah woodland near rivers; ** **de Winton's Long-eared Bat** *L. wintoni* (**287**) from Kenya, where all the specimens discovered so far have been found under tree bark; a single specimen from SW Cape Province of South Africa may prove to be another species. The * **Angolan Long-eared Bat** *L. angolensis* (**288**) is known from Zimbabwe, S Zaire and Angola.

**** BIG-EARED BROWN BAT**
Histiotus sp.nov. (**289**)
In the 1970s a new species of Big-eared Brown Bat was reported from Venezuela. Exact distribution and status unknown.

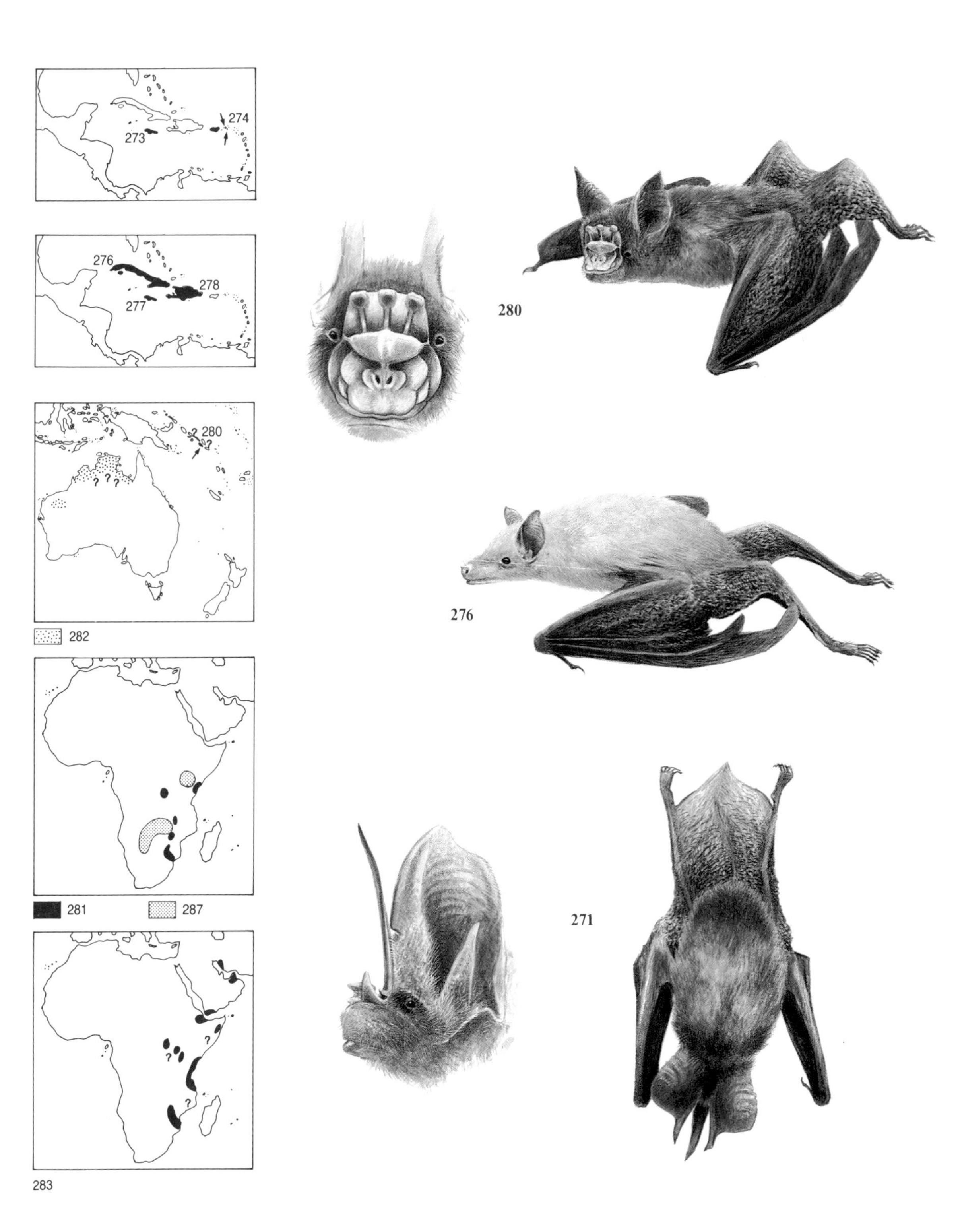
274
273
276
278
277
280
?
? ? ?
282
281
287
?
?
?
283
280
276
271

***** INDIANA BAT**
Myotis sodalis **(290)**
T 1.1–1.7ins; FA 1.4–1.6ins; WT 0.2–0.3oz.
Similar to the more common Little Brown Bat *M.lucifugus* but the Indiana Bat has a distinct keel on the calcar (spur). They roost colonially in winter, in caves and mines, and in summer in hollow trees. Their roosting requirements are very specialised, with a preferred temperature range of 4–8°C, and since few caves are cool enough, over 90% of the population occurs in only 10 roosts. Population once numbered well over 1 million, but fell to around 500,000 by 1978; decline is probably linked to the widespread use of persistent pesticides during the 1960s; in 1964, 85% of one population in Kentucky was killed in a flood. In summer, Indiana Bats are found in eastern USA from New Hampshire, west to Oklahoma and south to Florida. In winter they are more restricted. The populations are divided into two areas (*see* map). Fully protected and listed as Endangered in USA, and many roosts protected as a Critical Habitat, with public access restricted.

***** GREY BAT**
Myotis grisescens **(291)**
HB ca.2ins; T 1.3–1.7ins; FA 1.6–1.8ins; WT 0.2–0.3oz.
Similar to the Indiana Bat (290), but duller greyish-brown fur uniformly coloured to base of each hair. Is a colonial roosting species, found in limestone caves, mostly in central USA. About 95% of the population estimated to hibernate in 9 roosts, and one contains over 50%. Size of original population not accurately known, but numbers are estimated to have fallen by up to 76%. Distribution has always been patchy, but increasingly fragmented. Main cause of decline has been disturbance by vandals and cavers and, to a much lesser extent, biologists. Listed as Endangered and protected in USA. The most important summer roost, Souta Cave, Alabama, is owned by the US Fish & Wildlife Service, and several other sites now protected. ** **Miller's Myotis** *M.milleri* **(292)** is only known from La Grulla in Baja California, Mexico.

**** SAKHALIN BAT**
Myotis abei **(293)**
Only known since 1944 from the island of Sakhalin, USSR.

***** LARGE MOUSE-EARED BAT**
Myotis myotis **(294)**
HB 2.6–3.2ins; T 1.8–2.4ins; FA 2.3–2.6ins; WT 0.7–1.5oz.
One of Europe's largest bats. It occurs over most of Europe, east to Asia Minor and Israel. Apart from some 19th century records, it was unknown in England until the 1960s when small populations were discovered, which have since declined and are now virtually extinct. In other parts of Europe declines noted, and in Israel is nearly extinct, due to control programmes directed against fruit bats. In continental Europe migrations of up to 125mls recorded. Protected throughout all of their European range and proposed for special protection under the Bonn Convention.

*** NATHALINA BAT**
Myotis nathalinae **(295)**
HB 1.6–2ins; FA 1.3–1.4ins; WT 0.2–0.4oz.
Slightly smaller than the more common and widespread Daubenton's Bat, from which it was first distinguished in 1977; main differences are in genitalia and teeth. Only known from Spain, France, Germany, Switzerland and Poland.Often found in old mines and among old masonry. Doubts have been expressed on the validity of this species.

***** BECHSTEIN'S BAT**
Myotis bechsteini **(296)**
HB 1.7–2ins; T ca.1.5ins; FA 1.5–1.8ins; WT 0.2–0.4oz.
Ears relatively large (0.8–1in) and a wingspan of up to 11.8ins. Upperparts light brown, lower parts greyish to white. Recorded over most of Europe, and east to the Caucasus, but over most of range fairly rare, particularly in the west. Found mainly in woodlands and forests, and rarity is probably due to destruction of the European woodlands. Sub-fossil remains indicate were once abundant in Britain. Protected over most of their range, but are probably continuing to decline due to loss of woodland habitat and roosting sites, disturbance and pesticides. Many other *Myotis* bats declining in the northern temperate regions, and others are little known; * *M. insularum* **(297)** is known only from a single specimen from Samoa – which may be incorrectly labelled. All the European species are declining, at least locally; the *** **Lesser Mouse-eared Bat** *M. blythi* **(298)** is extinct in W Germany, and nearly extinct in Israel. It also occurs in apparently widely separated populations across Central Asia to Mongolia, and the *** **Long-fingered Bat** *M. capaccinii* **(299)** is extinct in Austria, Switzerland and Hungary, and declining elsewhere.

******* TANZANIAN WOOLLY BAT**
Kerivoula africana **(300)**
Discovered in the 1870s, and never encountered again, on the coast of Tanzania, opposite Zanzibar. Status, very rare, or perhaps extinct. May be closely related to *K. eriophora* from Ethiopia, but little is known about this species too. *K.papuensis*, although widespread in New Guinea, is rarer in Australia. Other *Kerivoula* spp. from SE Asia may be threatened, but little is known of their status.

*** NEW ZEALAND LOBE-LIPPED BAT**
Chalinolobus tuberculatus **(301)**
WT 0.28–0.35oz.
A forest-edge species, it is still reasonably widespread in a wide variety of habitats on both main islands on New Zealand, Stewart Island and a few smaller islands. However, they are vulnerable to introduced predators as well as changes to their habitat such as the introduction of exotic pine plantations.

***** HOARY BAT**
Lasiurus cinereus **(302)**
HB ca.3.5ins; T 1.7–2.6ins; FA 1.8–2.3ins; WT 0.7–1.2oz.
Has characteristic 'frosted' appearance. One of the largest and most widespread of all North American bats, from edge of the tundra, south through C and S America to Chile and Argentina. An isolated population on the Hawaiian Islands, *L.c. semotus* is threatened by habitat loss, but status unknown.

**** LAMINGTON FREE-EARED or LONG-EARED BAT**
Lamingtona (=Nyctophilus) lophorhina **(303)**
HB 2ins; FA 1.5–1.6ins.
A little-known species, described in 1968 from Mt Lamington, SE Papua New Guinea.

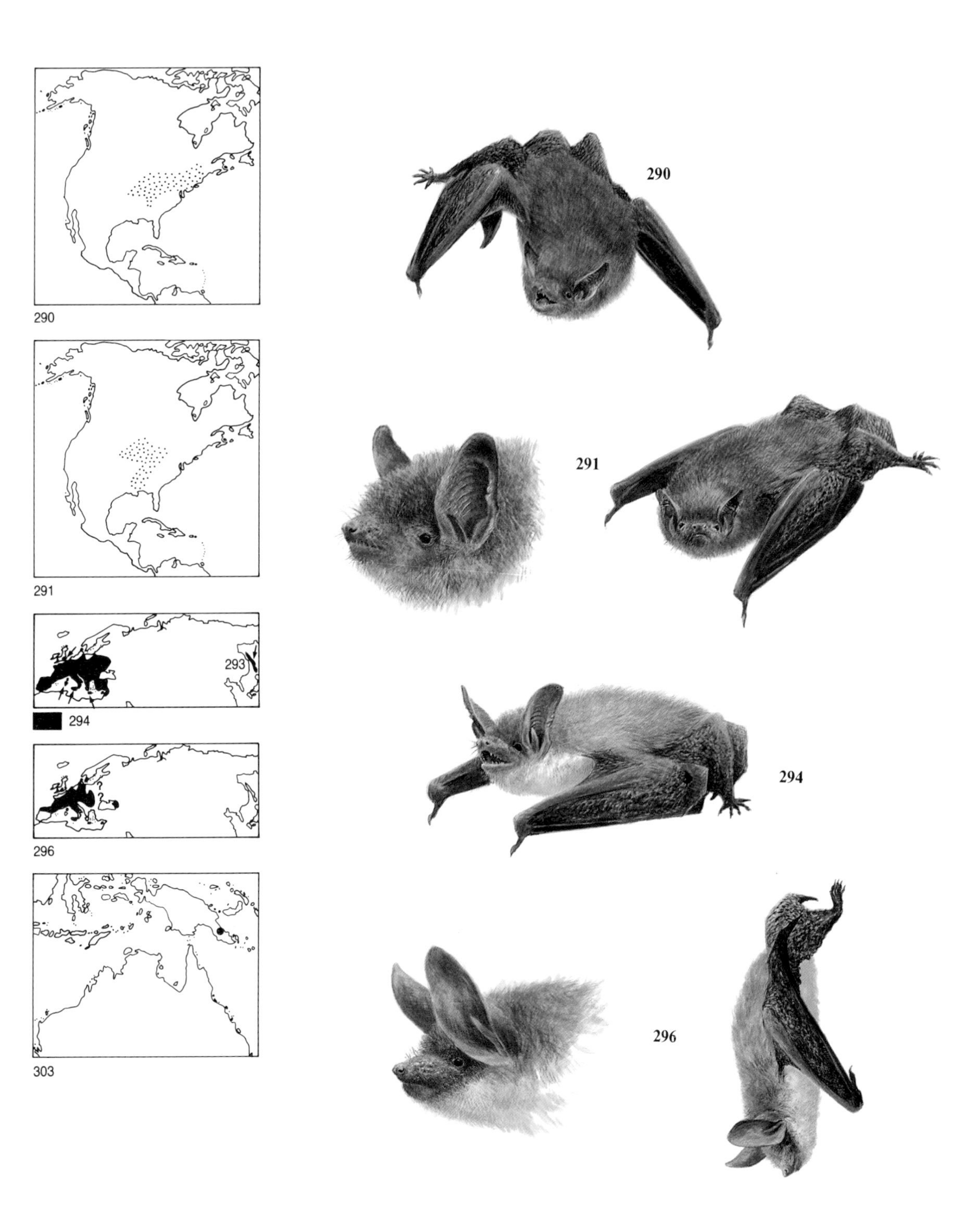
290
291
293
294
296
303
290
291
294
296

**** WING-GLAND or HAIRY BATS**
Cistugo (=Myotis) spp.
HB 1.6–1.9ins; T 1.6–1.7ins; FA 1.3–1.4ins.
C. seabrai (**304**) and *C. lesueuri* (**305**) have wing-glands, and are closely related. *C. lesueuri* is yellowish-brown above, and yellowish-white below; *C. seabrai* is brownish or greyish and slightly smaller. Wing-glands larger and more obvious in *C. seabrai.* Hairy bats are little known insectivores living in trees in arid regions. *C. seabrai* has been found in Angola, Namibia and New Cape Province, South Africa; *C. lesueuri* is confined to SW Cape Province, South Africa.

**** DISC-FOOTED BAT**
Eudiscopus denticulus (**306**)
HB 1.6–1.8ins; T 1.5–1.7ins; FA 1.5ins.
Upperparts cinnamon-brown, underside brighter. Superficially similar to Pipistrelles, but has flattened skull and adhesive pad on the foot. Little known of its biology. Only known from Phong Saly in Laos, and Pegu Yoma in C Burma.

PIPISTRELLES
Pipistrellus spp.
Although none is thought to be in immediate danger, the European Pipistrelles, * *P.pipistrellus* (**307**), * *P. nathusii* (**308**), * *P. kuhli* (**309**) and * *P. savii* (**310**), have all probably undergone declines since the 1950s, although in some areas *P. pipistrellus* is still very abundant and may even be increasing; *P. nathusii* may total only a few hundred. There are ca.50 species of Pipistrelles, many of which are poorly documented and may be threatened; these include: ** *P. societatis* (**311**); known only from one animal, described in 1972 from Pahang, Malaysia. ** *P. bodenheimeri* (**312**); described in 1960, only known from a handful of localities in Israel and Arabia. ** *P. aero* (**313**) known from NW Kenya, but may also occur in Ethiopia. ** *P.arabicus* (**314**) first described in 1979, only known from Oman. * *P. anthonyi* (**315**) and * *P. joffrei* (**316**) are little-known species from N Burma. * *P. permixtus* (**317**), a little-known species discovered in Dar-es-Salaam, Tanzania, in 1957. **P. maderensis* (**318**) is confined to the Madeira Islands. * *P. tenuis* (**319**) widespread from Thailand to N Australia, and occurs on small islands; some populations may be separate species. ** *P. kitcheneri* (**320**) is known from 2 localities in Borneo; ** *P. cuprosus* (**321**) from Sepilok in Sabah; and ** *P. vordermanni* (**322**) from Betitung Island and Sarawak.

***** GREATER NOCTULE**
Nyctalus lasiopterus (**323**)
HB 3.1–4ins; FA 1.8–2.2ins; WT 1.4–2.7oz.
The largest European bat, with a wingspan of 20ins, occurring from E France to the Urals and Iran. Very sporadic in occurrence; detailed distribution is not known, but probably one of Europe's rarest bats. * **Leisler's Bat** *N. leisleri* (**324**) is abundant only in Ireland, and rare throughout most of its European range; little-studied outside Europe; the * **Noctule** *N. nyctalus* (**325**) is declining in much of its European range.

**** EUPHRATES SEROTINE**
Eptesicus walli (**326**)
T ca.1.8ins; FA ca.1.5ins.
A small serotine sandy-buff above, buff-white below. Closely related to *E. nasutus*, with which it is sometimes regarded as conspecific. Only known from a very small area in Arabia around the Euphrates. No information on habits or status. ** *E. loveni* (**327**) is known only from the type specimen collected on the eastern slope of Mt Elgon, Kenya, at an altitude of 8000ft, described in 1924. The ** **Yellow-lipped Serotine** *E. douglasi* (**328**), described in 1976, is only known from the Kimberley area of N Western Australia; ** *E. platyops* (**329**) from single specimens from Nigeria and Senegal. * *E. guadeloupensis* (**330**) has only been found in Guadaloup, Lesser Antilles; ** *E. demissus* (**331**) is known from a single specimen from Thailand and has not been seen since the begining of this century. The * **Northern Serotine** *E. nilssoni* (**332**) and * **Serotine** *E. serotinus* (**333**) both occur in Europe where locally abundant, but declining in many areas; protected throughout Europe; little is known of their status outside Europe.

**** FALSE SEROTINE BAT**
Hesperoptenus doriae (**334**)
In general appearance like a serotine, *Eptesicus* spp., with light brown colouring. Only known from 2 specimens: from Sarawak, Borneo and Selangor, Malay Peninsula.

**** PIED BUTTERFLY BAT**
Glauconycteris superba (**335**)
Only known from a few specimens from Ghana, NE Zaire and Uganda. These and other butterfly bats may prove to be more widespread – but may also be threatened by habitat destruction in some areas. ** **Bibundi Bat** *Glauconycteris egeria* (**336**) only known from a single specimen from Cameroon.

**** TUBE-NOSED INSECTIVOROUS BAT**
Murina florium (**337**)
HB ca.1.9ins; T ca.1.3ins; FA ca.1.4ins; WT ca.0.3oz.
Has long brown fur, with scattered silvery hairs on the upper parts. The prominent nostrils are tubular and diverge. A single specimen was captured in Queensland, Australia in 1981; only 15 other specimens are known, from Sumbawa, Flores, Buru and Seram. They are believed to occur in montane tropical rainforest. They do not rest with the wings folded at the sides, but wrap them around the body, like fruit bats, but held away from the body to form an umbrella; this is thought to be an adaptation to its wet and misty environment. ** *Murina tenebrosa* (**338**) known only from one specimen from Tsushima Island, Japan. ** *Murina canescens* (**339**) is known only from the island of Nias, off the west coast of Sumatra. * *M. puta* (**340**) from Taiwan little known, and other *Murina* from islands in SE Asia poorly-known and may be threatened.

**** PYGMY or LITTLE TERRITORY LONG-EARED BAT**
Nyctophilus walkeri (**341**)
HB 1.5–1.8ins; T 1.2–1.4ins; FA 1.3–1.4ins; WT 0.14–0.16oz.
Fawn above, buff below, dark brown wings, pale tail. First discovered in 1891 near Adelaide River, Northern Territory, Australia, and then not seen again until 1975. 5 more specimens are known, from W Australia and N Territory.

**** BIG-EARED BAT**
Pharotis imogene (**342**)
HB 2ins; T 1.7ins; FA 1.5ins.
The only specimen known (female) is dark brown above and below. Similar to the more widespread *Nyctophilus* spp. big-eared bats, it has not been recorded since it was first described in 1914; it was collected from SE Papua New Guinea.

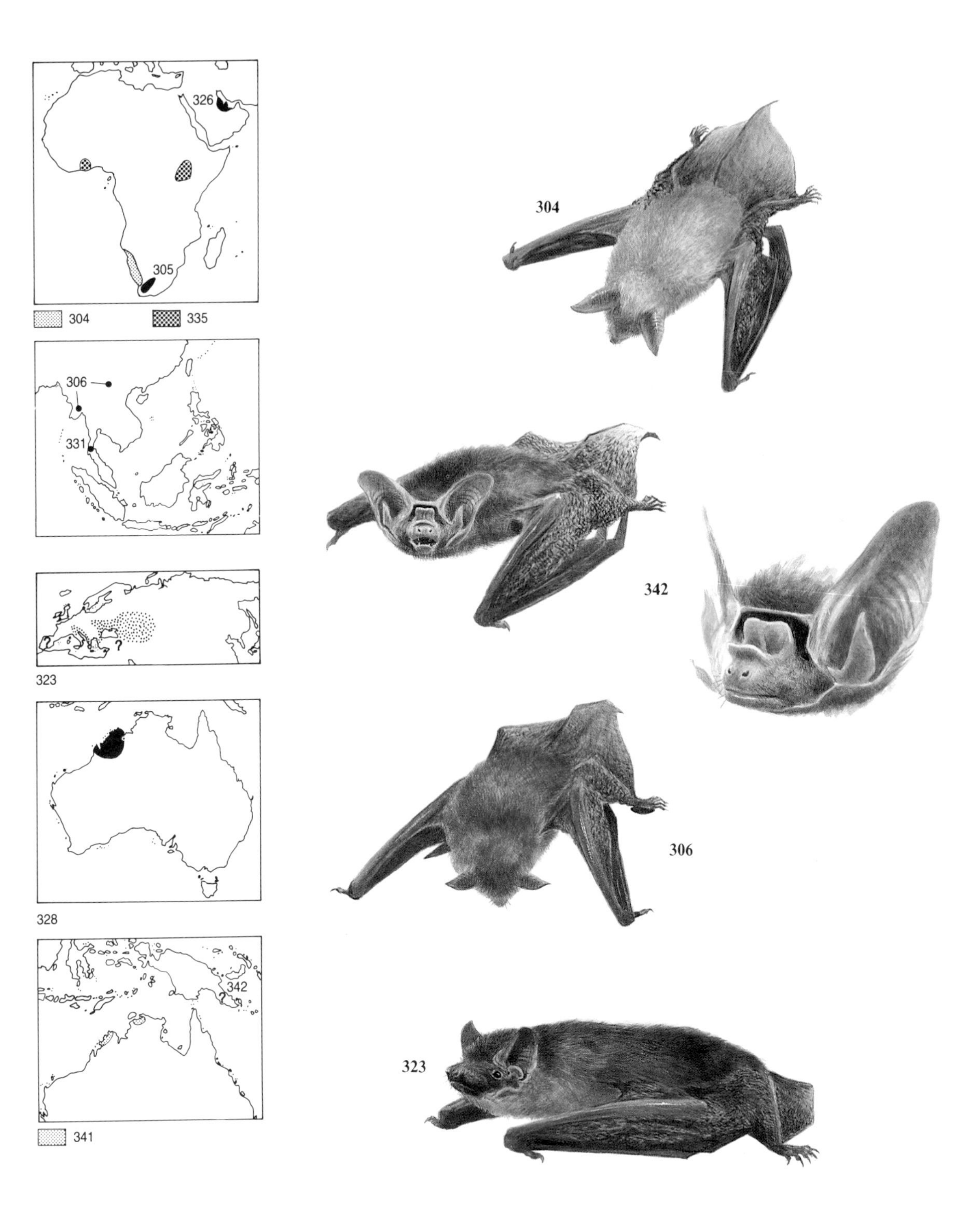
326
305
304
335
306
331
323
328
342
341
304
342
306
323

**** LITTLE YELLOW BAT**
Rhogeesa mira (**343**)
Known from one locality in Michoacan, C Mexico. ** *R. gracilis* (**344**) has a restricted range and is endemic to Mexico.

**** INDIAN HARLEQUIN BAT**
Scotomanes emarginatus (**345**)
Described in 1871 from a single specimen from an unknown locality in India.

***** TOWNSEND'S BIG-EARED BAT**
Plecotus townsendii (**346**)
HB 3.5–4.2ins; T 1.4–2.1ins; FA 1.5–1.9ins; WT 0.3–0.4oz.
A small, long-eared (up to 1.5ins) bat, living in variety of habitats, including deciduous forests; winters in caves. Widely distributed from southern British Columbia south through western USA to Oaxaco (Mexico) and east through W Virginia. Several populations threatened: *P. t. virginianus* from the Appalachian region, appears to be declining due to disturbance to caves where they roost, despite protection. *P. t. ingens*, found in caves in Arkansas, SW Missouri and E Oklahoma, also susceptible to disturbance. The European long-eared bats, * *P. auritus* (**347**) and * *P. austriacus* (**348**) declined locally, and in Britain *P. austriacus* only known from a few localities.

*** MEXICAN LONG-EARED BAT**
Plecotus mexicanus (**349**)
HB 3.5–4.1ins; FA 1.5–1.8ins; WT 0.3–0.4oz.
Two separate populations: in central Mexico and Yucatan. Formerly regarded as a subspecies of *P. townsendi* (346) but ranges overlap and differ in dentition and the inter-femoral (tail) membrane. Little known of their distribution or status.

*** SPOTTED (PINTO) BAT**
Euderma maculatum (**350**)
HB 2.4–3ins; FA 1.9–2ins; WT 0.5–0.6oz.
Dark reddish-brown or blackish with a white spot on each shoulder, and at the base of the tail. The ears grow to 1.95ins. Found in a wide variety of habitats, but mostly in dry and desert country. Generally solitary but may live in small groups. Discovered in 1890; only been found in a few localities in western USA and N Mexico. Occur in several protected areas including Yosemite NP, California and Big Bend NP, Texas.

*** SCHREIBER'S LONG-FINGERED BAT**
Miniopterus schreibersi (**351**)
HB 2–2.5ins; T 1.9–2.4ins; FA 1.7–1.9ins; WT 0.2–0.3oz.
One of the most widely distributed bats in the world, from Europe, east to Japan and south to Africa, SE Asia and Australia. Recent research suggests some populations may be separate species. Fragmented range in Australia, and cave habitats are under pressure. Near extinction in Israel and some other parts of its range.

***** LITTLE LONG-FINGERED BAT**
Miniopterus australis (**352**)
Slightly smaller than the previous species, with more reddish fur. Roosts communally in caves and tunnels; although locally abundant, its roosting sites are under pressure. Found in coastal Australia from Cape York, to NE New South Wales, and north through New Guinea, to the Philippines and Indonesia. * *M. robustior* (**353**) occurs on several Pacific Islands, including New Caledonia, Loyalty Islands, Lifu Island and Quepenee. * *M. tristis* (**354**) occurs more widely from the Philippines south to New Guinea, Solomons and New Hebrides; the status of both is unknown over most of range, where extensive changes to habitat take place.

*** NEW ZEALAND SHORT-TAILED BAT**
Mystacina tuberculata (**355**)
FA 1.6–1.8ins; WT 0.4–0.5oz.
The upperparts are greyish-brown to brown, and the underparts paler. Remarkably agile on the ground, they run and climb with specially modified wings and claws. Found in caves, crevices, trees and burrows, using their teeth to excavate holes. They feed on insects, fruit, nectar and pollen. Known from a few lowland and montane forests on North Island and Little Barrier Island, from a single locality on South Island, and Codfish Island off Stewart Island. A closely related, slightly larger species *M. robusta* was found over much of New Zealand, but not recorded since 1965 when rats invaded its last refuges; presumed extinct. *M. tuberculata* is threatened by introduced predators, such as rats, cats and stoats.

**** LARGE-EARED FREE-TAILED BAT**
Tadarida lobata (**356**)
FA 2.2–2.6ins.
The large ears and wings are both rather transluscent. Only one specimen found in Kenya, until the 1970s, when two more were found in Zimbabwe, 1490 mls away. Nothing is known of its habits or status. The ** **African Free-Tailed Bat** *T. africana* (**357**), closely related, has only been found in 5 widely separated localities in the eastern side of Africa; it may be more widespread. The ** **Natal Free-tailed Bat** *T. acetabulosus* (**358**) is known from Madagascar, Reunion, and Mauritius, and single specimens found in South Africa and Ethiopia; precise range or status on the African mainland little-known.

***** BRAZILIAN FREE-TAILED BAT**
Tadarida brasiliensis (**359**)
HB ca.2.6ins;T 1.2–1.4ins; FA 1.4–1.8ins; WT 0.4–0.5oz.
Variable habitat, roosting in buildings and caves. They live in the largest known mammalian aggregations, with nursery colonies of up to several million. During the 1960s, 13 caves in Texas were estimated to contain 100 million Brazilian Free-tailed Bats. The most famous colony, in Carslbad Cavern NP, New Mexico fell from 8.7 million in 1936 to only 200,000 in 1973; the population in Eagle Creek, Arizona, fell from 25–50 million to 600,000 by 1970. Decline usually attributed to the widespread use of persistent pesticides. DDT is now banned in the USA, but still used in Mexico, where the bats winter.

**** BINI FREE-TAILED BAT**
Myopterus whitleyi (**360**)
Only known from widely scattered localities in Ghana, Nigeria, Cameroon, Zaire and Uganda. Not gregarious, and found roosting in trees; current status unknown.

** *Molossops neglectus* (**361**)
Not described until 1980; only known from Powaka, Surinam.

*** PANAMANIAN DOMED-PALATE MASTIFF BAT**
Promops pamana (**362**)
Known only from a single imperfect specimen from Central Brazil; possibly the same as the more widespread *P. nasutus*.

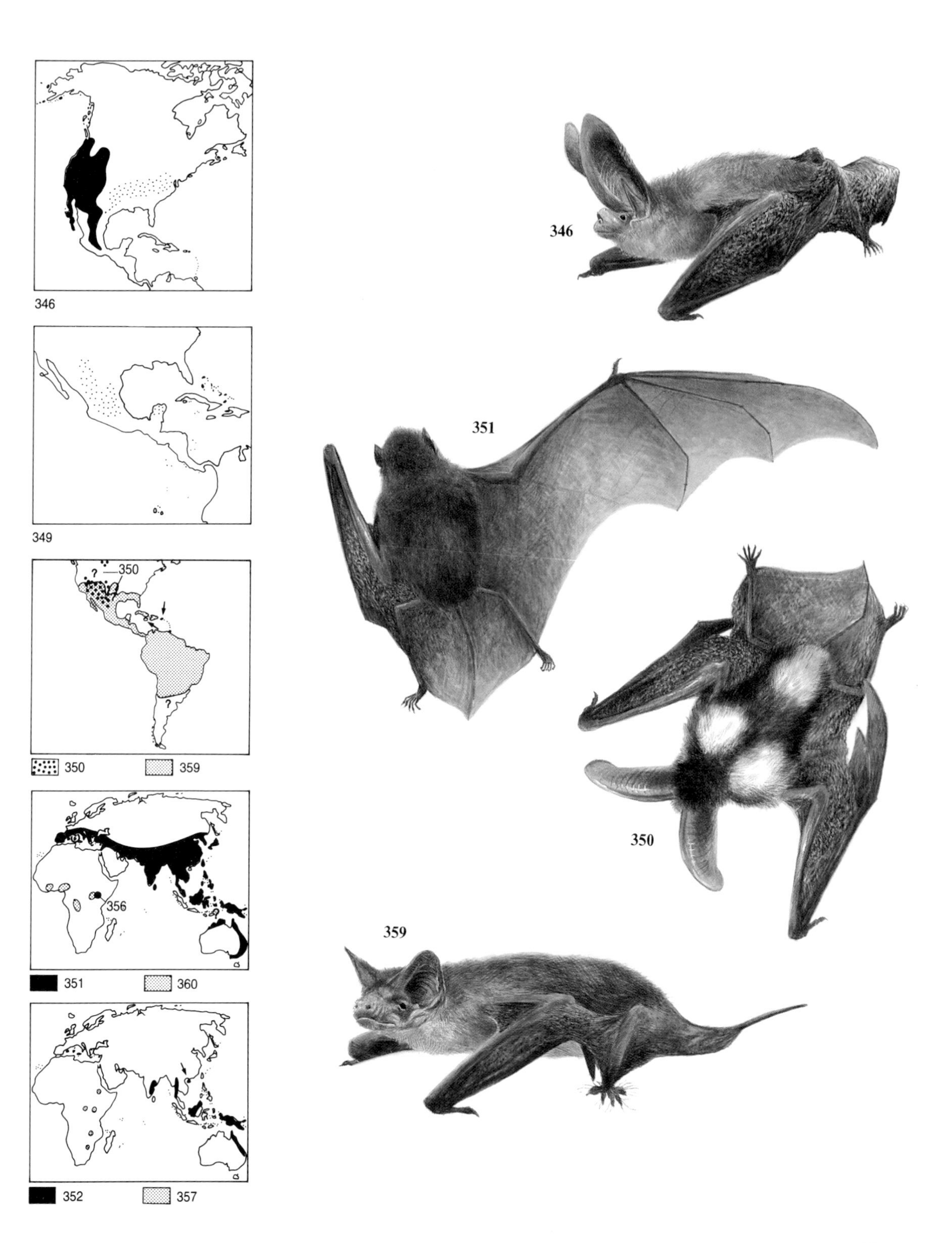
346
349
?
350
?
350
359
356
351
360
352
357
346
351
350
359

PRIMATES

The status of most primates is better documented than that of the majority of mammals, and there is an extensive literature describing them. Most of the primates are threatened, at least locally. More detailed accounts of their biology, distribution and status can be found in *The Primates of Madagascar* (1982), by Ian Tattersall, and *The Primates of the World* (1983), by Jaclyn H. Wolfheim, and in other books listed in the bibliography.

*** WESTERN or HORSFIELD'S TARSIER

Tarsius bancanus (**363**)

HB 4.7–6.1ins; T 7.1–8.9ins; WT 3–4oz.

Soft-furred, pale olive, reddish or greyish; the long tail only has fur at the end. The fingers and toes are slender, ending with pads and pointed nails, except on the second and third toes of the hind feet which have large claws. They are arboreal and nocturnal, feeding exclusively on small animals, including insects, lizards and amphibians. In addition to loss of habitat, they have suffered from the effects of pesticides in plantations. Can adapt to secondary habitats, and have been recorded at densities of up to 300 per sq.ml. In Sarawak they are reported to be captured as pets, but rarely survive for long. They occur in Kinabalu NP, Sabah and are protected in Sarawak, Sabah and Indonesia since 1931. Listed on CITES App.II. The * **Eastern** or **Sulawesi Tarsier** *T.spectrum* (**364**) is confined to three areas on Sulawesi, and Peleng, Sangihe, Savu and Saleyer, off Sulawesi. Little is known of its status, although it is likely to be threatened by habitat destruction. Occurs in the Tangkoko Batuangus Reserve, and has legal protection in Indonesia since 1931. The *** **Philippine Tarsier** *T. syrichta* (**365**) is declining through destruction of its forest habitat. In addition to three areas on Mindanao, it also occurs on the islands of Samar, Leyte, Dinagat and Siargao Bohol. Not found in any protected areas.

* SLENDER LORIS

Loris tardigradus (**366**)

HB 6.9–10.4ins; T absent; WT 3–12.2oz.

Smaller and more slender than the Slow Loris. Arboreal and nocturnal, generally found in dense forests, where they climb with slow deliberate movements and feed on a wide variety of plants and small animals. Usually solitary animals; the female gives birth to 1–2 young. Their range extends over southern India and Sri Lanka; it is thought that numbers have declined this century due to loss of habitat. Also, until recently they were trapped, both for use in biomedical research, and in traditional medicine (for eye diseases). Although they were often seen in zoos in the past there are no self-sustaining populations, and they are now comparatively rare.

*** SLOW LORIS

Nycticebus coucang (**367**)

HB 7.8–10.8ins; T up to 1in; WT 8–21.3oz.

The soft dense fur is greyish to reddish-brown, with a dark stripe along the mid-back, and usually a dark ring around large eyes. Found in a wide variety of habitats, mostly forested or wooded, but including secondary growth and plantations. Arboreal and nocturnal, they feed on small animals, including insects and also soft fruits. They have disappeared wherever forest cover has been destroyed, and they are also hunted in parts of Indonesia for use in traditional medicine. In Bangladesh they have vanished from most of their former range over the past 20 years. In peninsular Malaysia they are considered potentially threatened. They occur in several protected areas, and are occasionally exhibited in zoos. Their survival is threatened mainly by continued destruction of forest habitats and fragmentation of range. Legally protected in Indonesia, Malaysia, Thailand and Singapore. Almost nothing is known of the status of the closely related * **Lesser Slow Loris** *N. pygmaeus* (**368**) from Vietnam, Kampuchea and Laos, but the Indo-china war has damaged much of its habitat. Both are listed on CITES App II.

*** LESSER MOUSE-LEMUR

Microcebus murinus (**369**)

HB 4.9–5.9ins; T 4.9–5.9ins; WT 1.3–3.4oz.

Greyish or brownish above, with a slight dorsal stripe. The two populations are often considered as full species. *M. m. murinus* occurs in W and S Madagascar, *M. m. rufus* in E and N Madagascar. Confined to primary and secondary forests, building nests in foliage or in hollows, where they are nocturnal and arboreal. They accumulate fat in their tails and are less active in the dry season. They feed on insects, fruit, flowers, gums, lizards and frogs. Their habitat is being rapidly destroyed but they occur in several reserves, and the Montagne d'Ambre NP, Madagascar. Listed on CITES App.I and Class A of the African Convention, and fully protected in Madagascar. Small numbers are held in captivity, most of which have been bred there.

***** COQUEREL'S MOUSE-LEMUR

Microcebus coquereli (**370**)

HB ca.9.8ins; T ca.11ins; WT 9.8–11.7oz.

Dark greyish above, with reddish tinge, yellowish-grey below. One of the rarest lemurs, confined to dry deciduous forests of W Madagascar, where it is declining due to loss of habitat. Has been found at densities of up to 554 per sq.ml. Occurs in at least 3 protected areas, and listed on CITES App.I and Class A of the African Convention.

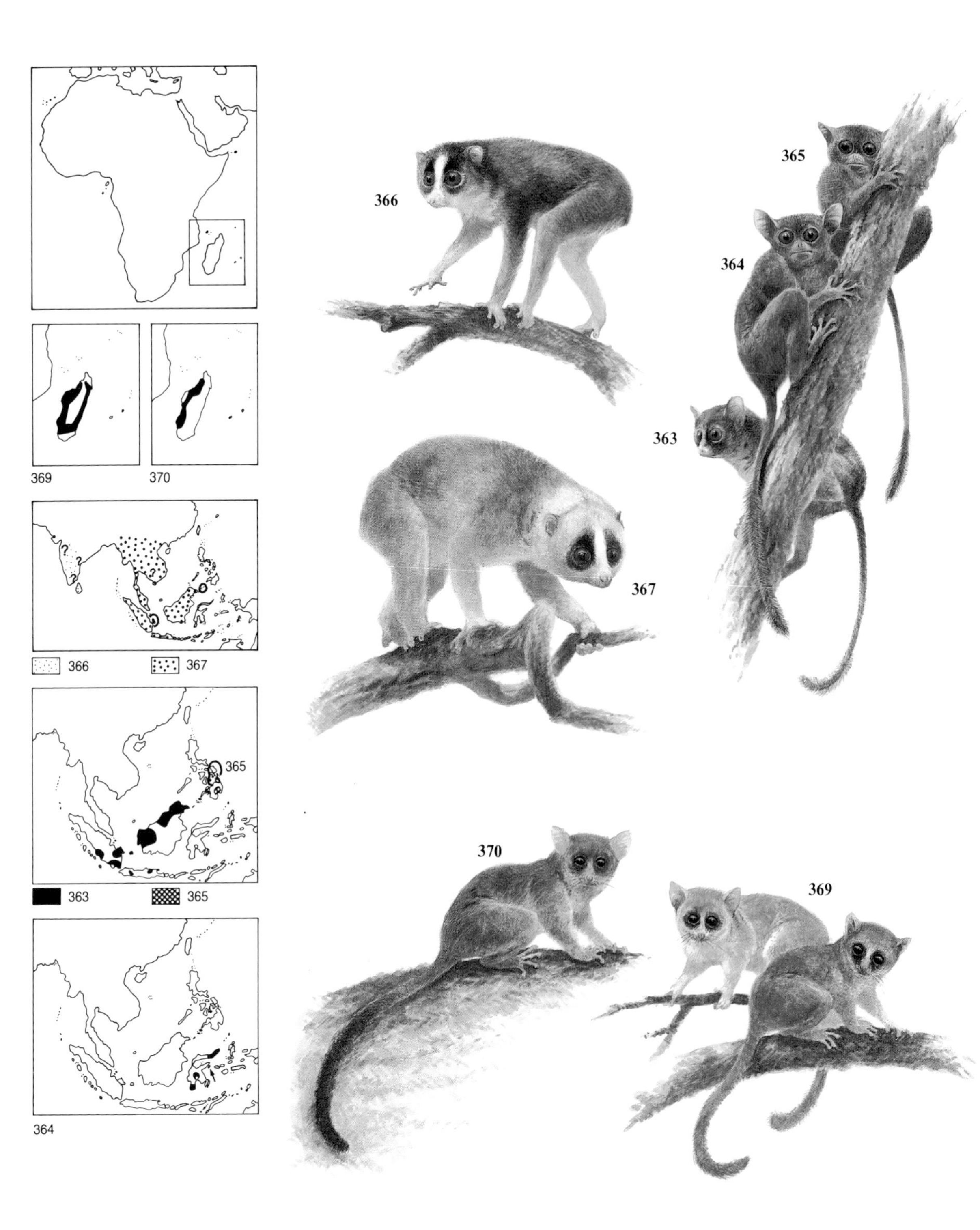
369
370
366
367
365
363
365
364
366
365
364
363
367
370
369

***** BLACK LEMUR**
Lemur macaco (**371**)
HB 13.8–15.8ins; T 17.7–19.7ins; WT ca.4.4lb.
Found in humid forests in the north of Madagascar and on the islands of Nosy Bé and Nosy Komba. Much of the forest in its range has already been destroyed, and it is threatened by the spread of agriculture, logging and fires. (The subspecies *L.m.flavifrons* may already be extinct.) Found in a few reserves and listed on CITES App.I and Class A of the African Convention. Captive populations of the typical Black Lemur *L.m. macaco* exist, which are probably self-sustaining.

******* HAIRY-EARED DWARF LEMUR**
Allocebus (=Cheirogaleus) trichotis (**372**)
HB ca.5.1ins; T ca.6.7ins.
Greyish-brown above, pale below, with reddish-brown tail. Ears short with long tufts of hair. One of the rarest lemurs, for a long time believed extinct; only two specimens known, found in 1874 and 1965. Agriculture and logging in their rainforest habitat continues to threaten their survival. Although legally protected they do not occur in any reserves. The closely related, but more widespread *** **Greater Dwarf Lemur** *Cheirogaleus major* (**373**) and *** **Fat-tailed Lemur** *C. medius* (**374**) are also declining through loss of forest habitat. Both species occur in several protected areas. Durham Zoo, USA, specialises in these and other rare lemurs, although they are rarely exhibited in zoos. All are listed on CITES App.I and Class A of the African Convention, and protected in Madagascar.

***** FORK-MARKED LEMUR**
Phaner furcifer (**375**)
HB 9.8–10.8ins; T 11.8–13.8ins; WT ca.12.2oz.
The dense fur is reddish or brownish, with black streaking on the head; the hands, feet and tail are darker. Found mainly in deciduous forests, also in woods in more open areas, including savannah; often associated with baobabs. They are nocturnal, spending the day in holes in trees or in dense foliage; very agile, leaping up to 33ft from tree to tree. They feed mostly on gums, by scraping off bark with their modified incisors and lapping the sap. Densities of up to 2254 per sq.ml have been recorded; they are generally solitary and territorial, the males' territories overlapping more than that of the females. The single young is carried by the mother. Like most other lemurs, numbers are declining due to burning and cutting of their habitat. They occur in the Montagne d'Ambre NP, Madagascar, and at least 4 other reserves, and are listed on CITES App.I and Class A of the African Convention. Not known to be kept in captivity at present.

*** RING-TAILED LEMUR**
Lemur catta (**376**)
Although still one of the most abundant Madagascan primates, they have undergone a massive decline, which is continuing. Found in at least 6 protected areas and breed well in many zoos.

***** BROWN LEMUR**
Lemur fulvus (**377**)
Similar in size to the Black Lemur (371), with which it is sometimes considered conspecific. Found in closed canopy forest, as well as deciduous forest with baobabs. On the island of Mayotte, in the Comoros, where they may have been introduced by man, they occur in secondary forest; this population was estimated at ca. 50,000 in the 1970s. Although among the more widespread lemurs, they are declining and at least two subspecies are particularly threatened: *L. f. fulvus* occurring within the Autsalo Reserve, and *L.f. sanfordi* in the Montagne d'Ambre NP. Other reserves which have populations of Brown Lemur include Nosy Mangabé.

***** MONGOOSE LEMUR**
Lemur mongoz (**378**)
HB 15.8–17.7ins; T 15.8–17.7ins; WT 4–4.8lb.
Found in a variety of dry forests, and on the Comoros Islands in secondary forest, as well as highland rainforest. With the widespread destruction of Madagascan forests, numbers are declining; they are protected here and occur in reserves. On the Comoros they are afforded some protection. Listed on CITES App. I and Class A of the African Convention.

***** CROWNED LEMUR**
Lemur coronatus (**379**)
Once thought to be a subspecies of the Mongoose Lemur, but now considered distinct. (They are similar in size.) Found in the north of Madagascar, where it has a very restricted range, occurring in dry forests, including the Montagne d'Ambre NP. Listed on CITES App. I and Class A of the African Convention.

******* RED-BELLIED LEMUR**
Lemur rubriventer (**380**)
HB 15.8–17.7ins; T 16.5–19.7ins; WT 4–4.8lb.
Their original range extended along the rainforest on the east of Madagascar. Now believed to be declining rapidly, and do not occur in any reserves. Already extinct in many forests of their former range, and on the brink of extinction in others. Listed on CITES App. I and Class A of the African Convention.

***** AVAHI or WOOLLY LEMUR**
Avahi laniger (**381**)
HB 11.8–17.7ins; T 12.6–15.8ins; WT 1.3–2.2lb.
The fur is thick and woolly, and even the face is covered with short hairs. Normally greyish, but colour varies from almost white to reddish. Arboreal and nocturnal, sleeping in hollow tree or hiding in thick vegetation, and feeding almost exclusively on vegetable matter, including leaves, buds and bark. They live in small groups in rainforest in E and NE Madagascar, and are declining through loss of habitat, particularly due to fire; they are also hunted for their meat. Rarely seen in zoos, they are found in a few reserves including Ankarafantsika.

371
372
373
374
375
377
378
379
380
381
♂
♀

******* AYE-AYE**

Daubentonia madagascariensis **(382)**

HB 14.2–17.3ins; T 19.7–23.6ins; WT ca.4.4lb.

The only living member of its family, found in forests, bamboo and mangrove, where they are arboreal and nocturnal, sleeping by day in nests. They feed mostly on fruit and insect larvae, by listening for insects in decaying wood, and biting them out with their powerful teeth, using the long third finger to extract the grubs. They extract the pulp of coconuts in a similar manner. Destruction of forest habitat has led to their decline: are now reduced to scattered individuals in NE and NW coastal Madagascar. Nine were released in 1967 and are reported to be breeding on Nosy Mangabé Island, a reserve off NE Madagascar. Although they have been known to live for over 23 years in captivity, they are no longer normally seen in zoos.

***** INDRIS**

Indri indri **(383)**

HB 24–35.5ins; T 2–2.4ins; WT 15.4–22lb.

Thick, silky fur rather variable in colour; can be patterned with grey, brown, white and black. Diurnal in coastal and montane moist forest, up to ca. 5900ft. Largely arboreal, leaping from tree to tree, feeding on a variety of leaves, flowers and fruit. Indris live in small groups (2–5); their song is audible to humans for up to 1.2 mls. Found in NE Madagascar, their range has been reduced by logging, but otherwise they do not appear to be threatened. Fully protected by law; in the past they were protected by local taboos. They occur in a few parks and reserves, including Perinet-Analamazaotra Faunal Reserve, Madagascar, which was mainly set up to preserve them. Rarely exhibited in zoos and are not being bred in captivity.

***** VERREAUX'S SIFAKA**

Propithecus verreauxi **(384)**

HB 15.4–18.3ins; T 19.7–23.2ins; WT 7.9–9.5lb.

Colour and pattern variable, but mostly white or yellowish-white with black or reddish patches on arms, legs and head. The arms are comparatively short, connected to the body by a small, gliding membrane. An extremely agile species, leaping up to 33ft. They feed mostly on fruit, flowers and leaves, and live in groups of 2–13 led by a male. Found in W and S Madagascar, where they are threatened by the spread of human habitations, which are isolating many populations. *P. v. coronatus* occurs in the Ankarafantsika Nature Reserve (240 sq.mls); *P. v. deckeni* occurs in the Tsingy de Namoroka Reserve (88 sq.mls). The closely related ***** Diademed Sifaka**, *P. diadema* **(385)**, which occurs in E and N Madagascar, is also threatened in the north of its range. *P. diadema perrieri* was thought to be reduced to under 500 by 1972. Verreaux's Sifaka has been kept and even bred in captivity occasionally, but the Diademed has never been bred, and rarely survives long in captivity.

***** RUFFED or VARIEGATED LEMUR**

Varecia variegata **(386)**

HB ca.23.6ins; T 23.6ins; WT 6.6–11lb.

The largest living lemur, with long, soft fur, is variable in colour, with an irregular pattern, often markedly asymmetrical. Family groups consist of a pair and up to 3 offspring. They are arboreal, living in rainforest, feeding on fruit, leaves, buds and other vegetable matter. Threatened by loss and fragmentation of their habitat, but fully protected by law and found in several reserves within their range, including Nosy Mangabé. Exhibited in many zoos, where they breed freely; however, most of the zoo stock are of rather mixed ancestry, and relatively few are identified to particular populations.

***** WEASEL LEMURS**

Lepilemur spp.

HB 11.8–14.8ins; T 10–12ins; WT 1.1–2lb.

The 7 species are generally rather similar in appearance, and until recently were usually treated as a single variable species. Colour ranges from greyish to rufous or brownish above, paler below. They are nocturnal forest dwellers, spending the day asleep in a hollow tree or dense foliage, and feed mostly on leaves and blossoms. All are threatened to some degree by habitat loss. The **Greater Weasel Lemur** *L. mustelinus* **(387)** is found in the northern forests of E Madagascar; the **Island Weasel Lemur** *L. dorsalis* **(388)** is restricted to the island of Nosy Bé and the adjacent NW coastal region; the **Light-necked Weasel Lemur** *L. microdon* **(389)** in the southern part of E Madagascar, from Tamatave to Fort Dauphin; the **Dry-bush Weasel Lemur** *L. leucopus* **(390)** in the extreme south between Fort Dauphin and Ambovombe; the **Lesser Weasel Lemur** *L. ruficaudatus* **(391)** in forests of SW Madagascar; **Milne-Edwards' Weasel Lemur** *L. edwardsi* **(392)** in W Madagascar and the **Northern Weasel Lemur** *L. septentrionalis* **(393)** in the northern tip of Madagascar. The Dry-bush, the Lesser and the Island Weasel Lemurs are probably the rarest and most threatened. Weasel lemurs are rarely exhibited in zoos; they occur in Ankarafantsika Nature Reserve, Lokobe Nature Reserve on Nosy Bé, and several others.

*** GREY GENTLE LEMUR**

Hapalemur griseus **(394)**

HB 10.6–12.2ins; T 12.6–15.8ins; WT up to 2.2lb.

A short-legged, thickset lemur, with a long, bushy tail. Found in humid and swampy forests and bamboo thickets. There are three populations – *H.g.griseus* which is found in SE Madagascar from Fort Dauphin to Fianarantson; *H.g. occidentalis* found in N and W Madagascar; and *H. g. alaotrensis*, restricted to the Lake Alaotra region in mountains up to 3250ft. Although still locally abundant, they are declining due to habitat loss and fragmentation, through burning and drainage for rice cultivation, as well as extensive hunting. Small numbers are kept in zoos and occasionally bred. They occur in the Montagne d'Ambre NP and a few other protected areas.

**** BROAD-NOSED GENTLE LEMUR**

Hapalemur simus **(395)**

HB 17.7ins; T 17.7ins; WT up to 5.5lb.

A short-legged, thickset lemur, with a bushy tail. Found in reed-beds, bamboo, sugarcane and similar habitats. They are diurnal, feeding on plants, including grass, sugarcane, bamboo and reed. Found only in a single population in E Madagascar.

NB. All species above are listed on CITES App. I and Class A of the African Convention.

382
383
384
385
386
387
394
395
388
389
390
391
392
393
?
386
383
382
387
394
384

MARMOSETS and TAMARINS

Many of the marmosets and tamarins are threatened, at least locally, by forest clearance, and the species described are only those most at risk and/or better documented (*see Living New World Monkeys* (1977), by P. Herschkovitz and *The Primates of the World* (1983), by Jaclyn H. Wolfheim in the Bibliography). They breed readily in captivity and there is no reason why self-sustaining populations of all species and subspecies could not be maintained.

TAMARINS

Saguinus spp.

HB up to 11.8ins; T up to 17.3ins; WT up to 2lb.

Rather variable in appearance; often considerable variation within the species, with many subspecies described. Found mainly in tropical forests, they are diurnal and arboreal, feeding on fruits, blossoms and small animals.

*** COTTON-TOP or CRESTED TAMARIN

Saguinus oedipus **(396)**

Found in forest edges, and also in secondary forests from Costa Rica south to NW Colombia. They can benefit from some selective logging which opens up forest habitat; it is possible that since the arrival of European colonists, some populations have increased, due to the decline of the intensive agriculture practised by the Amerindians. However, deforestation is causing declines in most populations. The western population *S.o. geoffroyi* is sometimes treated as a full species, and the eastern Colombian populations are the most threatened. Both subspecies are frequently kept in captivity and breed freely, with self-sustaining populations.

*** EMPEROR TAMARIN

Saguinus imperator **(397)**

Found mainly in the vine-canopy of forests in SE Peru, W Brazil, and NW Bolivia. Although still common in some areas, believed to have declined in parts of Peru; they are hunted for food and their teeth are used as ornaments. They occur in the Manu NP, Peru, and are listed on CITES App.II. Small numbers are kept in captivity, which may be self-sustaining.

* BARE-FACED TAMARIN

Saguinus bicolor **(398)**

Found in a small area of Brazil, north of the River Amazon, in a variety of forested habitats. Appear to be adaptable and survive in disturbed forests, even within the outskirts of the city of Manaus. However the effects of the extensive habitat changes within their range, and the rapid growth of Manaus in particular, may cause a decline.

*** WHITE-FOOTED TAMARIN

Saguinus leucopus **(399)**

Confined to a small area in N Colombia, its range has been reduced by forest clearance, and although apparently able to survive in disturbed areas and secondary forest, its range is now much fragmented. It is protected by law, and listed on CITES App.I. Does not occur in any protected areas.

MARMOSETS

Callithrix spp.

HB up to 11.8ins; T up to 15.8ins; WT up to 1lb.

Like tamarins, marmosets are forest-dwelling, diurnal and arboreal, feeding on small animals and plant matter; they have modified lower canines which are used for cutting tree bark to gather sap and gums. They live in groups of up to 12 and have 1–2 young.

*** SILVERY MARMOSET

Callithrix argentata **(400)**

Found in both primary and secondary forest, in two widely separated areas. Still common in some areas and found in protected areas. However, *C.a.leucippe* which is confined to a small area in Brazilian Amazonia, is threatened by forest clearance to create cattle pasture, and the construction of the Trans-Amazonian Highway. There are 2 other subspecies: *C.a.melanura* which occurs in C Brazil and adjacent Bolivia and is believed to be common, and *C.a.argentatus*, which occurs further north, and is not believed to be in any danger. Although Silvery Marmosets are bred in captivity, there are no captive *C.a.leucippe*. Listed on CITES App.II.

*** COMMON or TUFTED MARMOSET

Callithrix jacchus **(401)**

There are 5 well-marked subspecies which are often treated as full species; *C.j.jacchus*, *C.j.penicillata*, *C.j.geoffroyi*, *C.j.flaviceps* and *C.j.aurita*. The Common Marmoset is still common locally, has a large range and is adaptable in a wide variety of mostly forested habitats, although clear-cutting, burning and the spread of agriculture are increasingly fragmenting and isolating many populations. *C.j.aurita* has a very small range in Atlantic coastal forests of SE Brazil; *C.j.flaviceps* is confined to a small area of SE Brazil, where much of the forest has already been destroyed. Although Common Marmosets occur in national parks and reserves they have disappeared from some – possibly due to epidemic disease. Listed on CITES App.II; *C.j.aurita* and *C.j.flaviceps* are listed on App.I and are considered severely threatened. They have, nevertheless, been used in biomedical research and also extensively traded as pets.

*** SANTAREM MARMOSET

Callithrix humeralifer **(402)**

Little is known of this species, but it is thought to be seriously threatened; the Trans-Amazonian Highway runs through its range. There are no reserves within its range but it is protected in Brazil and is listed on CITES App.I.

* PYGMY MARMOSET

Cebuella pygmaea **(403)**

HB up to 5.9ins; T up to 11.8ins: WT up to 4.9oz.

The smallest marmoset and one of the smallest living primates. Although not generally threatened, and occurring in a few well-protected areas, they may be vulnerable in the future.

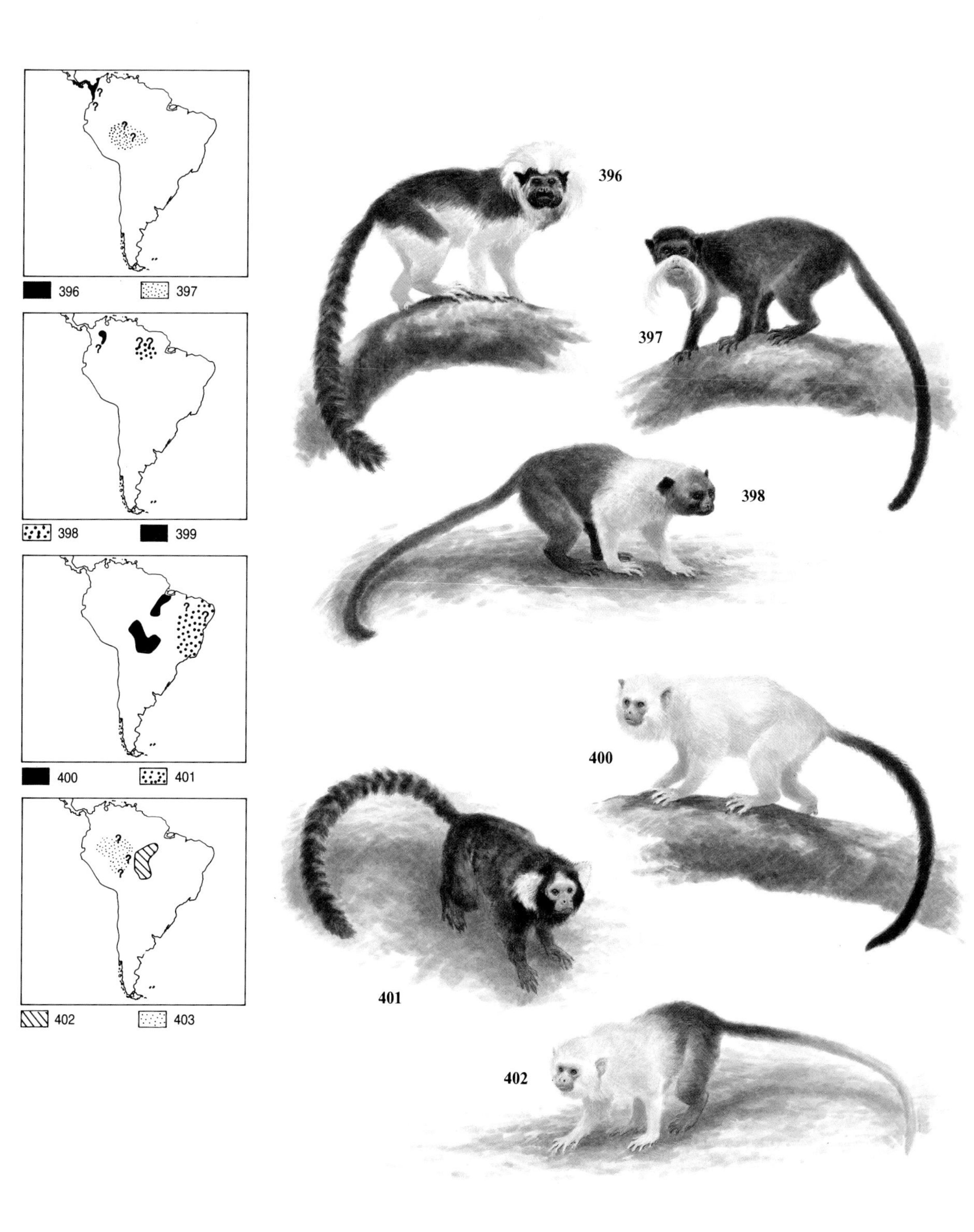
396
397
398
399
400
401
402
403
396
397
398
400
401
402

*** GOELDI'S MARMOSET
Callimico goeldii (**404**)

HB 8.3–9.3ins; T 10–12.8ins; WT 13.7–30.1oz.

Dark brown above, with lighter markings, and paler rings on the tail; although superficially similar to marmosets and tamarins, they have many affinities with the South American monkeys (Cebidae). Found mostly in thick forest where they are naturally rather sparsely distributed, feeding on invertebrates and other small animals, as well as fruit. They live in small groups and normally have a single young, which the males may help rear. Rare throughout their range, and slash-and-burn agriculture has destroyed much of their habitat in parts of Colombia. They occur in the Manu NP, Peru, and are listed on CITES App.I.

*** RED-BACKED SQUIRREL MONKEY
Saimiri oerstedii (**405**)

HB ca.11.8ins; T ca.15.8ins; WT up to ca.2.2lb.

Closely related to, and sometimes treated as, a subspecies of the more widespread and abundant Common Squirrel Monkey *S. sciureus.* Their ranges are separated by several hundred kilometres; the Red-backed Squirrel Monkey is confined to Panama and Costa Rica. They live in troops of up to 35 individuals. The single young is carried on the mother's back for the first few weeks. They feed on insects and other small animals, and also fruit and other vegetable matter. Their range has been considerably reduced by forest clearance, they have also been extensively traded, and insecticide spraying against malaria and yellow fever is reported to have harmed them. They occur in Cuenco Corcovada Reserve in Costa Rica, and are listed on CITES App.I. There are small numbers in captivity, and they are now being bred regularly.

***** LION TAMARINS
Leontopithecus (=Leontideus) spp.

HB 7.9–13.4ins; T 12.4–15.8ins; WT 12.6–24.8oz.

The fur is long and silky, particularly on the shoulders, giving a 'lion'-like mane. There are three well-marked populations, variously regarded as full species or well-marked subspecies of a single species; the only obvious differences are in the proportions of gold and black colorations. They are confined to tropical forest in SE Brazil, sometimes occurring in secondary forest and cultivated areas. They usually live 3–10m up in trees covered with vines and epiphytes; they are diurnal, feeding on insects and small vertebrates and fruit. The **Golden Lion Marmoset** *L. rosalia* (**406**) was estimated to number less than ca.100 by 1980, with the majority in the Poco d'Anta Reserve; however there were about 211 in captivity in 1981, including a small population in the Rio de Janeiro Primate Centre, Brazil. In 1984 the first 10 animals were reintroduced into the wild from captive-bred stocks. A revised estimate in 1984 put the world population at 331, with 115 in the Reserve. The **Golden-headed Lion Tamarin** *L. chrysomelas* (**407**) was estimated at 400–500 in 1972, ca.200 in 1977, and even less by 1981. In 1980 most of the area in which they are found was declared a reserve, but it is poorly protected. The rarest lion tamarin is the **Golden-rumped** or **Black Lion Tamarin** *L. chrysopygus* (**408**), which is confined to 2 areas, both reserves in SE Brazil: Morro do Diabo and Caitetus; it had been considered extinct, but was rediscovered in 1970 with a population of ca.100. By 1985 the wild population was estimated at 75–100, but a small captive colony of 25 at the Rio de Janeiro Primate Centre was increasing. The main cause of the decline of all 3 populations was undoubtedly the destruction and fragmentation of their habitat; but in recent years the main threat has come from collecting for zoos and private collections. The captive populations are probably the most closely managed of any species of threatened mammal, with a high degree of international cooperation. Listed on CITES App. 1.

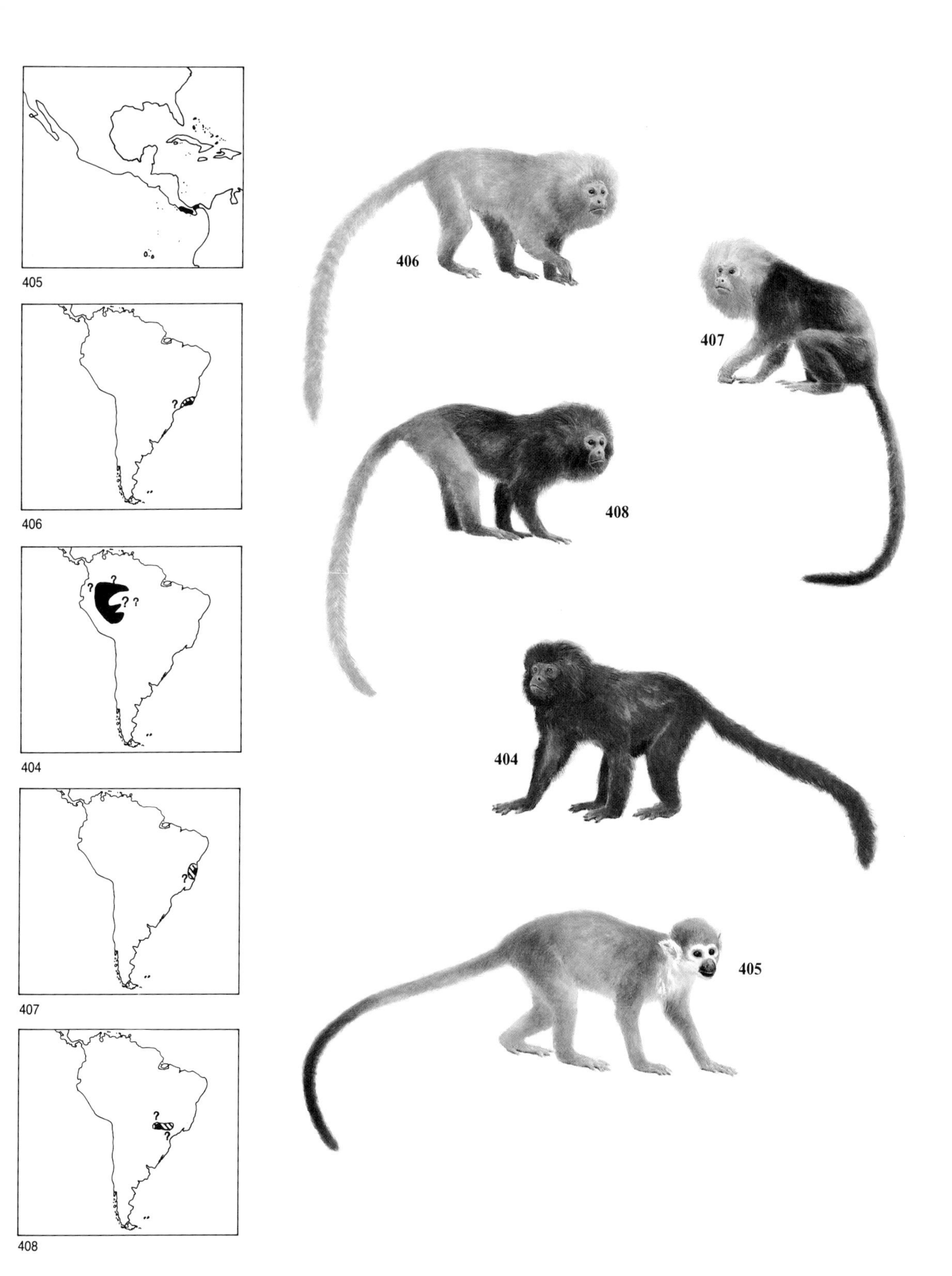
405
406
408
404
407
406
407
408
404
405

*** MASKED TITI

Callicebus personatus **(409)**

HB 11.8–19.7ins; T 16.2–21.7ins.

Confined to Atlantic coastal forests and swamp forests in Brazil, where widespread habitat destruction is causing their range to become increasingly fragmented. They are also extensively hunted for food. Already extinct in many parts of their former range, although they do occur in several protected areas where they appear to be thriving.

*** MANTLED HOWLER

Alouatta palliata **(410)**

HB 18.7–25.2ins; T 19.7–26.4ins.

Their range stretches from Tabasco, Mexico, south through Central America to coastal Colombia, Ecuador and possibly NW Peru. Found in forests, including mangroves, swampy coastal forests and cloud forest up to altitudes of 6550ft. Over much of their range, but particularly lowland areas, their habitat is being destroyed by logging and other forms of development. They are also hunted in many parts of their range, and this is thought to be responsible for local extinctions. Occur in national parks and reserves in Nicaragua, Colombia, Costa Rica, and protected by law in Mexico. Rarely seen in captivity and there are no self-sustaining populations. Listed on CITES App.I.

*** GUATEMALAN HOWLER

Alouatta pigra **(411)**

Live in undisturbed, dense forest from sea level to 800ft. They are found in one of the relatively few areas of extensive undisturbed rain forests in Central America, in the Tikal NP, Guatemala. In Mexico, and other parts of their range, they are declining since they do not adapt well to disturbed habitats. Became extinct in most of the Yucatan Peninsula before 1917. Listed on CITES App.II.

*** BROWN HOWLER

Alouatta fusca **(412)**

HB ca.23.5ins; Tca.23.5ins.

Males darker than females. Found in forests of SE Brazil, NE Argentina, and possibly N Bolivia, at altitudes of 1300–3750ft; these forests are now extremely fragmented, and continue to be destroyed. Also hunted for meat. They occur in several national parks and reserves and are protected by law in Brazil. Like other howlers they are rare in zoos and do not thrive. Listed on CITES App.II.

*** WHITE-NOSED SAKI

Chiropotes albinasus **(413)**

HB ca.15ins; T ca.16.5ins.

Shiny black fur with stiff whitish hairs on the nose and lip, and bare red skin on the face. Found in tropical forests, including those that are seasonally flooded, where they feed mostly on seeds, as well as fruits, flowers and other vegetable matter. They occur in C Brazil, south of the River Amazon; throughout their range they are rather sparsely distributed, and decreasing locally, partly due to destruction of habitat. They are also extensively hunted for food, for use as bait for cat traps, and the tail is also used as a duster. They occur in at least one national park, are totally protected and listed on CITES App.I. Rarely seen in zoos.

*** BLACK-BEARDED SAKI

Chiropotes satanas **(414)**

HB ca.17.7ins; T ca.13.8ins; WT ca.3.3lb.

The colouring varies through shades of reddish, chestnut, and blackish. Like the White-nosed Saki, it has a thickly-furred tail. Found mainly in humid tropical rain forest, usually near streams or rivers, where they live in troops of up to ca.30. There are two populations: *C.s. chiropotes* which occurs north of the Amazon, and *C.s. satanas* which is confined to Brazil, south of the River Amazon, in an area which has an extremely dense human population, and where much of the forest has been destroyed. They have protection of some sort in most countries in their range, and occur in several protected areas; *C.s. satanus*, which is already extinct over much of its former range, occurs in several reserves. They are very rare in captivity.

*** RED AND WHITE UAKARIS

Cacajao calvus **(415)**

HB 12.2–16.2ins; T 3.9–6.3ins; WT 2.2–3.3lb.

Uakaris are the only short-tailed New World primates. This species has two well-marked subspecies, often regarded as full species. The **Red Uakari** *C.c. rubicundus*, is bright reddish-brown or chestnut, with a vermilion face; the **White Uakari** *C.c. calvus* is whitish with a scarlet face. They are confined to forests in the floodplains of rivers and live in troops of up to 50 or more. Their habitat appears to have suffered relatively little damage, and the main reason for their decline is probably over-hunting, mostly for food, but they have also been trapped for pets and for the export trade. They are totally protected by law in Brazil and Peru, and listed on CITES App.I. They do not occur in any protected areas. Very few are kept in captivity, but they do breed regularly.

*** BLACK-HEADED UAKARI

Cacajao melanocephalus **(416)**

HB ca.18ins; T ca.6.3ins.

The general colouring is chestnut brown on the upperparts, with the underside, head, feet and hands black. Found in forests up to an altitude of 1650ft in Brazil, Venezuela and Colombia. Although in the Brazilian part of their range they are often very abundant, elsewhere they are considered seriously threatened. They are extensively hunted in parts of Colombia, despite being protected. Also protected by law in Brazil, and listed on CITES App.I. Occur in La Macarena NP. They are only rarely seen in zoos.

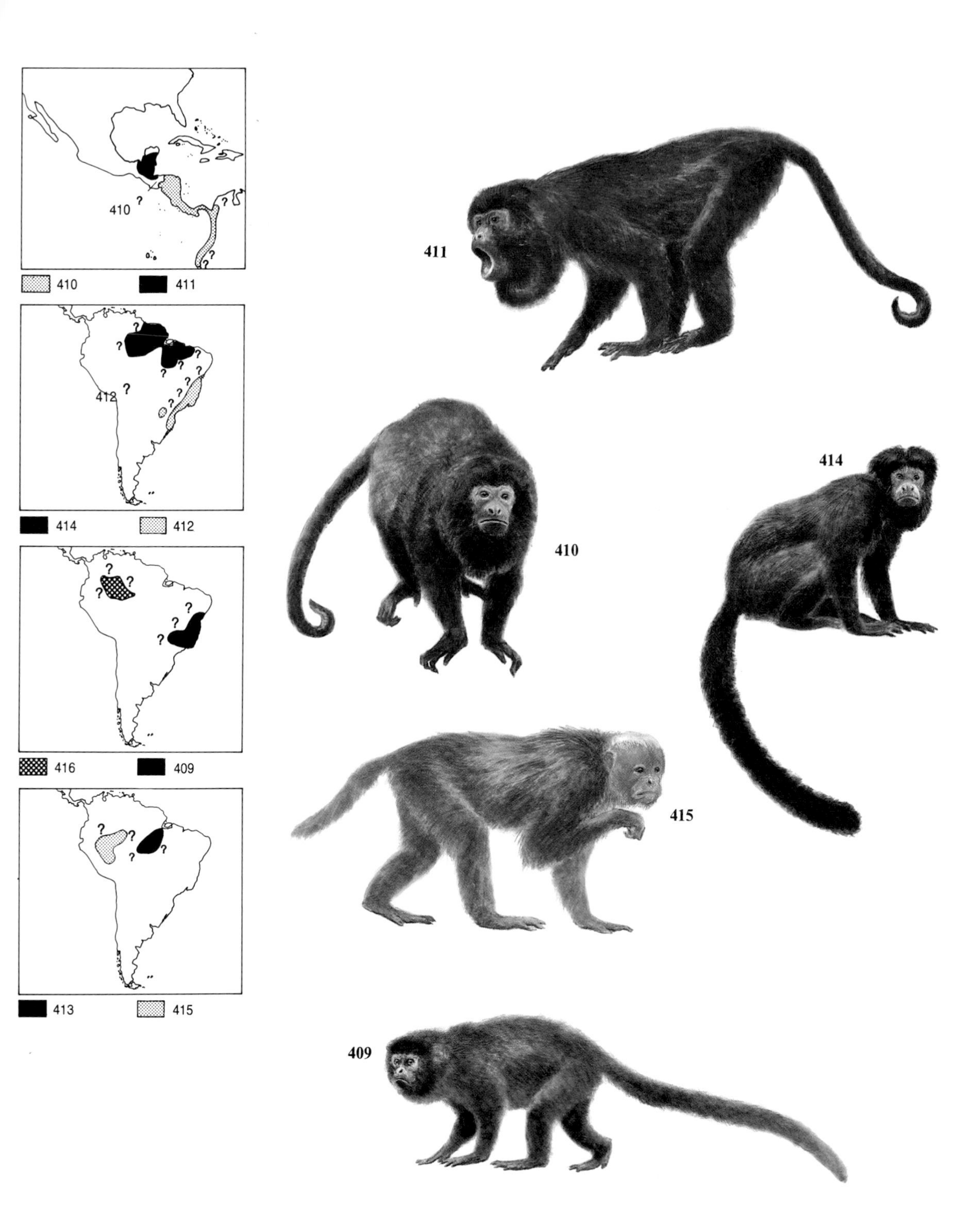

410
410
411
414
412
412
416
409
413
415
411
410
414
415
409

*** COMMON or HUMBOLDT'S WOOLLY MONKEY

Lagothrix lagothricha (**417**)

HB 19.7–27.2ins; T 23.6–28.4ins; WT up to 24.2lb.

One of the largest species of monkey in South America, with woolly, short, dense fur usually greyish, or dark brown, but sometimes yellowish-buff; the face is naked, usually black. The tail is bare on the underside near the tip, and strongly prehensile. Found in a variety of forests, at altitudes of up to 9850ft, but rarely in secondary forests. They live in groups of 20–70 moving about half a mile a day, feeding on fruit together with some leaves, seeds and occasionally insects. Occurs in W Brazil, north to Colombia and Ecuador, Peru and Bolivia, with a small population in Venezuela. The population from SW Brazil and SE Peru (*L. l. cana*) is sometimes regarded as a full species. One of the main causes of their decline has been hunting; their flesh is highly prized in many areas, and their fat provides a grease for cooking. In the mid-1960s they were already being noted as declining due to overhunting. They have also been popular as pets and in zoos; to obtain the young the mother is normally killed. Their habitat is vanishing in many areas. Found in several national parks and reserves throughout their range, but legal protection for them is often weak.

*** YELLOW-TAILED or HENDEE'S WOOLLY MONKEY

Lagothrix flavicauda (**418**)

Similar in size to the Common Woolly Monkey, they are deep mahogany red with a yellow stripe around the naked patch on the prehensile tip to the tail, and the males have a tuft of yellow fur on the scrotum. Restricted to N Peru in montane rain forests at altitudes of 1650–9850ft. Much of their habitat has been destroyed to provide land for agriculture and cattle ranching, the new roads providing easy access to the area. They are hunted extensively for their flesh, and the mothers are shot in order to obtain babies and juveniles for pets. Between its original discovery in the early 1800s and the 1930s, only five specimens were known, and by the 1970s it was believed extinct. However, in 1974 captive animals and small wild populations were discovered and a conservation programme has now been developed by WWF. A captive breeding programme has been proposed in Lima Zoo, and a campaign to protect its habitat has been launched.

*** WOOLLY SPIDER MONKEY

Brachyteles arachnoides (**419**)

HB 18.1–24.8ins; T 25.6–31.5ins; WT 13.2–22lb.

Intermediate in appearance between woolly monkeys and spider monkeys; the fur is generally greyish, sometimes with a rufous tinge, particularly around the conspicuous genitalia. Occur at altitudes of up to 4900ft, in Atlantic rainforest on the coast of Brazil, which since the beginning of the present century has been almost totally destroyed; they have also been hunted for their meat. Its original population, before the arrival of European colonists, is thought to have been ca.400,000; in 1971 it was ca.3000, dropping to 1,000 in 1977. They have been protected since 1967, and occur in 5 national parks as well as several other protected areas; the largest population occurs in the Serra da Bocaina NP, Brazil. They are listed in CITES App.I. Rare in captivity; there are no self-sustaining captive populations.

SPIDER MONKEYS

Ateles spp.

HB 15–24.8ins; T 19.7–35.1ins; WT up to ca.13.2lb.

There is some disagreement over the classification of spider monkeys, some zoologists lumping them together as a single species, and others splitting into four or more species. They are very variable, and do hybridise; for the purposes of conservation it is important to identify as many discrete populations as possible, and here they are treated as four species, distinguished mostly by coloration and range. They are forest dwellers, found in both rain forest and dry forest, and also in other habitats including mangroves. All *Ateles* spp. lack thumbs. Their main threat is deforestation, extensive in many parts of their range, and many populations are now isolated and fragmented. They are also extensively hunted for their meat, which is the major threat to their survival in some areas. They were formerly collected in large numbers for the pet and zoo trade; over 12,000 were imported into the USA between 1968–1973.

*** BLACK-HANDED or GEOFFROY'S SPIDER MONKEY

Ateles geoffroyi (**420**)

Rather variable in colour, but usually grey or buff above, paler below, with dark face, hands and feet. Nine subspecies are usually recognised, but they most probably intergrade. Found in Central America from S Mexico to S Panama and NW Colombia. Although several subspecies are seriously threatened, others are still locally abundant and some are breeding well in captivity. *A. g. frontatus* and *A. g. panamensis* listed on CITES App.I; all others on App.II.

*** LONG-HAIRED SPIDER MONKEY

Ateles belzebuth (**421**)

Rather variable, but usually fairly dark above, often paler below, but can be almost entirely blackish; usually has pale eyebrows. Three subspecies occur in two widely separated areas: *A. b. marginatus* occurs south of the Amazon in Brazil, and *A. b. belzebuth* in W Brazil, Peru, Colombia, Ecuador and Venezuela; in N Colombia and W Venezuela *A. b. hybridus* occurs. Found in several parks and protected areas, and kept in a few zoos.

*** BROWN-HEADED SPIDER MONKEY

Ateles fusciceps (**422**)

Generally dark brown or black. Two poorly-defined subspecies occur from S Panama to SW Colombia and W Ecuador. Found in several protected areas, and also kept in many zoos, where they breed freely.

*** BLACK SPIDER MONKEY

Ateles paniscus (**423**)

There are two populations, widely separated: *A. p. paniscus* in the east, and *A. p. chamek* in the west. Probably the best protected of the spider monkeys, occurring in at least 20 national parks and other protected areas, and also in substantial numbers in captivity.

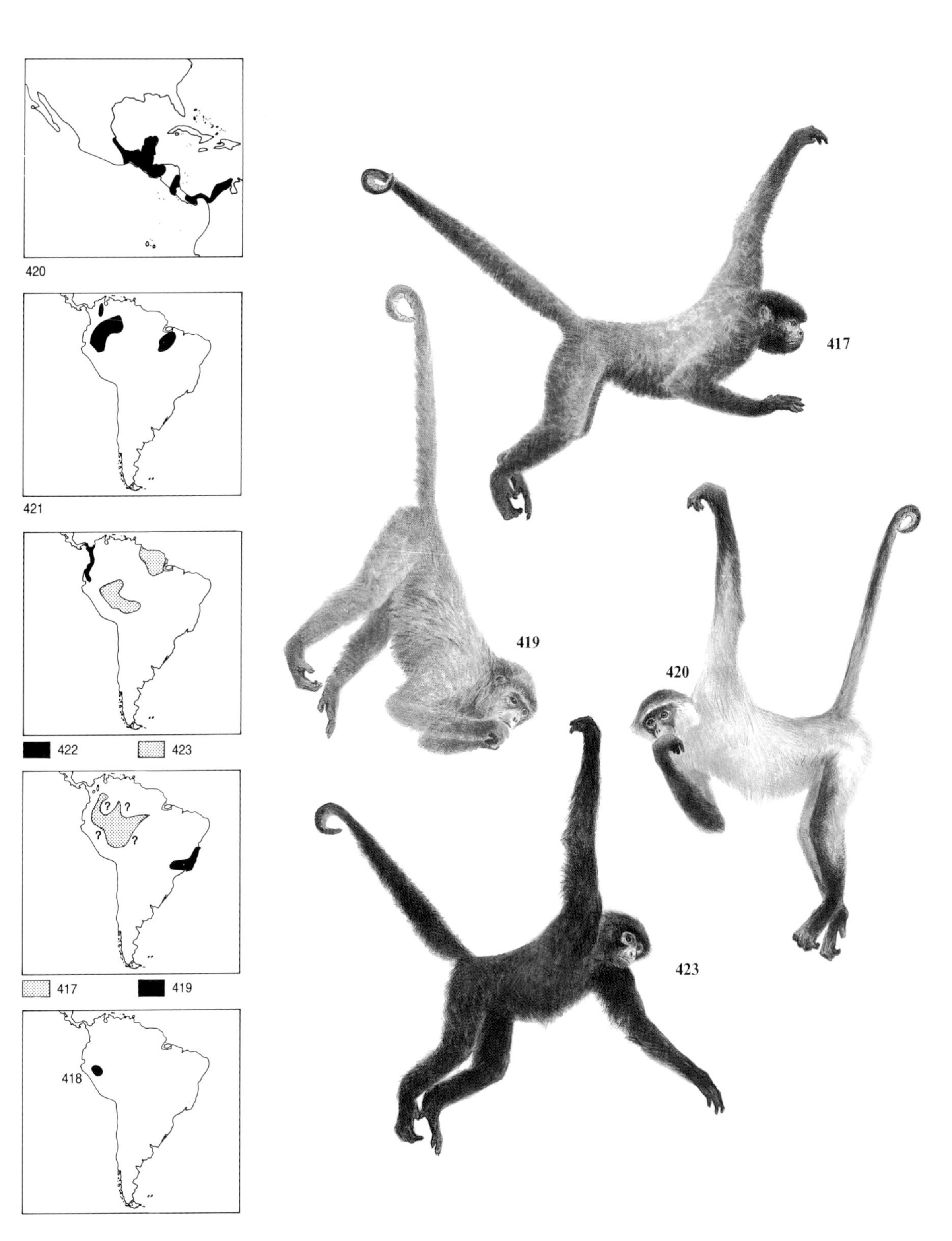
420
421
422
423
417
419
?
418
417
419
420
423

*** DIANA MONKEY
Cercopithecus diana (**424**)
HB 15.8–22.5ins; T 20.5–32.3ins; WT 4.8–16.5lb.
Male larger than female. A slender monkey that carries its tail in a wide curve. Live in primary rain forest, where they are arboreal, in troops of up to ca.50. By the 1950s had already become rare in Ghana, Liberia, and Ivory Coast; their decline continues in these countries mainly due to destruction of forests, but they are also hunted for their flesh. Not known to damage crops but treated as pests, and due to striking coloration, are popular as pets. Occur in the Tai NP, Ivory Coast ; Bia Forest Reserve, and other reserves in Ghana. Listed on CITES App.II and Class A of the African Convention; they are still fairly common in zoos and breed freely.

** RED-EARED MONKEY
Cercopithecus erythrotis (**425**)
HB 15.8–21.7ins; T 18.1–30.3ins; WT 5–9.4lb.
Male larger than female. Rather variable with 3 distinct subspecies. *C.e. erythrotis* from the island of Fernando Po and adjacent mainland has a brick-red nose patch and white throat; *C.e. camerunensis* from N Cameroon has a greyish throat and red nose; *C.e. sclateri* from S Nigeria has a whitish nose patch. Little is known of its status in its primary and secondary forest habitat. It is believed to be declining in Cameroon due to forest destruction, and is also hunted. Has restricted range so must be considered potentially endangered. It occurs in at least 4 protected areas in Cameroon.

*** L'HOEST'S MONKEY
Cercopithecus lhoesti (**426**)
HB 17.7–27.6ins; T 18.1–31.5ins; WT 6.6–17.6lb.
Male larger than female. The tail is slightly prehensile. Rather variable colouring but generally dark, with a white chin; male has a bright mauve scrotum. They are found in troops averaging 17, in montane forests up to 8200ft, feeding on a wide variety of plant matter and occasionally raiding crops. Although locally abundant in Central Africa, their range is shrinking in the west because of habitat destruction. The western population. *C.l. preussi* is in a critical state outside reserves. Listed on CITES App.II and Class B of the African Convention. Comparatively rare in zoos.

***** RED-BELLIED MONKEY
Cercopithecus erythrogaster (**427**)
HB ca.17.7ins; T ca.23.6ins; WT 4.4–6.6lb.
Only known from SW Nigeria and possibly Benin.

*** BLUE MONKEY
Cercopithecus mitis (**428**)
HB 17.3–26.4ins; T 24.8–33.5ins; WT 13.2–26.5lb.
An extremely variable species, with several well-marked subspecies; the colour on the upperparts varies from bluish-grey to bright golden, greenish or rufous; there are varying amounts of white on the face and underparts. They feed on a wide variety of vegetable matter depending on habitat, and are found in a range of habitats including high altitude bamboo forest, swamps and lowland forests. Their distribution, once mostly continuous, is increasingly fragmented and isolated, particularly around the edges and in South Africa, and several of the subspecies may be threatened. The race found in the Virunga Volcanoes, *C.m. kandti*, the **Golden Monkey**, numbers a few hundred, and appears to coexist with subspecies, *C.m. doggetti*, which suggests it may in fact be a full species. Earlier this century many skins were used in the fur trade, leading to local extinctions, at least in Uganda.

Other Cercopthecine monkeys:
* **De Brazza's Monkey** *C. neglectus* (**429**) may be declining in eastern parts of its range, but elsewhere, in its extensive C African distribution, it is probably safe. The * **White-nosed Monkey** *C. nictitans* (**430**) and the * **Spot-nosed Monkey** *C. petaurista* (**431**) have declined in parts of W Africa where their rainforest habitat has been cleared.The * **Dryas Monkey** *C. dryas* (**432**) from Zaire, and the * **Owl-faced Monkey** *C. hamlyni* (**433**) from Nigeria and the Cameroon are both little studied species, status unknown. The ** **Salongo Monkey** *C. salongo* (**434**) only been found in a single location in Zaire; it was first recognised in 1977, but may not be a valid species.

* ALLEN'S MONKEY
Allenopithecus nigroviridis (**435**)
HB ca.17.7ins; T ca.19.7ins.
Greyish or blackish face, with lighter colouring around eyes; body fur grey or blackish. Little is known of status of this species, which is found in NW Zaire and in NE Congo.

** AGILE OR CRESTED MANGEBEY
Cercocebus galeritus (**436**)
HB 17.3–25.6ins; T 17.7–31.1ins; WT 9.9–28.7lb.
There are three separate subspecies: the **Agile Mangabey** *C.g. agilis* (sometimes regarded as a full species) with rather short fur, greyish above, speckled with golden yellow, occurs from S E Nigeria through Zaire; the **Golden-bellied Mangabey** *C.g. chrysogaster* appears somewhat greenish, from south of the Zaire River; and the **Tana River Mangabey** *C.g.galeritus*, rather greyish-brown, occuring in a totally isolated population in the riverine forests of the Tana River, on the east coast of Kenya. This latter population is enclosed by arid semi-desert and the Indian Ocean, in an area of under 8 sq.mls. It is fully protected and occurs in the Tana River Primate Game Reserve. Listed on CITES App.I and in Class A of the African Convention. (The other subspecies are not known to be in immediate danger, although in the west of their range, habitat loss is causing declines.)

* WHITE-COLLARED MANGABEY
Cercocebus torquatus (**437**)
HB 18.1–26.4ins; T 15.8–31.1ins; WT 9.9–27.6lb.
Very variable but generally greyish above, with white underside and around neck, with brownish cap. Found in forests and cultivated lands, often near swamps, living in troops of up to 20 led by old males. They feed on fruits, nuts, seeds and also raid crops. The western population, the **Sooty Mangabey** *C.t. atys* is often considered a separate species. Although their main threat is destruction of their forest habitat, are also hunted for food in some areas, and killed as pests where they raid crops, particularly in the Sierra Leone cacao growing areas. They occur in the Tai NP, Ivory Coast; Bia Forest Reserve, Ghana; Omo Forest Reserve, Nigeria; Korup Reserve, Cameroon, and several other protected areas. Listed on CITES App.II and in Class B of the African Convention. They are regularly bred in zoos. The **Black Mangabey** *C. aterrimus* from Zaire and Angola is a little studied species which may be threatened.

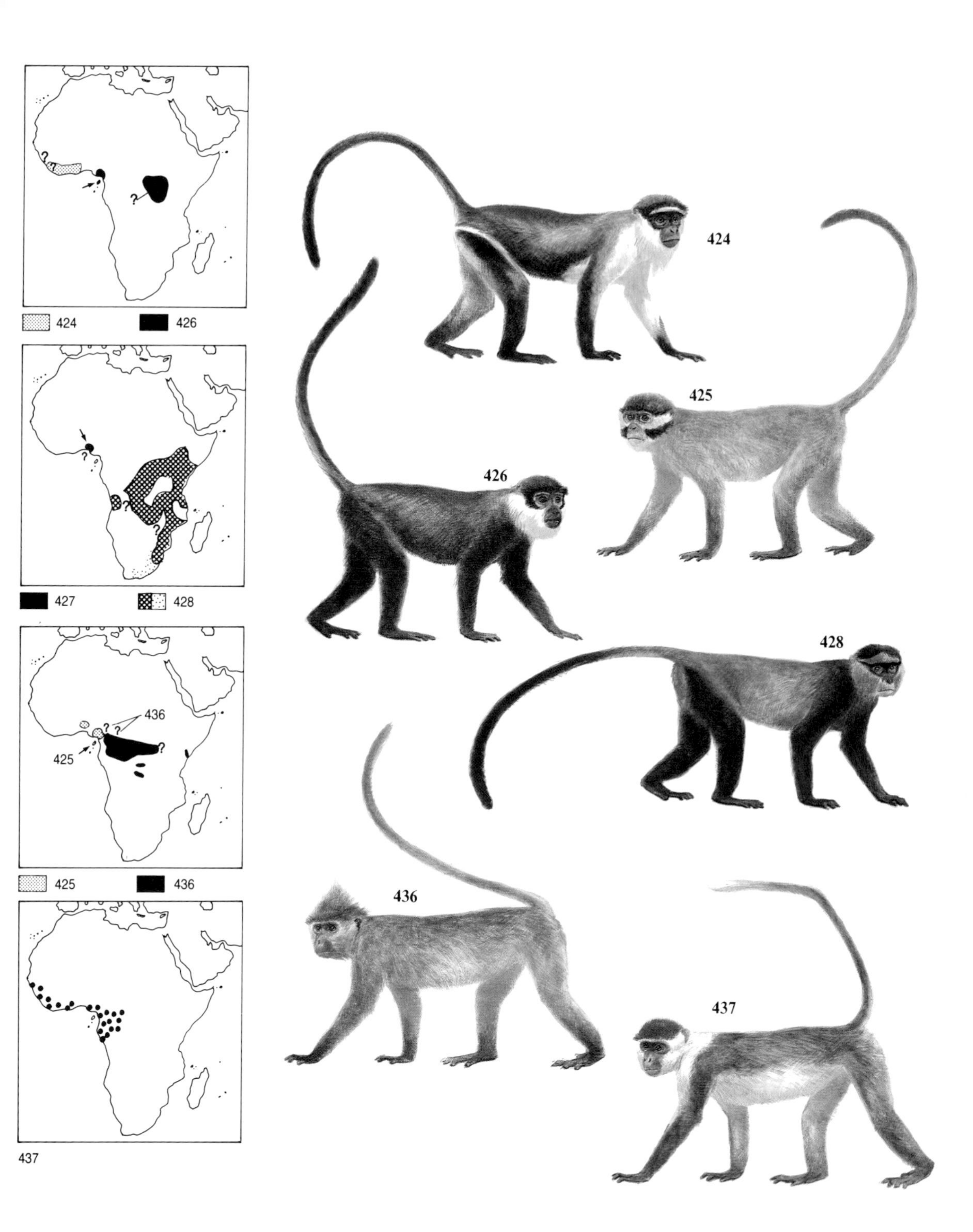
424
426
427
428
436
425
437
424
425
426
428
436
437

***** LION-TAILED MACAQUE or WANDEROO

Macaca silenus (**438**)

HB 19.7–23.6ins; T 9.8–15ins.

Found in tropical forests in the Western Ghats of S India. Habitat has been drastically reduced by the spread of agriculture and teak, coffee, tea and other plantations. Because they appear neither to use plantations, nor even travel through them, their range has become increasingly isolated and fragmented. By 1968 the total population was estimated at less than 1000, less than 800 in 1974 and under 400 by 1975. The minimum size for a reserve is 50 sq.mls, which would support 500 individuals. They were formerly extensively traded, both as pets and also for zoos and research. Now totally protected, they occur in at least four protected areas. One of their strongholds in Silent Valley (Bhavani River) has been threatened by hydro-electric schemes, but has also been proposed as a reserve. Listed on CITES App.I and kept at present in many zoos where they breed freely and are self-sustaining. Should a reintroduction programme be developed there are ample stocks.

* STUMP-TAILED MACAQUE

Macaca arctoides (**439**)

HB 19.1–25ins; T 1.4–3.2ins; WT 17.6–26.5lb.

Occurs in forest from sea level to 7850 ft, from E India and Bangladesh, east to Indochina and south to Peninsular Malaysia. It was noted as being endangered in China in 1937, and in the late 1960s in India and Thailand; more recently it has been considered rare in Malaysia. Little is known of its status elsewhere. It has been used for biomedical research – in particular for manufacture of polio vaccine. Found in a few protected areas and listed on CITES App.II.

* ASSAM MACAQUE

Macaca assamensis (**440**)

Although little is known of its status in many parts of its range, it is not thought to be in any immediate danger. A recently discovered giant form of Assam Macaque from China may prove to be a separate species; it has a very restricted range. The closely related * **Pere David's Macaque** *M. thibetana* (**441**) has declined, but little is known of its current status.

*** PIG-TAILED MACAQUE

Macaca nemestrina (**442**)

HB 18.5–23ins; T 5.5–9.1ins; WT 7.7–19.8lb.

Very variable fur colour, but the short tail is characteristic. Diurnal, living in troops of 15–40, feeding on fruit and small animals. They are the only primates that, when alarmed, descend from trees and flee on the ground. Found in forests and also agricultural areas, where they can do considerable damage. Although locally abundant, in many parts of their range they are declining. They live at low densities and it has been calculated that ca. 1286 sq.mls is needed for a single population of 5000. They occur in an area of 26 sq.mls on Mentawi Island, Indonesia and many other reserves throughout their range. Listed on CITES App.II and bred in substantial numbers in captivity.

*** BARBARY APE

Macaca sylvanus (**443**)

HB 22.1–27.6ins; T absent; WT up to 37.5lb.

One of the few monkeys to lack a tail. Occurs in dry rocky habitats, cedar and holm oak forests. The introduced population on Gibraltar has been augmented with fresh stock from N Africa, but in recent years has been culled and the surplus supplied to zoos. By the 1970s the world population was estimated at 12,840 – 21,340 and continues to decline. Habitat destruction still occurs and they are hunted where they raid crops. They are found in several protected areas in Algeria, but lack proper protection in Morocco. Listed on CITES App.II and Class A of the African Convention.

* CRAB-EATING MACAQUE

Macaca fascicularis (**444**)

HB 12.2–24.8ins; T 12.6–26.4ins; WT 6.6–18.3lb.

Generally similar to the Pig-tailed Macaque, which has a shorter tail. Found in a wide variety of habitats, almost invariably near water; they are active by day in troops of up to 30, led by 2–4 adult males. They feed on fruit, as well as small animals, including insects, amphibians and crabs, and also raid crops. Still locally abundant, but declining in many parts of their range. Research in Malaysia found that 193 sq.mls is the minimum size for a reserve to support a viable population. They occur in several parks and reserves, and are protected around temples in Thailand. Have been exported in large numbers (up to 70,000 a year from Malaysia) but this trade has declined in recent years.

* SULAWESI BLACK APE

Macaca nigra (**445**)

HB 17.5–21.7ins; T 17.3–25.6ins; WT ca.13.2lb.

There are 7 separated populations, sometimes treated as full species: *M.n. tonkeana* from C Sulawesi; *M.n. maura* from SW Sulawesi; *M.n. ochreata* from SE Sulawesi; *M.n. brunnescens* from Muna and Butung Islands, SE Sulawesi; *M.n. hecki* from NW Sulawesi; *M.n. nigrescens* from N Sulawesi and *M.n. nigra* from NE Sulawesi. Virtually nothing is known of the precise status of any of these. Protected since 1970.

Other Macaque species:

* **Japanese Macaque** *M.fuscata* (**446**) may number up to 100,000;occurs on Shikoku, Kyushu and Honshu and several smaller islands, but is believed to be declining rapidly. * **Rhesus Macaque** *M.mulatta* (**447**) is among the most abundant of the macaques, and probably increasing in urban areas; however, in many rural parts of its range it is thought to be declining rapidly; there is little precise data on its status or range. In the past it has been heavily exploited with ca.250,000 exported from India in 1938. * **Toque Monkey** *M.sinica* (**448**) is confined to Sri Lanka, and declining in many areas; in 1976 it was estimated at ca.600,000. * **Formosan Rock Macaque** *M.cyclopis* (**449**) is confined to Taiwan, where it has been forced into the higher altitudes by pressure from human populations, occurring up to 9850 ft (most common from 1970 – 5900 ft).

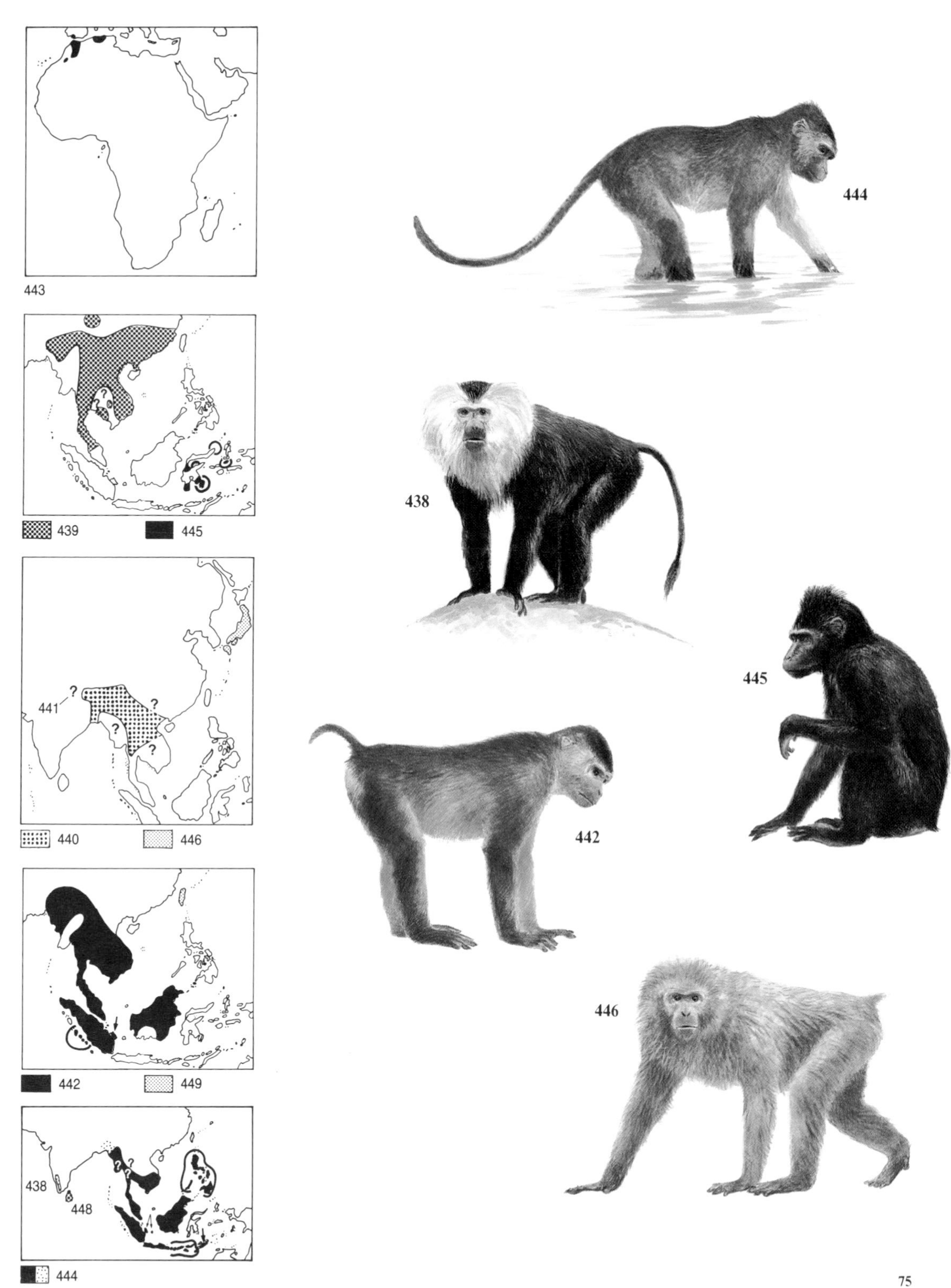
443
439
445
441
440
446
442
449
438
448
444
444
438
445
442
446

***** GELADA BABOON**
Theropithecus gelada (**450**)
HB 19.7–29.6ins; T 15.8–21.7ins; SH 15.8–25.6ins; WT 22–44lb.
Male is noticeably larger than female, with a thick mane and naked patch on the chest. Found in NE Africa in montane habitats at altitudes of 6550–16,400ft, in alpine meadows, rocky gorges, and other generally treeless areas. Live in troops led by an old male, sometimes gathering together into packs of up to 600. The world population was estimated to be ca.600,000 in the late 1970s. Although there is no immediate cause for concern, their range is under increasing pressure from expanding human populations. In addition to being displaced by intensive agriculture, their habitat is being altered by *Eucalyptus* plantations and they are also hunted for the capes of the males, which are used in traditional costumes as well as for sale to tourists. They occur in the Simien NP, Ethiopia. Small numbers are held in zoos, and they are bred regularly.

***** HAMADRYAS or SACRED BABOON**
Papio hamadryas (**451**)
HB 19.7–37.4ins; T 28.8–25.6ins; SH 15.8–25.6ins; WT 22–44lb.
Male is noticeably larger than female with a shoulder mane of long hair (up to 9.8ins). They live in troops consisting of an old male and several females and their young; troops may gather into packs of up to 500. Found in open dry habitats including savannah, but invariably with rocky cliffs and hillsides. They were considered sacred to Thoth, a moon god, in ancient Egypt, and so were more widely distributed. They are now extinct in Egypt, but elsewhere are not thought to be in any immediate danger. However, their limited range and rather specialised habitat requirements, and the spread of agriculture and grazing, means they are increasingly vulnerable. They occur in the Simien NP, Ethiopia. Exhibited in some of the larger zoos; a few are bred each year.

***** MANDRILL**
Papio sphinx (**452**)
HB 21.7–37.4ins; T 2.8–3.9ins; WT 22–66lb.
Male is larger than female, who lacks his brilliant colouring. Restricted to forest habitats in West Africa, and although they are able to adapt to regenerating forests and secondary growth, they are also extensively hunted and trapped. They do not survive in areas where there is intensive agriculture. Groups of up to 250 individuals have a range of up to 20sq.mls. depending on group size. They are declining in Cameroon and Equatorial Guinea, and considered to be one of the most threatened monkeys in Africa. Found in the Wong-Wonue NP, Gabon. They have been involved in international trade both as laboratory animals, and more generally for zoos, because of their spectacular face and posterior. Substantial numbers are kept and bred in captivity.

***** DRILL**
Papio leucophaeus (**453**)
HB 17.7–35.5ins; T 2.4–4.7ins; SH 17.7–23.6ins; WT 22–44lb (max 79.5lb).
A forest species found in lowland rain forest, coastal forest and riverine forest at altitudes of up to 3300ft. They live in troops of up to about 20, consisting of a male, several females and their young; these troops may gather into packs of up to 200. Their habitat is particularly vulnerable to destruction in coastal areas, and they are already extinct in many areas where they formerly occurred; they are considered one of the most threatened monkeys in Africa. They occur in reasonable numbers in the Wonga-Wonue NP, Gabon, and in Korup NP, Cameroon. There are small numbers in many of the larger zoos, where they are bred regularly.

*** CHACMA BABOON**
Papio ursinus (**454**)
The Chacma Baboon is widespread and often extremely abundant in many parts of its range, which extends over most of southern Africa. However, there are reports that it is declining in parts of Zimbabwe, South Africa, and Botswana. The recent droughts in the Sahel may have had an adverse effect on the *** Guinea Baboon** *P. papio* (**455**), formerly abundant and widespread in West Africa; little is known of its current status or precise distribution.

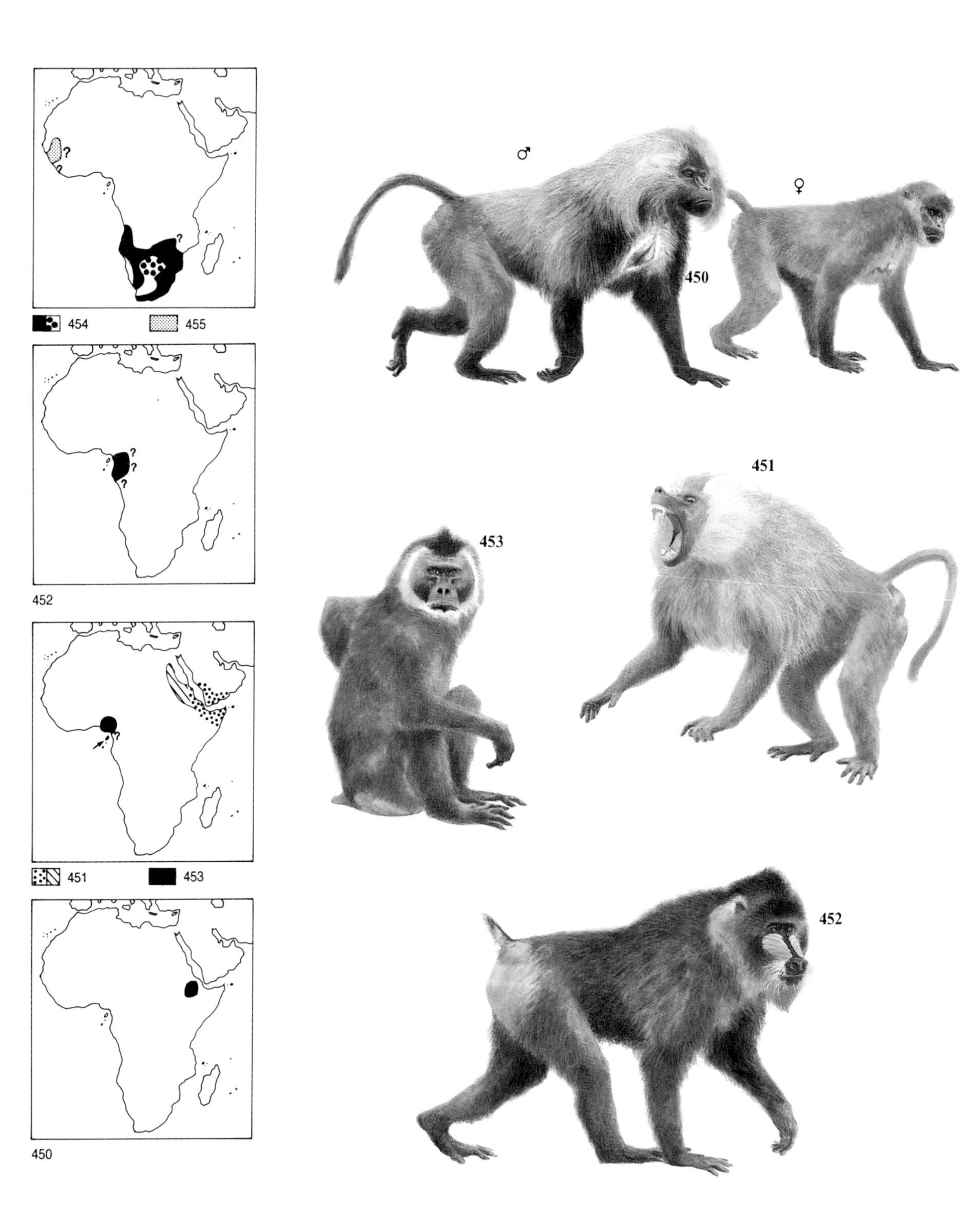
454
455
452
451
453
450
♂
♀
450
451
453
452

LANGURS
Presbytis spp.
HB 16.9–31.1ins; T 0.2–0.4ins; WT 11–39.7lb.
All rather slender-bodied with long tails and long hands. Mostly greyish or brownish above, paler below. Of the 15 species, 10 give cause for conservation concern, but they are probably all declining in some parts of their range. Diurnal forest dwellers, they are generally arboreal, feeding almost entirely on vegetable matter, including leaves, flowers and fruits; they also raid crops. Among those most threatened are as follows:

***** DUSKY or SPECTACLED LANGUR**
Presbytis obscura (**456**)
Prominent pale rings around the eyes, and extremely variable body colouring, which varies, as with most other langurs, both seasonally, regionally and individually. Confined to Malaysia, Peninsular Thailand, and Tenasserim, Burma and a few nearby islands. They live in a wide range of habitats, but generally prefer tall forests, and occasionally raid rubber plantations. They eat up to 4lb of leaves and other vegetable matter per day, together with some insects. Groups consist of 3–20, and it has been suggested that an area of at least 193sq.mls is needed for a population of 5000. Although still locally common in Thailand and possibly Burma, they have declined sharply in Malaysia; the decline is likely to continue throughout their range as deforestation is widespread.

***** PHAYRE'S LEAF MONKEY**
Presbytis phayrei (**457**)
Rather similar to the Dusky Langur, but the eye-rings are bluish-white, and the crown is the same colour as the rest of the back. Found from E India and SW China south to N Thailand and C Vietnam. Little is known of their status, but in Burma they were exterminated in areas where they congregated at salt licks; they were hunted for their gallstones, used in traditional medicine. In Vietnam the recent war has affected much of their habitat, with forest destruction elsewhere. They occur in a few reserves and are protected in Thailand.

***** SILVERED LANGUR**
Presbytis cristata (**458**)
The individual hairs are tipped white, giving a silvery appearance. The new-born young are bright orange with black hands and feet, and do not assume adult colouring until after 3 months. Found in forests and mangroves in S Burma, Indo-China, E Thailand, Peninsular Malaysia and Sumatra east to Borneo Lombok. They live in groups of up to about 50. Although apparently still abundant in S Sumatra and Borneo in the late 1970s, they suffered an estimated 33.3% decline to 4000 between 1958–1975 in Peninsular Malaysia. Protected in Malaysia and Thailand, and occur in a few reserves. Frequently exhibited in zoos, and bred in small numbers.

**** GOLDEN LANGUR**
Presbytis geei (**459**)
First described in 1956, with a very restricted range in NW Assam, India, and Bhutan; the limits of its distribution and status in the wild are not known. Like other langurs they are forest-dwelling, feeding mainly on leaves and other vegetation. Fully protected in India, occurring in Ripu Forest Reserve, and also in Manas Wildlife Sanctuary in Bhutan.

***** COMMON LANGUR or HANUMAN**
Presbytis entellus (**460**)
Generally the most common and widespread langur within its range, and unlike other langurs very adaptable and often found in modified habitats, even in and around villages. In India the population has been estimated at around one million, in Bangladesh at under 500; elsewhere little is known of numbers or status. Even in India, where they are protected and considered sacred over most of their range, they are declining in many areas because of increasing habitat destruction. They cause considerable damage to agriculture, so they are often persecuted; they are also hunted in some areas. Found in many national parks and protected areas, are protected in most of their range, and listed on CITES App.I.

***** NILGIRI LANGUR**
Presbytis john (**461**)
Glossy black or brownish with a yellowish head; female has a distinct white patch on the inside of the thighs, even when young. Found at altitudes of 2950–6550ft or more in the Western Ghats. Although population is stable or possibly increasing in a few areas, their range is extremely fragmented; they continue to decline, sometimes to extinction, in some areas. They are extensively hunted, both for their flesh and also for various organs, blood, etc. used for medicinal purposes. They occur in several protected areas and are protected by law, both at national and state level.

*** BANDED LEAF MONKEYS**
Presbytis aygula/melalophos group (**462**)
This is a complex group about which there is little taxonomic agreement, with a number of species/subspecies occurring from S Burma through Peninsular Thailand and Malaysia to Sumatra, Java, Borneo and some smaller islands. Those of Thailand and Malaysia have declined in recent years, but are protected and occur in several reserves. In Java, they are particularly threatened. They are also found in several reserves in the Indonesian part of their range.

**** FRANCOIS'S LEAF MONKEY**
Presbytis francoisi (**463**)
Found in a relatively small area from SW China to C Laos and N Vietnam. No detailed information is available on its distribution or status.

*** MENTAWAI ISLAND LANGUR**
Presbytis potenziani (**464**)
Confined to the Mentawai Islands off Sumatra, Indonesia, where they are scarce and declining except possibly in the interior of Siberut.

*** CAPPED LANGUR**
Presbytis pileata (**465**)
Is possibly still abundant in some parts of its range which extends from E Bangladesh and Assam to N Burma, but there is very little recent data.

***** PURPLE-FACED LANGUR**
Presbytis senex (**466**)
Confined to Sri Lanka since 1934; populations declining; now protected and occur in some protected areas.

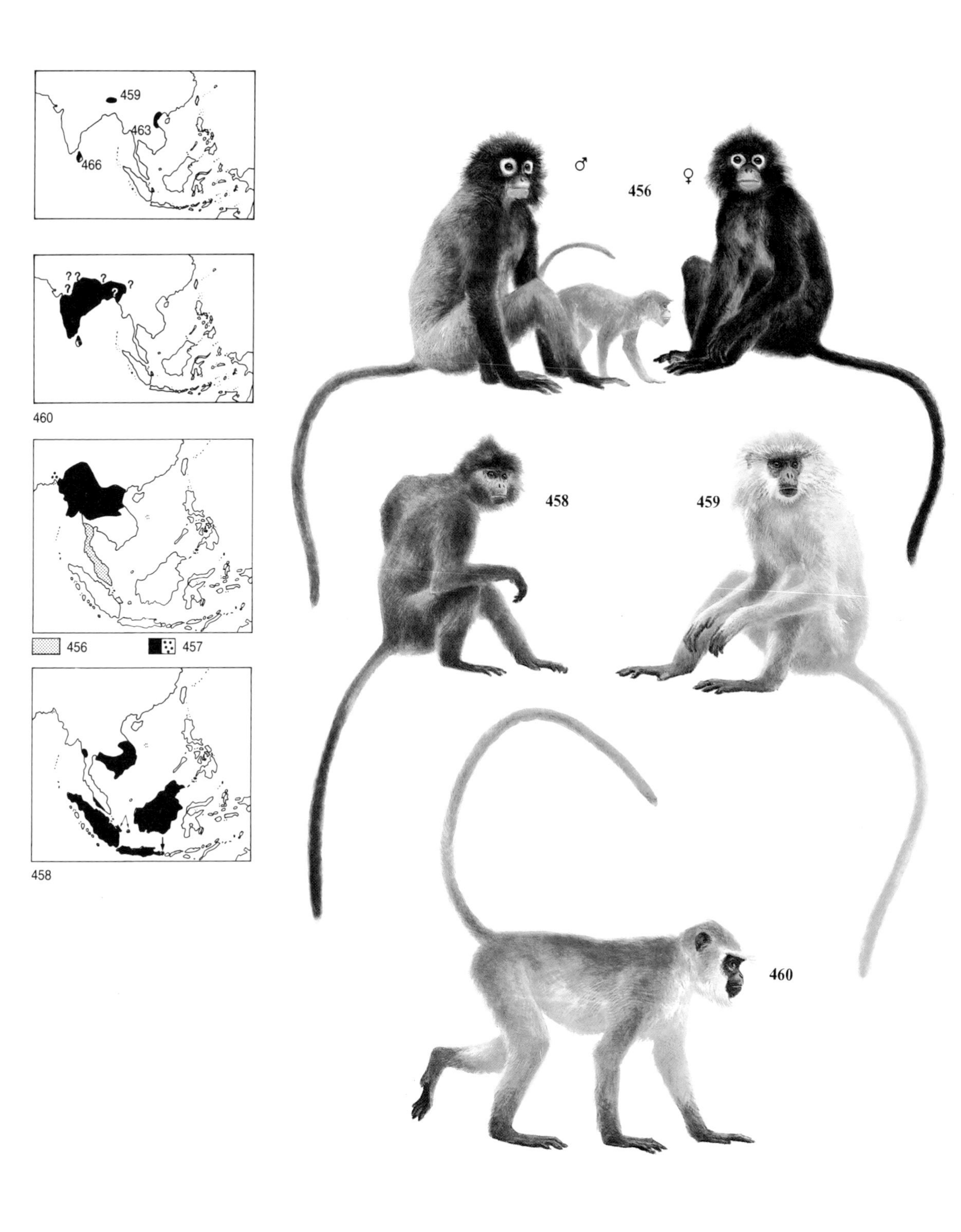
459
463
466
460
456
457
458
456
458
459
460

**** PIG-TAILED LANGUR or SIMAKOBU MONKEY**
Nasalis (=Simias) concolor (**467**)
HB ca.20.3ins; T ca.6.1ins.
Found mainly in primary forest, and confined to the Mentawai Islands off the west Coast of Sumatra, Indonesia. They once occurred on several nearby small islands, but are now restricted to Siberut, Sipora, Pagai Utara, Pagai Seletan and a tiny islet near the Katurei Peninsula, Siberut. In the late 1970s the maximum total population was estimated at ca.25,000 but believed to be much lower. Its main threat comes from selective logging, which, in addition to habitat destruction, allows easier access for hunting, a major factor in their decline. They are almost arboreal, feeding on leaves, and some fruit and berries. They occur in the Teiteibatti Wildlife Reserve, Siberut, have been protected by Indonesian law since 1972 and are listed on CITES App.I.

***** DOUC LANGUR**
Pygathrix nemaeus (**468**)
HB ca.23ins; T ca.27ins.
Forest dwellers, living in both primary and secondary forests at altitudes of up to 6550 ft. The precise limits of their former range are not known, and little is known of their habits in the wild. The widespread use of defoliants, as well as bombing in Indochina, undoubtedly had a serious effect on many populations, and they were also hunted for meat by military personnel. Protected in Vietnam, and listed on CITES App.I.

***** GOLDEN SNUB-NOSED MONKEY**
Pygathrix roxellanae (**469**)
HB ca.29.5ins; T ca.28.5ins.
A large, robust monkey, found in bamboo forests, rhododendron thickets and coniferous forests at altitudes of up to 9850 ft, descending to lower altitudes to avoid snow during winter. Little is known of its precise distribution, particularly the westerly limits of its range, which extends into SE Tibet, and may go as far as Arunachal Pradesh, India. In 1937 it was considered threatened in China; it is now protected there, and known to occur in the Wanglang Nature Reserve in Sichuan, China. A few are exhibited in Chinese zoos.

**** GUIZHOU SNUB-NOSED MONKEY**
Pygathrix brelichi (**470**)
Known only from 2 specimens described in 1903, from the Van Gin Shan Range, Guizhou, China; very similar to the Golden Snub-nosed Monkey, but apparently distinct.

*** TONKIN SNUB-NOSED MONKEY**
Pygathrix avunculus (**471**)
HB ca.23.2ins; T ca.34.5ins.
Confined to a relatively small range, apparently entirely within N Vietnam. Little is known of its current status or precise distribution, although the recent war in the region has probably had an adverse effect on its habitat.

***** PROBOSCIS MONKEY**
Nasalis larvatus (**472**)
HB 21.7–25.6ins; T 24.4–29.5ins; WT 22–50.7lb.
Male is larger than female. Greyish-yellow to reddish-brown, paler on rump and tail. The distinctive nose is larger in adult male. Diurnal, mostly arboreal, usually feeding near water in mangroves and other swampy forests, where they eat leaves, shoots and fruit. Restricted to Borneo, and a few coastal islands near the Sabah/ E Kalimantan border. Formerly most widespread and abundant in coastal and lowland areas, but have disappeared from many parts, and elsewhere are reduced in numbers, although still locally abundant. Their main threats are from hunting and loss of habitat, particularly mangroves. Have been protected in Indonesia since 1931. Do not thrive in captivity, and are relatively rare in zoos. Listed on CITES App.I.

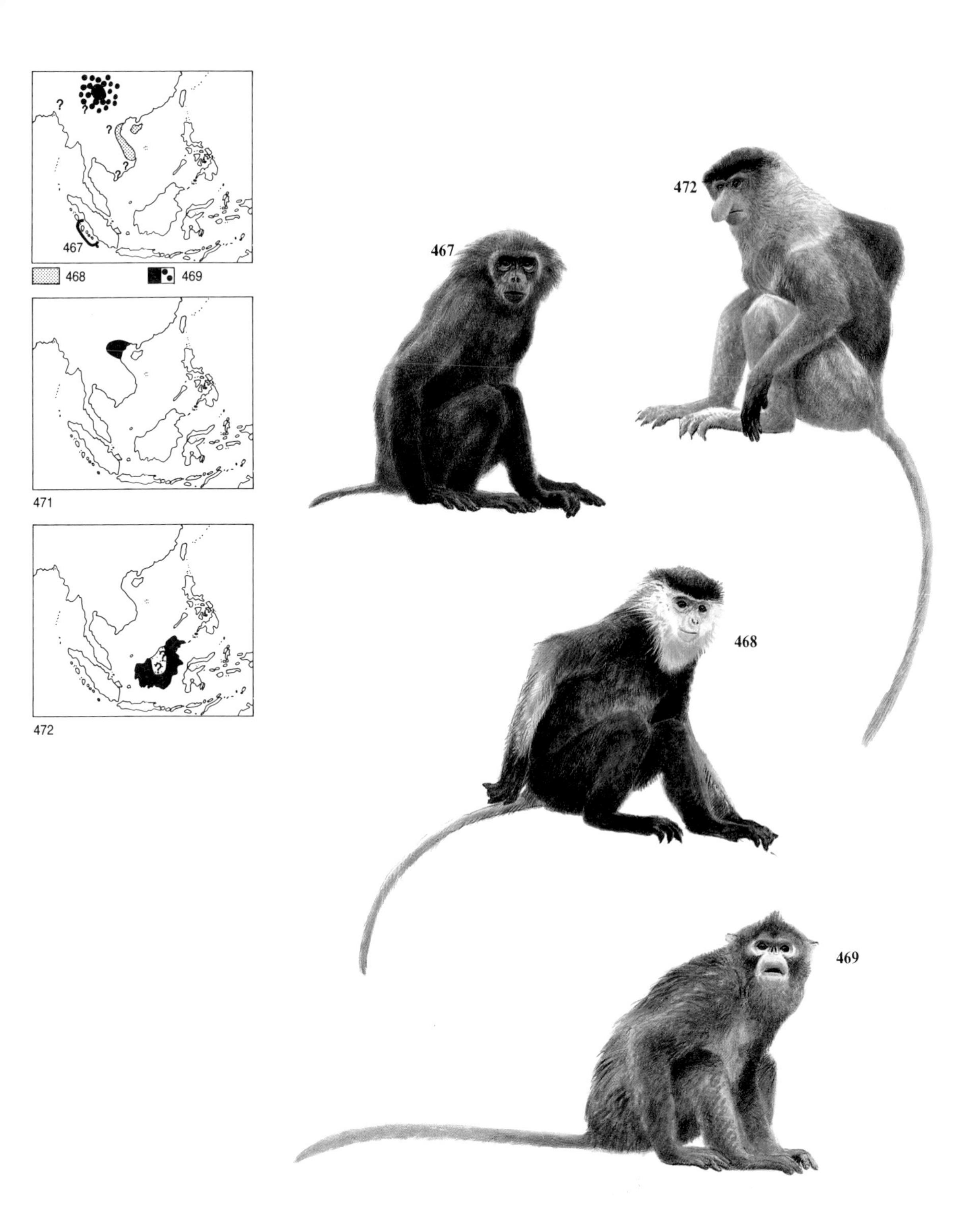

467
468
469
471
472
467
472
468
469

BLACK AND WHITE or PIED COLOBUS MONKEYS
Colobus spp.
HB 23.6–29.5ins; T 29.2–32.7ins; WT 28.7–50.7lb.
The proportions of black and white vary considerably, and there are usually fringes of long hair on the flanks and tail. There is no general agreement on the taxonomy of the colobus monkeys, and the species and subspecies described below will be variously found as species or subspecies in other publications. The important conservation feature is that most of the isolated populations are threatened, and their taxonomic status is more of academic interest than of practical conservation value.

***** WESTERN COLOBUS or WESTERN BLACK AND WHITE MONKEY**
Colobus polykomos **(473)**
Although still common in some parts of its range, it is extinct or declining rapidly in many parts of Ghana, Nigeria and Togo; virtually nothing is known of its status in the more westerly part of its range. Like other *Colobus*, they are arboreal, living in a wide range of forest habitats. Deforestation has had a catastrophic effect on their populations. They occur in many parks and reserves, including Tai NP, Ivory Coast; Bia Forest Reserve and several other protected areas, Ghana; Olokemeji Forest and other reserves, Nigeria. Listed on CITES App.II and Class B of the African Convention. Until very recently extensively exploited for international trade and skins are still in demand locally.

***** BLACK AND WHITE COLOBUS**
Colobus angolensis **(474)**
Although probably less threatened than other *Colobus*, and still common in some parts of Zaire, it is rare in most of its range, noteably Angola, Kenya, Rwanda,Uganda and Zambia. A forest species with an increasingly fragmented range.

***** ABYSSINIAN BLACK AND WHITE COLOBUS**
Colobus guereza **(475)**
Although still locally abundant in the Congo and Zaire, it is declining in Sudan, Ethiopia, Cameroon and Uganda. In addition to widespread habitat loss, it is extensively hunted for its spectacular pelts, which are in demand both locally within their range, and also for export as rugs, etc. Its extinction over most of E Africa is attributed to the fur trade, which was exporting 2 million skins in the late 1800s. In 1972 an estimated 27,000 skins were stocked by shops in Kenya. A further 200,000 were offered for sale in Ethiopia. Occurs in a large number of national parks and reserves within its range.

***** BLACK COLOBUS**
Colobus satanas **(476)**
Almost entirely sooty black with long whiskers and mantle. Like other *Colobus*, they are mainly found in forests, but on Fernando Poo (Macias Nguema) also occur in bushes at 8200–9850ft. Deforestation and hunting for meat is reducing their range and even in the reserves within their range they are not always adequately protected.

***** RED COLOBUS**
Colobus badius **(477)**
HB 18.1–27.6ins; T 16.5–31.5ins; WT 15.4–28.7lb.
There is no general agreement on the taxonomy and systematics of the Red Colobus Monkeys, but several populations have restricted ranges and are sometimes treated as full species: *C. b. ellioti* is confined to a small area of Bwamba, Uganda, and *C. b. gordonorum* is known only from the Uzungwa Mountains, Tanzania. *C. b. rufomitratus* (illustrated) is confined to the lower Tana River area, Kenya. Red Colobus are rare in captivity and are not bred regularly. All listed on CITES App.II.

***** KIRK'S or ZANZIBAR RED COLOBUS**
Colobus kirki **(478)**
Closely related and similar to the more widespread Red Colobus, it is found mostly in forest, and occasionally raiding gardens and agriculture, on the island of Zanzibar. Its total population of ca.1400 is divided into groups of between 15 and 300 individuals, which are usually subdivided into groups of up to 10. Populations have been translocated to Masingini and Kichwele Forest in Zanzibar, and to Ngezi in Pemba Island. They are threatened by timber felling and agriculture, and are also hunted for the pot. In the 1970s a large number were exported to zoos, but they are rarely seen in captivity, and there are no self-sustaining populations. Listed on CITES App.I.

***** OLIVE COLOBUS**
Procolobus (Colobus) verus **(479)**
HB 16.9–19.7ins; T 22.5–25.2ins; WT 6.4–11.9lb.
A small *Colobus*, olive above with a tinge of brown. Troops consist of an old male, several females and their offspring. They are arboreal, feeding mostly on leaves, and rarely descending to the ground. Found in a variety of forested habitats, usually with a dense, lush underbrush. There has been extensive deforestation in all countries within their range, and there are now very few areas of suitable habitat likely to be able to support sizable populations in the future. They occur in the Tai NP, Ivory Coast, Bia and Nini-Suhien NP, Ghana, and other protected areas. Listed in Class A of the African Convention.

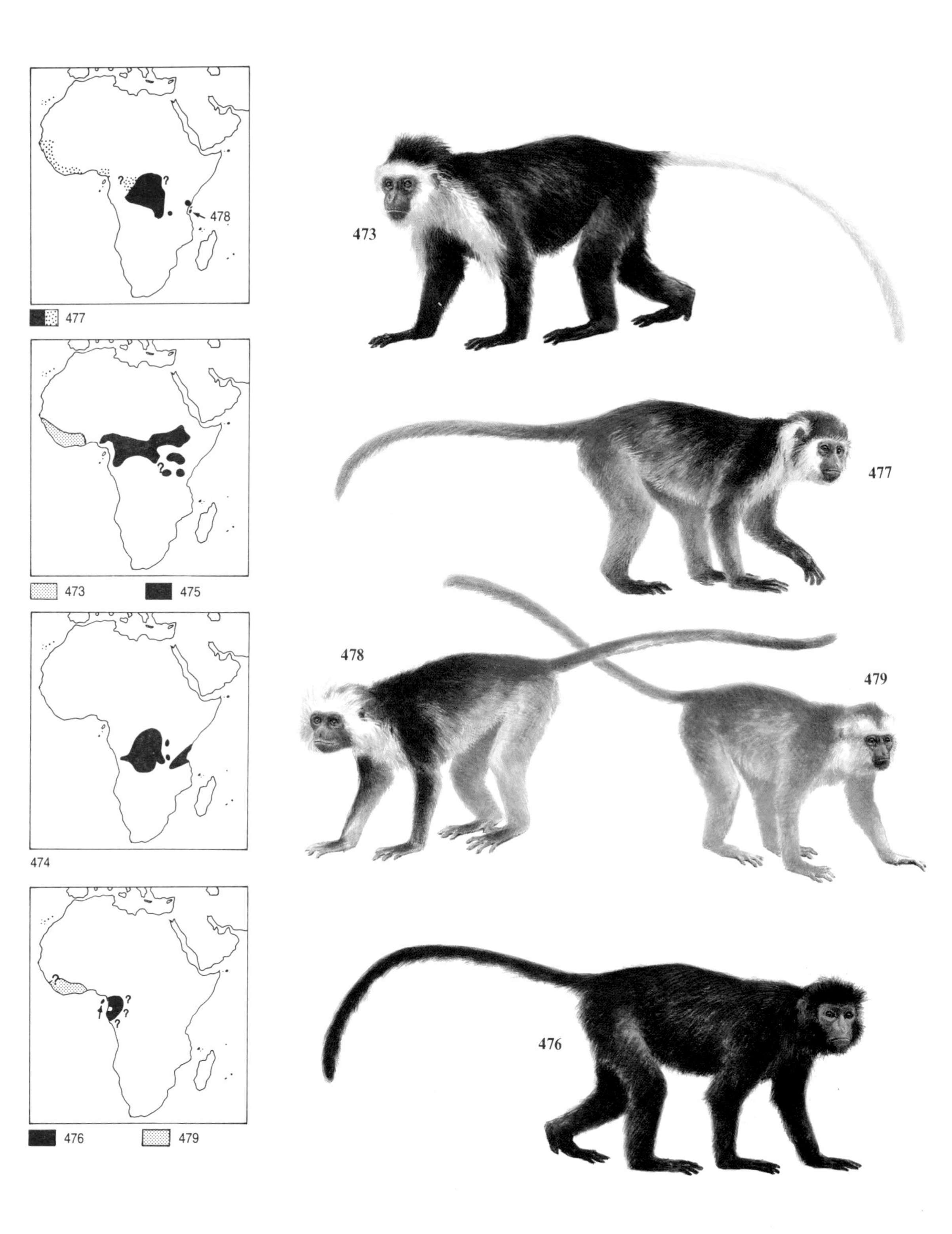

478
477
473
475
474
476
479
473
477
478
479
476

GIBBONS

Hylobates spp.

Gibbons are forest-dwelling, and almost exclusively arboreal, using their long arms to 'brachiate' or swing from one hand to the other; they can also leap 30ft or more. All are diurnal and extremely vocal, making a wide variety of calls, including hoots, trills, and booms, which carry for up to 2 miles. They feed mostly on plant matter such as buds, shoots, flowers and fruit, and also insects and other small animals. The single young is carried by the mother until it is old enough to follow the family. All the gibbons are declining to a greater or lesser degree, and all are listed on CITES App.I.

There is considerable confusion in the nomenclature, and many biologists split the species as described here into several separate species.

***** KLOSS'S or DWARF GIBBON**

Hylobates klossi (**480**)

HB ca.17.3ins; T absent; WT 10.6–13.2lb.

Found in rain forest in the Mentawai Islands, off Sumatra, Indonesia (Siberut, Sipora, Pagai Utara and Pagai Seletan). Although still common locally, they are declining throughout most of their range due to rapid deforestation. They are most abundant on Siberut, where there has been less logging, and they occur in the Tei-Tei-Batti Nature Park. In the mid-1970s the total population was estimated at no more than 84,000, predicted to drop to ca.3000 by the 1980s. Legally protected throughout their range.

***** COMMON or LAR GIBBON**

Hylobates lar (**481**)

HB 17.7–23.6ins; T absent; WT 8.6–15.4lb.

Rather variable in appearance, with several subspecies described; some of the subspecies are often treated as full species, notably *H.l.agilis* from the Malay Peninsula, Sumatra and Kalimantan; *H.l.moloch* from Java; *H. l. muelleri* from Borneo; the type subspecies *H.l.lar* occurs in the northern parts of the species's range. They are dimorphic, with some chocolate brown or blackish, others grey or whitish. Found in a wide variety of forest habitats where they live in family groups of 2–12 individuals. Throughout their range they appear to be declining; 1000 years ago their range was considerably more extensive in China. In the mid-1970s their total population was estimated at 2.8 million, predicted to fall to less than 400,000 by 1980, and to isolated relict populations by the late 1990s. In most parts of their range they are legally protected. They occur in at least 30 parks and reserves, including Khao Yai NP, Thailand, Kinabalu NP, Kalimantan, Udjung Kulon Reserve, Java, and Gunung Leuser Reserve, Sumatra. Commonly kept in zoos, and bred freely in captivity, especially *H.l moloch*; unfortunately many of the captives are of unknown ancestry.

***** PILEATED GIBBON**

Hylobates pileatus (**482**)

HB 18.5–19.7ins; T absent; WT 8.8–15.4lb.

Closely related to, and sometimes considered conspecific with *H.lar*, with long white hairs pointing outwards on the side of the head; the female is pale, the male dark-coloured. Found in a wide variety of forest habitats in Thailand, SW Laos, W Cambodia, and possibly S Vietnam, but has declined rapidly throughout its range. In the 1970s the total population was estimated at ca.100,000, and predicted to fall to 22,500 by the 1980s. In addition to habitat loss, other factors affecting them include human hunting for food. They occur in at least six protected areas in Thailand, including Khao Yai NP. They are protected in Thailand. Small numbers are kept in captivity, and a few are bred.

***** SIAMANG**

Hylobates (=Symphalangus) syndactylus (**483**)

HB 29.5–35.5ins; T absent; WT 17.6–28.7lb.

Generally black with a large greyish or pinkish throat sac, inflated when calling. Confined to rain forest, up to altitudes of 5900ft, where they live in family groups of 2–10 individuals. Their range is being drastically reduced by logging, and although believed to have once occurred as far north as S Burma and Thailand, there is no up-to-date information to suggest they still occur there. In the mid-1970s the total population was estimated at 167,000–196,000 and predicted to fall to 110,000 by 1980 and to isolated relict populations by the 1990s. Found in several protected areas in both Malaysia and Sumatra, and protected in both countries.

*** CRESTED or BLACK GIBBON**

Hylobates concolor (**484**)

Similar in size to the Pileated Gibbon (482). A little-known species found in forests of S China, Hainan, south to S Kampuchea and Vietnam. Little is known of its status, but extensive military activities within its range has probably had a detrimental effect. Six subspecies are recognised and several of these may be seriously threatened. Although exhibited in many zoos, comparatively few are being bred, and many zoo populations are of mixed ancestry; *H.c.leucogenys* is the only discrete subspecies being maintained in reasonable numbers in captivity.

***** HOOLOCK GIBBON**

Hylobates hoolock (**485**)

Similar in size to the Pileated Gibbon (482). Found at altitudes of up to 4500ft and formerly widespread from E India, through Bangladesh to China and south to the Irrawaddy in Burma. They have declined drastically, particularly in Bangladesh and India, where much of their habitat is being destroyed by the expanding human population; they are now confined to the east of Bangladesh. Found in a few protected areas in Bangladesh, India and Burma. Relatively few are kept in captivity, and they are only rarely bred.

480
482
485
484
481
483
480
483
481
484
485

**** ORANG UTAN

Pongo pygmaeus (**486**)

HB 49.2–59.1ins; T absent; WT 77.2–220.5lb.

Male is larger than female. Orange or reddish-brown colouring darkens with age. Diurnal and generally solitary, but young remain with the mother until they are 5 or 6 years old. They feed mostly on leaves, fruit and insects and are arboreal, building nests of broken twigs and branches. Found mainly in forested areas, occurring at densities of up to 5 per sq.ml. in lowland dipterocarp forest. Now confined to Sumatra and Borneo, but once found throughout Indo-China, Malaysia and also north to China, wherever there was suitable habitat. Retreated from man's presence, but malaria may be the reason they have always been absent from some areas. In Sumatra, range has declined by 20–30% since the 1930s. In the past, collection for the pet and zoo trade contributed to their decline, but at present habitat loss is the most important factor. They occur in several protected areas, notably Mt Kinabalu NP, Sabah, and Gunung Leuser in Sumatra. Two subspecies are recognised: *P. p. abelii* from Sumatra and *P. p. pygmaeus* from Borneo. There are large numbers in zoos and they are now being bred regularly. Listed on CITES App.I, and protected throughout their range.

*** GORILLA

Gorilla gorilla (**487**)

HB up to 69 ins (standing); T absent; WT up to ca.331lb, occasionally up to 606lb.

Male is larger and heavier than female. They occur in two widely separated areas: the **Western Lowland Gorilla** *G.g.gorilla* is found in rainforest in West Africa from SE Nigeria to S Congo; 621 miles to the east the **Eastern Lowland Gorilla** *G.g.graueri* occurs in EC Zaire, and the **Mountain Gorilla** *G.g.beringei* occurs in the Virunga range on the borders of Zaire, Rwanda and Uganda, at altitudes of up to 11,500ft in bamboo forest and montane rain forests. The subspecies can be distinguished by differences in fur colour. They are diurnal, building a nest each night. Family groups are led by mature males; the groups usually contain around 15 including infants, but occasionally there may be more than 40. They are vegetarian, feeding on a wide variety of herbage, buds, shoots and stems. In the early 1980s, the Western Lowland Gorilla was estimated at less than 14,250, the Eastern population at under 5000, and that of the Mountain Gorilla at under 400. Throughout their range they have declined during the past century partly due to habitat loss, but also due to hunting. Mountain Gorillas were discovered by Europeans at the beginning of this century and have probably always been rare. Gorillas often thrive in secondary forest, and in some areas with adequate protection, their populations have stabilised or even increased. They occur in several protected areas, notably in the Virunga Volcanoes in Zaire and Rwanda; Campo Reserve, Cameroon; Bwindi Forest, Uganda; they are also widespread in Gabon, which has one of the lowest human population densities in the world. In Rwanda a conservation programme, allowing tourist visits to the Gorillas has proved successful, and numbers appear to be increasing. There are many breeding groups in captivity, but until recently they were still being captured for the zoo trade. They are protected in most countries in their range. Listed on CITES App.I, and Class A of the African Convention.

*** CHIMPANZEE

Pan troglodytes (**488**)

HB up to 37.4ins; T absent; WT up to 176.5lb, occasionally more.

Male larger than female. The face is usually bare, and generally blackish, with flesh-coloured nose, ears, hands and feet. Found in tropical forests and also more open habitats, including wooded savannah, at altitudes of up to 9850ft. They are good climbers, but move between feeding areas on the ground. They are diurnal, and usually spend the night in nests, constructing a new one each night. They are one of the few animals to use tools, using sticks to extract termites etc. (This behaviour is learnt, not inherited.) They are omnivores feeding on a wide variety of vegetable matter but mostly fruit, birds' eggs, insects and small mammals; they also occasionally kill the young of animals up to the size of antelope and baboons. Bands of 30–80 live together, and these usually contain several subgroups, with no consistent leader. Their range is now extremely fragmented; of the three subspecies *P. t. verus* is the most threatened, and is confined to scattered populations in the west of its range; *P. t. troglodytes* is the most widespread, with numerous subspecies occurring in the central area; *P. t. schweinfurthii* occurs in the east, and the precise limits of its distribution are poorly known. They have all declined, largely through habitat loss, and also through hunting, particularly in areas close to agriculture because they raid crops. Have also been exported in large numbers in the past for exhibition in zoos, and for use in biomedical research; infants were also sold as pets. In the 1980s small numbers were still being exported. The effects of export are considerable since it normally involves young animals, which are obtained by killing the mothers and often others in the band. Chimpanzees are protected in many countries and found in a large number of protected areas; they are listed on CITES App.I and Class A of the African Convention.There are large numbers in captivity, including several successful breeding colonies. Captive animals have been successfully reintroduced into the wild, but this is an extremely slow and expensive process.

*** PYGMY CHIMPANZEE

Pan paniscus (**489**)

HB up to 23.6ins; T absent; WT up to 99.2lb.

Very similar to the Chimpanzee, but smaller and more lightly built. They are very little-studied in the wild, and although much of their forest habitat remains intact, their range has now been fragmented and they do not occur in any major protected areas. They are listed on CITES App.I and Class A of the African Convention. Only small numbers are being kept and bred in captivity.

486
489
487
488
486
489
488
487

*** GIANT ANTEATER

Myrmecophaga tridactyla (**490**)

HB 39.4–47.3ins; T 25.6–35.5ins; WT 39.7–86lb.

Very distinctive and easily recognised by its long snout and long bushy tail. The large powerful claws are used in feeding; it tears open termite mounds and collects termites, larvae and eggs with its very long, sticky tongue, consuming up to 30,000 a day. It also feed on ants and other insects and occasionally fruit. Found in a wide range of habitats, from forests and swamps to dry open savannah. Mainly diurnal, except in areas close to human habitation when it often becomes nocturnal. The single young is carried on the mother's back, and remains with her until the next one is born. Giant Anteaters were once widespread from S Belize to N Argentina, but have declined throughout most of their range, and are now only found in scattered, isolated populations; hunting has probably been one of the main causes of their decline. Extinct in many of the northern parts of their former range. Exhibited in many of the larger zoos, but only very rarely bred in captivity. Listed on CITES App.II, and found in a few national parks within its range, including Serra de Canastra NP (Brazil). Legally protected in Brazil and French Guiana, and has varying degrees of protection in other parts of its range, but often hunting bans are poorly enforced.

*** GIANT ARMADILLO

Priodontes maximus (=giganteus) (**491**)

HB 29.5–39.4ins; T 19.7ins; WT up to 132.3lb.

There are 11–13 bands on the back, and 3–4 on the neck, with powerful claws on the forefeet. Easily distinguished from all other armadillos by its much larger size. Found mostly in undisturbed savannah or forest, usually close to water, where largely nocturnal. They dig extensively, both searching for food, and to make burrows for shelter. They feed mainly on termites, but also eat other insects and invertebrates, snakes and carrion. Although overall range has probably not changed recently, within that range numbers have declined drastically, and many populations are now fragmented and isolated; it is locally extinct in areas which have been disturbed by man. In many places the flesh is prized by humans, and the opening up of roads has had a serious effect. Listed on CITES App.I. Very few are held in captivity and it has not been bred there. Protected in most of its range states, and also occurs in national parks including several in the Mato Grosso (Brazil), Manu NP (Peru), Sierra de la Macarena (Colombia) and several reserves in Suriname.

** BRAZILIAN THREE-BANDED ARMADILLO

Tolypeutes tricinctus (**492**)

HB 11.8ins; T 2.6ins.

Although most have 3 moveable bands on the 'shell' some may have 2, others 4. The sides of the bands which form the 'shell' hang free of the body along their lower parts. Distinguished from the closely related *T. matacus* by having 5 claws on the forefeet, instead of 4. Like many other armadillos they are believed to feed mainly on termites, but probably also eat other invertebrates and some fruit. Found in dry, open country, in C and NE Brazil. Like other armadillos they are hunted for food. Very little is known of their current distribution status. Protected in Brazil, and may occur in a few reserves. Only occasionally exhibited in zoos, and there are no breeding colonies in captivity. Other armadillos may be threatened, but little is documented; the 11–Banded Armadillo *Cabassous centralis* from Honduras, south to northern S America is locally threatened, and *C. tatouay* from Paraguay and SE Brazil, south to N Argentina is considered theatened in Uruguay.

** PINK FAIRY ARMADILLO

Chlamyphorus truncatus (**493**)

HB 4.9–5.9ins; T 1in.

The smallest armadillo with pale pink armour, and soft white fur covering the rest of the body. The shell is almost completely separated from the body, only attached at the pelvic region. A very elusive mammal, endemic to Argentina, where it was found along the south-eastern edge of the Pampas, the north of Patagonia, and the edge of Buenos Aires province. Lives mainly on light soils, feeding on insects and also some plant matter. Its disappearance over the last 40 years is linked with ploughing of its habitat. Only rarely seen in zoos and does not thrive in captivity.

** BURMEISTER'S FAIRY ARMADILLO

Chlamyphorus (=Calyptophractus) retusus (**494**)

HB 5.5–6.9ins; T 1.4ins.

Whitish or yellowish-brown carapace and fine whitish woolly hair. Similar to the smaller Pink Fairy Armadillo (493) but the carapace is attached over its entire length, and the tail is partly protected with plates. Like the Pink Fairy Armadillo it is found in open grassland, and is threatened by the spread of agriculture. Has also been collected for museums and may be predated by domestic dogs. Its range extends from the extreme north of Argentina to the Chaco of Bolivia and to Paraguay. Although it was apparently originally described from a Bolivian specimen, it is not known to occur there now; little is known of its precise distribution. Like the previous species it does not flourish in captivity.

*** THREE-TOED SLOTHS

Bradypus spp.

HB 16.2–27.6ins; T vestigial; WT 8.8–9.9lb.

The long, coarse fur is greyish-brown, with a finer, softer under fur; it often appears greenish because of algae growing on the fur. There are three species: the **Brown-throated Sloth,** *B. variegatus* (**495**), found from Honduras, south to N Argentina; the **Pale-throated Sloth,** *B. tridactylus* (**496**), found from S Venezuela and the Guianas, south to N Brazil; and the **Maned Sloth,** *B. torquatus* (**497**), only found in E Brazil. Confined to forests, they are almost entirely arboreal, hanging from branches where they feed on leaves, buds and shoots. They are generally solitary and they give birth to a single young. Less adaptable than the Two-toed Sloths; all Three-toed species, particularly the Maned Sloth, are probably decining due to loss of their forest habitat. All occur within parks and reserves, and the Maned Sloth has been introduced into the Tijca NP, in Rio de Janeiro, Brazil. They do not thrive in captivity and are rarely seen in zoos. Protected in most parts of their range, and *B. variegatus* is listed on CITES App.II. The **Two-Toed Sloths** *Choloepus* spp. are generally more widespread, except * *C. hoffmanni* (**498**), which is threatened in many parts of Central America.

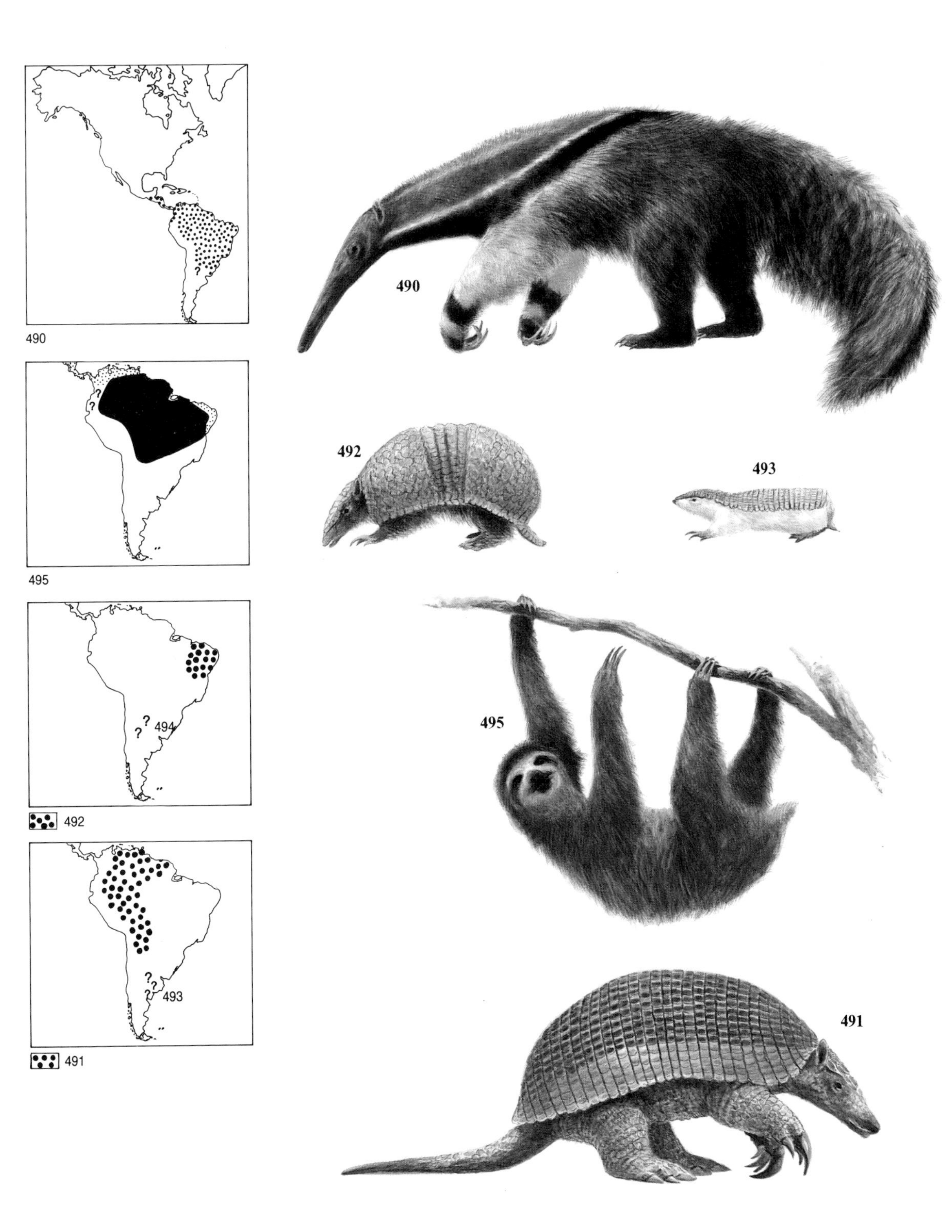
490
495
?
?
494
492
?
?
?
493
491
490
492
493
495
491

*** TEMMINCK'S GROUND PANGOLIN**
Manis temmincki (**499**)
HB 19.7–23.6ins; T 15.8–19.7ins; WT 33.1–39.7lb.
Also known as the **Cape Pangolin** or **Scaly Anteater**, it is covered with brownish, overlapping scales. The underside is almost naked, and the first and fifth claws on the hands are almost vestigial. Found in dry bush country, it sleeps by day in burrows, emerging at night to feed on ants and termites, breaking open the nests with powerful claws. Although their range is extensive, extending over most of East and South Africa, it is believed to be declining in most parts. Listed on CITES App.1. The other African species of pangolin, the * **Giant Pangolin** *M. gigantea* (**500**) and the * **Tree Pangolin** *M. tricuspis* (**501**) all suffer heavy mortality in fires, and increasingly from road casualties. Their scales are also very popular as charms and they are hunted for meat. There is little information on the effects of this mortality but the West African species are particularly likely to be affected.

ASIAN PANGOLINS
Manis spp.
The three species are very similar. All are covered with overlapping horny scales which are formed from compressed hair – similar in structure to the horn of rhinoceroses. They all have a prehensile tail, used when climbing, and carried clear of the ground when walking. They are nocturnal, passing the day in a burrow which they usually dig themselves. They have a very long sticky tongue (about 50% longer than the head and body put together) and feed almost exclusively on ants and termites; ca.200,000 together have been counted from a single stomach. They are preyed upon by larger carnivores such as bears and leopards, and are extensively trapped because of the alleged value of their scales; some people also believe that the blood cures internal bleeding. They are substantially traded in many parts of the Far East; between 1958–1964 over 66 tons of scales were exported from Sarawak. They are all listed on App.II of CITES.

*** MALAYAN PANGOLIN**
Manis javanica (**502**)
HB 16.7–21.7ins; T 13.4–18.5ins; WT 11–15.4lb.
Similar in general appearance to the other Asiatic pangolins but with shorter claws on the forefeet. Found mostly in forested areas, but also in more disturbed habitats. Little is known of their precise status, distribution, and life cycle.

*** CHINESE PANGOLIN**
Manis pentadactyla (**503**)
HB 17.3–18.9ins; T 6.3–13ins.
Similar to the Malayan Pangolin, with longer foreclaws and fleshy ears. Its range extends from Nepal earthwards to S. China and Hainan. Virtually nothing is known of its current status.

*** INDIAN PANGOLIN**
Manis crassicaudata (**504**)
HB 23.6–29.5ins; T ca.17.5ins.
Distinguished from the Chinese Pangolin by having 11–13 rows of scales around the body (the Chinese has 15–18 rows).They are found in peninsular India, Sri Lanka, Pakistan, China, and Bangladesh. Little is known of their status, but have apparently disappeared from many parts of Bangladesh.

*** AARDVARK**
Orycteropus afer (**505**)
HB 39.4–63ins; T 17.7–23.6ins; SH 23.6–25.6ins; WT 110.2–154.3lb (max.181lb).
One of the most distinctive mammals, having a long pig-like muzzle, with hairs (up to 0.2ins) growing from the nostrils; the tubular ears are 0.6–0.8in long. Aardvarks occur in a wide variety of mainly dry habitats, wherever ants and termites are abundant. They are exceptionally powerful diggers, and excavate extensive burrows. The Aardvark formerly had a wide range in Africa, and 8 subspecies have been described. However its status is poorly known, and in many parts of Africa it has disappeared or is very much rarer than in the past, particularly in the north and west of its range. Egyptian tomb paintings suggest that it once ranged north to the Mediterranean. Often displaced by the spread of agriculture and irrigation, and also hunted in some areas for its hide and meat; the claws are sometimes used as amulets. Exhibited in some of the larger zoos and occasionally bred, but not self-sustaining in captivity. Listed on CITES App.II. Legally protected in most parts of their range and found in many national parks and reserves, particularly in East and South Africa.

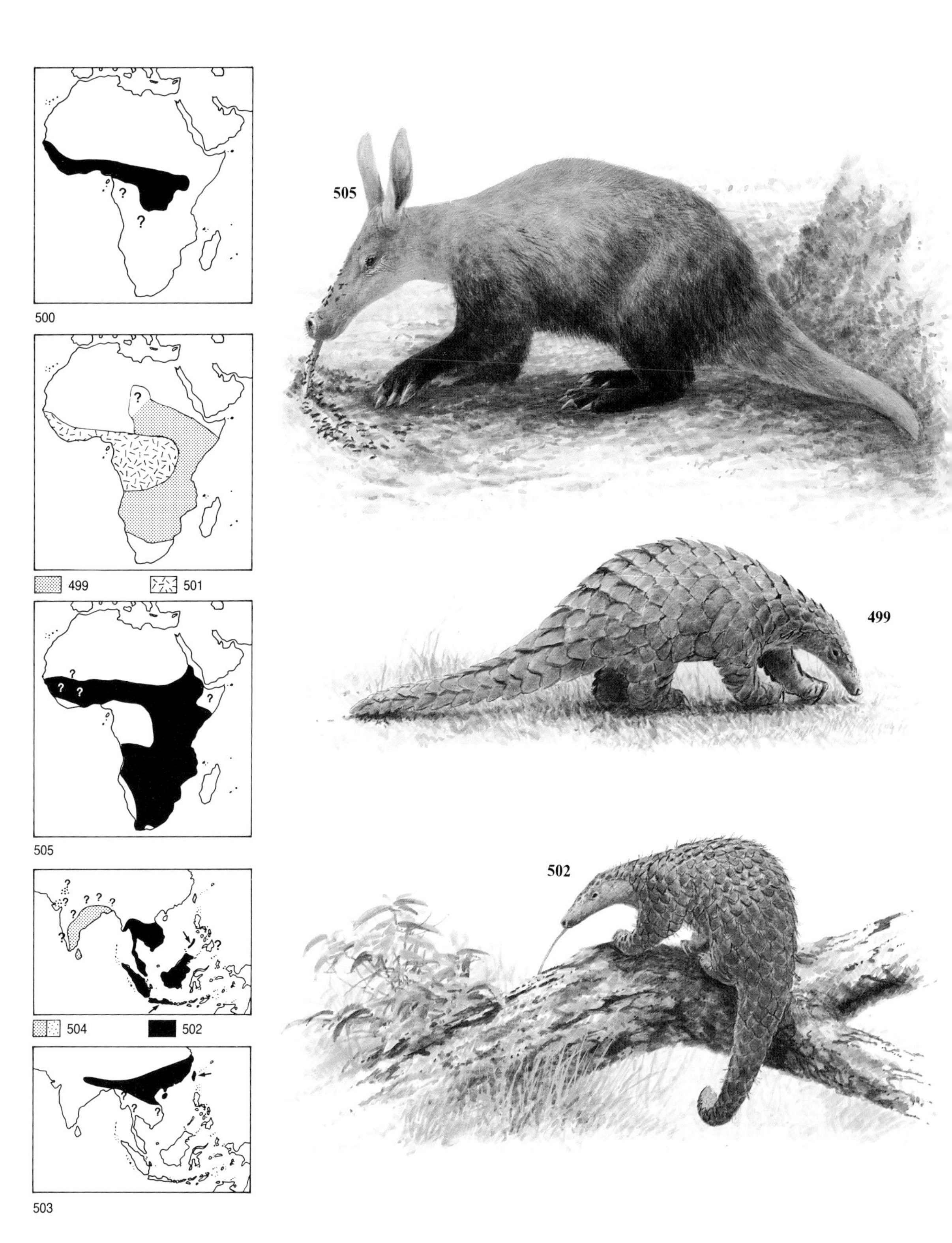

500
499
501
505
504
502
503
505
499
502

***** AMAMI RABBIT**
Pentalagus furnessi **(506)**
HB 16.9–20.1ins; T 0.6in.
The Amami Rabbit has short woolly fur, dark brown above, reddish-brown on its sides, and paler below. It has rather long claws (up to 0.8in), and ears up to 1.8ins. Found in dense forests feeding on berries, shoots, leaves, bamboo and other vegetation. It digs tunnels, where several litters of 2 or 3 young are born each year. Confined to the islands of Amami Oshima and Toku-no-Oshima, south of Japan. By 1900 their numbers were considerably reduced by over-hunting and they were given full protection in 1921. In 1979 the population was estimated at ca.5000 and declining, due to habitat loss and predation by feral dogs and introduced mink. None currently in captivity, and it is not known if they would breed easily. Because of their rarity, and unique status, they are protected as a 'Special Natural Monument'.

******* SUMATRAN RABBIT**
Nesolagus netscheri **(507)**
HB 13.8–15.8ins; T 0.6in.
Probably one of the world's rarest mammals. The Sumatran (or **Short-eared**) Rabbit, is small with short ears and soft fur, with the most distinctive pattern of any wild rabbit. Generally greyish with brown stripes, and with a bright reddish rump and tail. They are found in forested areas at altitudes of 1950–4600ft, where they are largely nocturnal, hiding in burrows by day. In the wild they feed on the forest undergrowth, but apparently thrive in captivity on rice, bananas and maize bread, but refuse vegetables. Very little is known of their present distribution or status. Confined to the Barisan range of mountains, in SW Sumatra, where forest clearance has destroyed much of the habitat. Only a few specimens are held by museums, and the species has rarely been encountered by naturalists; a possible sighting in 1978 is the only record after 1916. Protected since 1972. Listed on CITES App.II.

*** NEW ENGLAND COTTONTAIL**
Sylvilagus transitionalis **(508)**
HB 15–16.9ins; T ca.1.6ins; WT 1.6–3lb.
A medium to large species of cottontail; dark brown or buff above, overlaid with blackish. The back edge of the ears is covered with black hair. Lives in a wide variety of habitats, usually with grasses and often dense undergrowth. There are several litters a year, and the 3–5 young are born in nests in dense brush, or in burrows made by other animals. In recent years the New England Cottontail has declined drastically throughout most of its range. The main reason appears to be the competition from the introduced Florida Cottontail *S.floridanus*, but they are also less able to adapt to changes in habitat in areas thickly settled by people. Can be raised in captivity, but there do not appear to be any self-sustaining populations. Protection is hampered by its similarity to the Florida Cottontail, one of the most important game animals in the USA.

**** OMILTEME RABBIT**
Sylvilagus insonus **(509)**
The Omilteme Rabbit is a cottontail known only from the mountains of Guerrero in southern Mexico.

*** SAN JOSÉ RABBIT**
Sylvilagus mansuetus **(510)**
Found only on San José Island, off SE Baja California, Mexico. It is closely related to the Brush Rabbit *S. bachmani*, which is more widespread in western USA.

***** VOLCANO RABBIT**
Romerolagus diazi **(511)**
HB 8.9–13.8ins; T not visible; WT ca.154lb.
A short-eared (1.4–1.8ins) rabbit with dark brown fur above, dark grey below. In general appearance it is more like a pika than other rabbits. Found in small colonies of open pine forests with thick grass cover, where they live in runs and burrows among the tussocks. The burrows are complex, and include nests lined with grass and fur. Mostly nocturnal and crepuscular, they are also active by day, particularly when the sky is overcast. The breeding season is from March to early July and 1–4 (usually 3) young are born in an underground nest. The Volcano Rabbit has a very restricted range, being confined to the middle slopes of Popocatepetl and Ixtacihuatl and adjacent mountains in the south and east of the Valley of Mexico. Small colonies are being maintained in captivity, notably at Jersey Zoo.

******* BUSHMAN RABBIT**
Bunolagus monticularis **(512)**
HB 13–18.5ins; T 2.8–3.9ins; WT 2.2–3.3lb.
The Bushman Rabbit (or **Riverine Rabbit**) is harelike, with long ears, woolly feet and a short, thick tail. The body fur is generally greyish, but with contrasting rich reddish on the nape and a yellowish-white chin, with a black line above. It was discovered in 1902, and subsequently seen again in 1929, 1937, 1947, 1948, and then not until 1979. Found in dense riverine bush, notably around the Rhinoceros and Fish Rivers of the Karoo, where the habitat is disappearing with the spread of agriculture. Within their range they occur in two separate populations: near Calvinia and near Deelfontein, Cape Province. Virtually nothing is known of the Bushman Rabbit's habits, such as feeding or reproduction. It has never been exhibited in zoos.

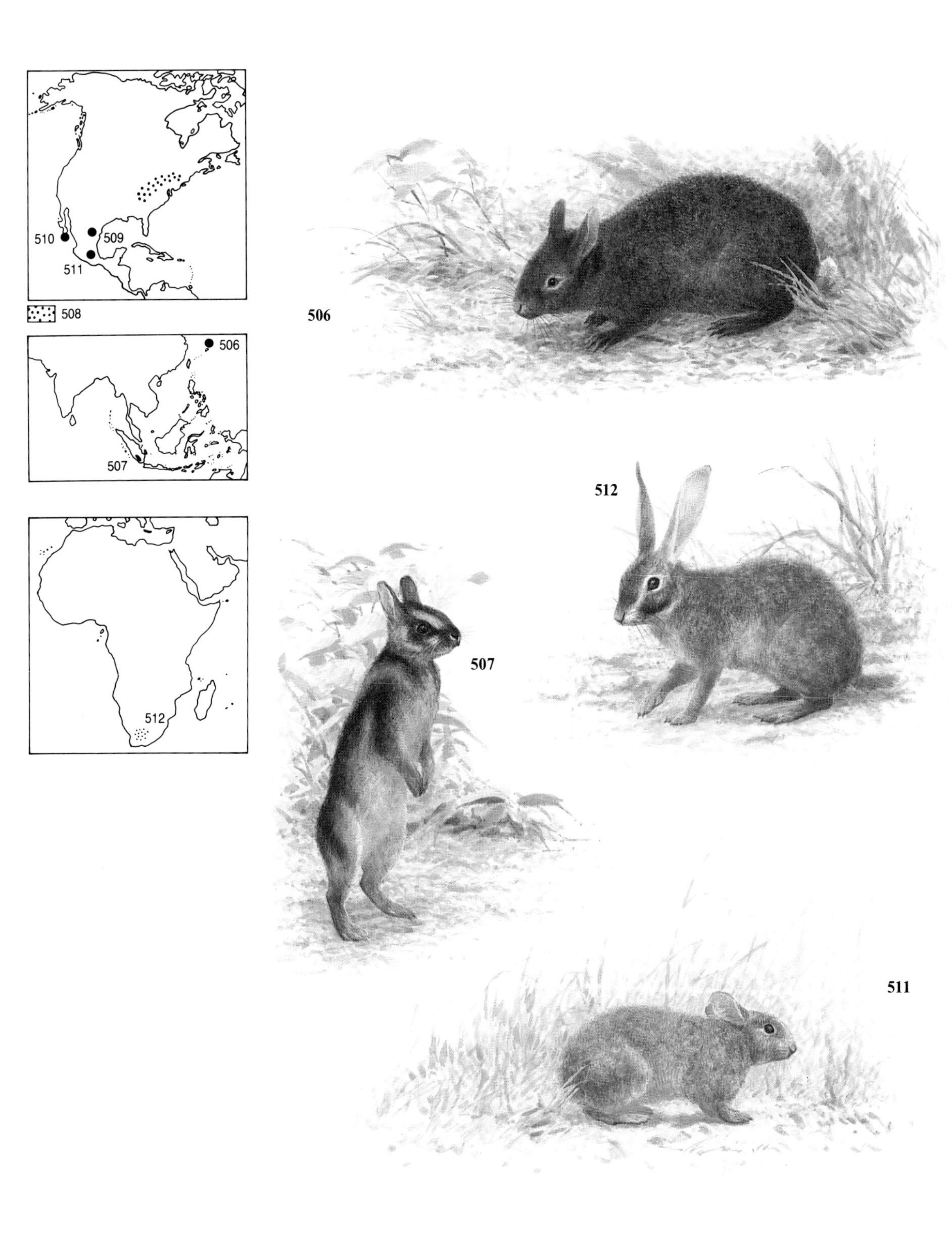
510
509
511
508
506
507
512
506
512
507
511

* WHITE-SIDED JACK RABBIT

Lepus callotis (inc. mexicanus) (**513**)

HB 18.5–20.7ins; T 2.4–2.8ins; WT ca.5.9lb.

Colour varies from pale buff to blackish, with a dark nape, white sides and underparts, grey rump and black on upper tail. Normally live in pairs in short-grass habitats, where they are almost completely nocturnal. Has a restricted range in extreme SW New Mexico, and parts of N and C. Mexico. In areas where its habitat has been degraded by overgrazing (which is widespread), the White-sided Jack Rabbit is being displaced by the Black-tailed *L. californicus*, which is better adapted to degraded habitats. A related species, * *L. insularis* (**514**) is confined to Espiritu Santo Island off Baja California, Mexico; it is a melanistic form most closely related to *L. californicus.*

* SNOWSHOE HARE

Lepus americanus (**515**)

HB 14.2–20.5ins; T 1–2.2ins; WT 2.2–3.5lb.

In summer brownish with darker nose and ears, and white-edged nostrils; the underside is white. In winter pure white except for ear tips. Common over most of Canada and south along the Rockies into Utah, Nevada and also south to North Carolina, USA. Although Canadian populations are thriving, some US populations are declining, due to habitat changes, and the introduction of species such as the European Hare *Lepus capensis*. The Virginian population may be endangered, due to logging of its habitat.

* ORIENTAL HARE

Lepus sinensis (**516**)

Is considered threatened in Vietnam; little is known of its status elswhere in its range which extends into SE China, with discrete populations in S Korea and Taiwan.

*** HISPID HARE or BRISTLY RABBIT

Caprolagus hispidus (**517**)

HB ca.18.5ins; T ca.2ins; WT ca.5.5lb.

Unusually coarse and bristly outer fur, with short fine underfur; short ears and relatively short hindlegs for a hare. Dark brownish above, brownish below. Found in tall thatch grasslands (up to 12ft); during the monsoon season, when the grasses are too waterlogged, they move into the foothills, and when the dry grasslands are burnt by local villagers, they move into the forested foothills and agricultural lands. They feed mainly on grasses, shoots, bark and roots and occasionally crops. Not gregarious, but may occasionally be found in pairs. Formerly recorded along the southern foothills of the Himalayas from Uttar Pradesh, through Nepal, Sikkim, Bengal and Bhutan to Assam, but now rare or completely extinct over most of this area, and probably confined to NW Assam and a few areas in Nepal. In addition to burning of their main habitat, they are also threatened by hunting, and are killed by dogs. They are the subject of a conservation programme by the Jersey Wildlife Preservation Trust. Protected in India and listed on CITES App.I.

***** VANCOUVER MARMOT

Marmota vancouverensis (**518**)

HB 26.4–29.6ins; T 7.1–9.5ins; WT ca.6.6lb.

A large, heavy-bodied rodent occurring only on Vancouver Island, BC, Canada. Found in the mountains, in forested clearings and above the tree-line, between 3300–6550ft. They live in burrows in dry, rocky ground, and since their habitat is snow-covered for much of the year, they hibernate for longer than most other marmots. Reduced to 100–150 individuals in four areas where they are threatened by logging and ski development. Protected under British Columbia law; there are proposals for reserves. The closely related ** **Olympic Marmot** *M. olympia* (**519**) is confined to the Olympic Mountains in Washington DC, USA, with a total range of 680 sq.mls.

*** BOBAK MARMOT

Marmota bobak (**520**)

The Bobak or **Steppe Marmot** once occurred from the steppes of eastern Europe, eastwards to Kazakhstan, Mongolia and western China. It is completely extinct in Hungary, Romania, Poland and other parts of Europe, and is thought to be declining in other parts of its range, wherever agriculture is encroaching on steppe habitat.

*** MENZBIER'S MARMOT

Marmota menzbieri (**521**)

Superficially similar to other marmots, Menzbier's is confined to two discrete areas in the Tien Shan Mountains of Uzbek and eastern Kazakhstan, USSR, where they occur at altitudes of up to 6550ft and are considered theatened.

* EUROPEAN SOUSLIK

Spermophilus citellus (**522**)

HB 7.1–9.5ins; T 1.8ins; WT 8.4–11.9oz.

A short-eared, large-eyed ground squirrel with a fairly mottled back. Always terrestrial, they live in colonial burrows and are active by day. They feed mainly on seeds, which they carry to underground stores in their cheek pouches, and during the coldest months they hibernate. They were until the mid 19th century very abundant in the open steppe country of E and SE Europe, as far east as the Crimea. However, the spread of agriculture into the steppes has drastically reduced their range in Hungary, although in more upland areas (up to 7200ft in Yugoslavia, 4250ft in Czechoslovakia) they are sometimes still locally abundant. There are several little-known species of *Spermophilus* occurring in Mexico, similar in general appearance, some populations of which may be threatened.

* BLACK-TAILED PRAIRIE DOG

Cynomys ludovicianus (**523**)

HB 11–13.8ins; T 2.8–4.5ins; WT 2–3lb.

Tawny-coloured, with short legs and tail tipped with black. They live in 'towns', in short-grass habitats in the prairies. A once common and widespread species, which in 1927 spread to Canada. However, also much reduced in many areas: the Texas population, estimated at 800 million in 1905, was reduced to 2.25 million by the 1970s. Similarly, they once occupied ca. 4000 square miles in Kansas, but are now reduced to 60 sq.mls. A single 'town' in Texas is reputed to have once contained 400 million prairie dogs, and occupied an area of 247,040 sq.mls. In S Dakota a range of 3000 sq.mls is now reduced to 96 sq.mls, and elsewhere similar declines have been noted. One of the main causes is poisoning, because they compete with sheep and cattle-grazing. They may also eat crops such as alfalfa beans, potatoes, corn, and sorghum wheat, and will bark young fruit trees. They apparently thrive in captivity, but the most urgent conservation measure is to set up colonies on suitable reserves within former range.

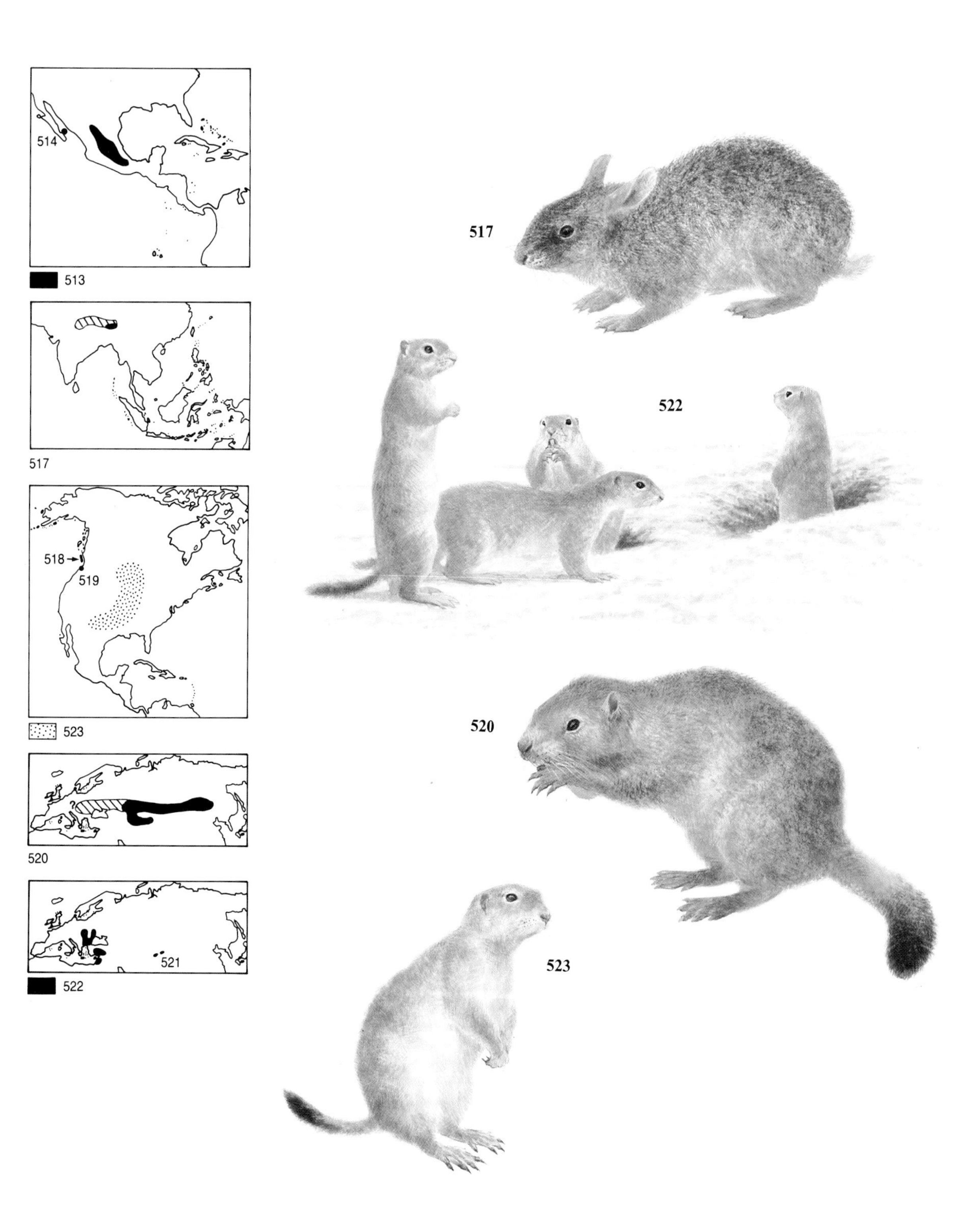
514
513
517
517
518
519
523
520
521
522
522
517
520
523

*** UTAH PRAIRIE DOG

Cynomys parvidens (**524**)

Uniformly brown with short legs and a short, white tail. Live colonially, in 'towns'. Believed to have always been confined to Utah; as recently as 1935 it was found in 9 counties, but by 1962 restricted to only 5. In 1972 a programme was initiated to establish colonies on protected land and by the 1980s the decline had been halted. In the 1920s its population may have been around 95,000, but by 1979 a census gave a total of 84 towns inhabited by 2887 individuals. Classified as Endangered under US legislation, and a recovery plan is well under way. A small number are being captive-bred.

* MEXICAN PRAIRIE DOG

Cynomys mexicanus (**525**)

HB ca.11.8ins; T ca.3.2ins; WT up to 2.6lb.

Like other prairie dogs, they are losing habitat to agriculture and are now confined to an area of less than 310sq.mls, S Coahuila and N San Luis Potosi in N Mexico. They live at 5250–7200ft in 'towns' often isolated by desert. Listed on CITES App. I.

** WOOLLY FLYING SQUIRREL

Eupetaurus cinereus (**526**)

HB 20.3–24ins; T 15–18.9ins.

The entire body is covered with soft thick woolly fur, greyish above, paler below. Found in rocky terrain and believed to feed on lichens and moss. Only known from a few specimens, collected in Pakistan and Gilgit, N Kashmir, at altitudes of up to 9850ft.

* NORTHERN FLYING SQUIRREL

Glaucomys sabrinus (**527**)

HB 9.7–14.6ins; T 4.3–7.1ins; WT 1.6–2.4oz.

Upperparts are greyish-brown, underparts whitish. The flying membrane extends between the legs, but not the tail. Many subspecies have been described, and some populations are now isolated from each other. In the southern parts of its range it overlaps with the Southern Flying Squirrel *G. volans*; both are found in woodland, mostly coniferous, where they occupy a similar niche. Flying squirrels do not hibernate, but in cold weather remain in their nests; *G. sabrinus* appears less restricted than *G. volans* in its winter activity. In areas where the Northern species is threatened by habitat destruction, it is also being displaced by the Southern, which also transmits a harmful parasite and is more adaptable and aggressive. *G.s. fuscus* and *G.s. coloratus* are among the subspecies most threatened, and listed in the US as Endangered.

** FLIGHTLESS SCALY-TAILED SQUIRREL

Zenkerella insignis (**528**)

HB ca.7.9ins; T 6.5ins.

Although closely related to flying squirrels (*Idiurus* spp. and *Anomalurus* spp.) this species has no trace of a flying membrane. The soft, dense fur is greyish above and paler below with an almost black tail. At the base of the underside of the tail there are 13 sharp-tipped scales. Confined to forests in Cameroon, Central African Republic, Rio Muni and possibly Gabon. It lives in holes in ancient trees and, with the increasing rate of destruction of its forest habitat, it is likely to become threatened. Only known from a handful of museum specimens, and virtually nothing is known of its habits.

** GREY BELLIED SCULPTOR SQUIRREL

Glyphotes canalvus (**529**)

HB ca.5.1ins; T ca.3.9ins.

Only known from 3 specimens collected on Mt Dulit, Sabah, Malaysia. A closely related species, *G. simus* is also known from the mountains of Sabah and Sarawak.

* AFRICAN PALM SQUIRREL

Epixerus ebii (**530**)

HB 10–12ins; T 11–12ins.

Eastern population *E.e.wilsoni* is rufous and black above, and yellowish rufous below; western populations *E.e. ebii* and *E.e.jonesi* are more rufous; the lower parts of the limbs bright red, underside orange, and tail sometimes banded with white. Found in moist forests, where they are largely terrestrial, at altitudes of up to 3300ft. Little-known in the wild, and they have rarely been seen alive. Their range stretches from Sierra Leone, Ghana and Cameroon to Gabon and W Congo. This species has probably always been scarce, and is almost certainly threatened by forest destruction.

* TASSEL-EARED SQUIRREL

Sciurus aberti (**531**)

HB 6.7–13ins; T 7.7–10ins; WT 1.3–1.5lb.

Characterised by long black ear tufts (up to 1in). Mainly found in Ponderosa pines, in foothills of Rocky Mountains, and Colorado Plateau. The Kaibab Squirrel *S.a. kaibabensis*, sometimes regarded as a full species, is confined to the Kaibab National Forest and Grand Canyon NP. Although numbers have at times dropped to 500, they are probably at carrying capacity for their habitat and range.

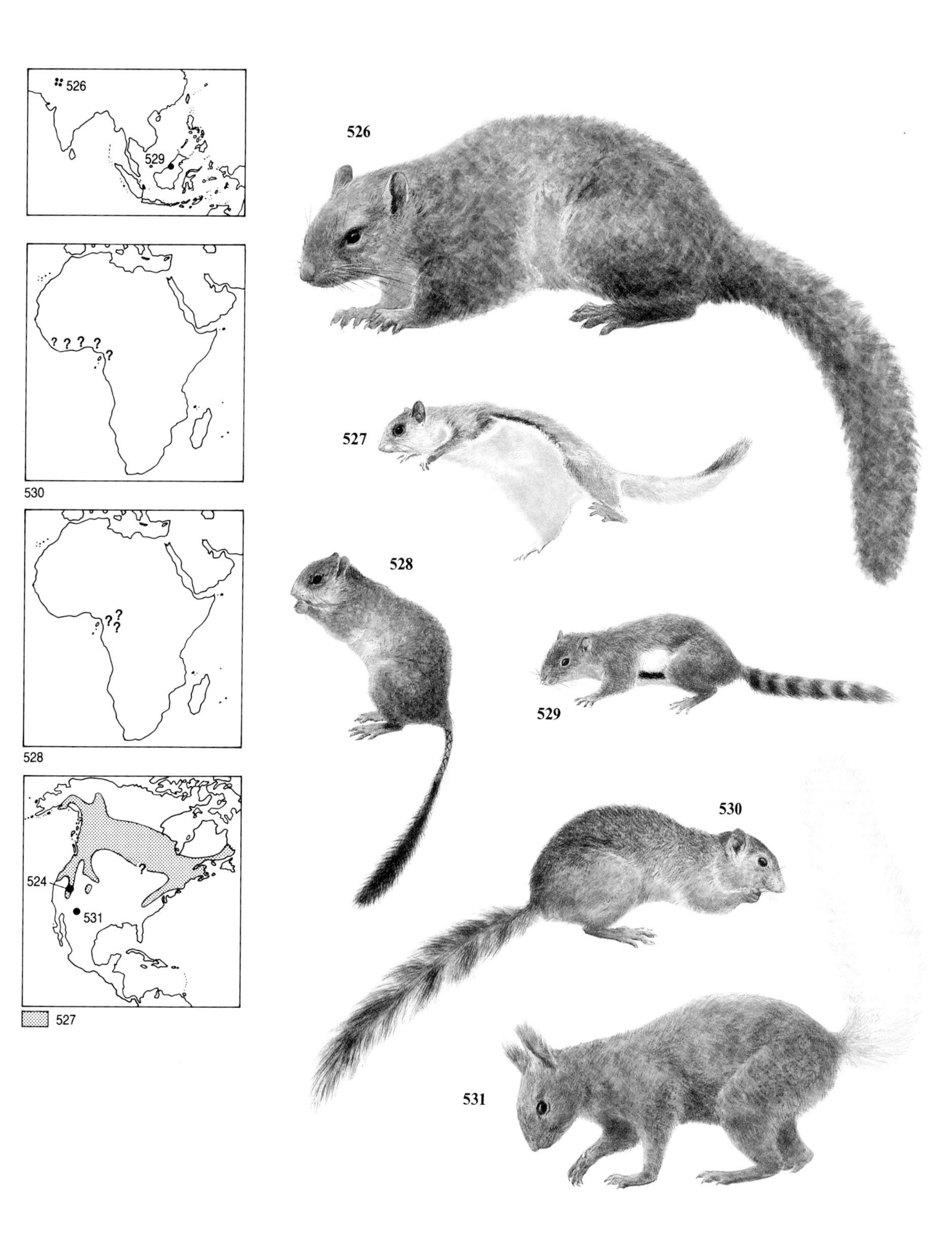
526
529
530
528
524
531
527
526
527
528
529
530
531

*** PETER'S SQUIRREL**
Sciurus oculatus **(532)**
HB 20.9–22.1ins; T 9.8–10.6ins; WT 1.2–1.6lb.
Closely related to the Fox Squirrel *S. niger* (541), which is widespread and common in the eastern USA, Peter's Squirrel has buff ear patches and eye-rings. Decline is due to loss of forest. Status not known. Only found in C Mexico.

*** ALLEN'S SQUIRREL**
Sciurus alleni **(533)**
Closely related to the Fox Squirrel (541). Confined to NE Mexico, where it is declining due to deforestation. Little is known of its precise status. Only found in temperate montane forests of Sierra Madre Oriental, of SE Coahuila, Nuevo Leon and Tamaulipas.

*** NAYARIT SQUIRREL**
Sciurus nayaritensis **(534)**
Related to Fox Squirrel (541). Found in WC Mexico north to SE Arizona. Occurs in forests at altitudes of 6550–9200ft in W Sierra Madre Occidental, in S Durango, Nayarit, Zacatecas, and Jalisco; declining due to deforestation. Precise status not known.

*** ARIZONA GRAY SQUIRREL**
Sciurus arizonensis **(535)**
Found from C Arizona to W New Mexico, NE Sonora, and NW Chihuahua, Mexico, in pine forests as well as more open oak woodlands; probably declining due to loss of habitat. NB: other *Sciurus* with restricted ranges, about which little is known include ** *S. richmondi* **(536)**, from Nicaragua; ** *S. sanborni* **(537)** from SE Peru and ** *S. stramineus* **(538)** which occurs on the borders of NE Peru and SE Ecuador; *S. colliaei* **(539)** from W Mexico and *S. deppei* **(540)** from E Mexico to N Costa Rica, where is is possibly threatened.

***** FOX SQUIRREL**
Sciurus niger **(541)**
HB 9.8–14.2ins; T 7.9–13ins; WT 1.5–2.6lb.
Black to silvery or rusty brown above and orange below, with much variation in a single population. Found mainly in woodland edge habitats, and deciduous woodlands. Their range extends over most of USA, east of the prairies with a few populations in S Canada. Whilst numerous and increasing in parts of their range, in eastern states several subspecies have considerably reduced ranges; *S. n. cinereus* is now restricted to Maryland; *S. n. vulpinus* has lost 50% of its range in New England; in Florida *S. n. shermani* and *S. n. avicennia* have both declined through loss of mature forest, and in Mexico most populations are suffering habitat loss.

*** EURASIAN RED SQUIRREL**
Sciurus vulgaris **(542)**
HB 7.1–9.5ins; T 5.5–7.9ins; WT up to 1.1lb.
Usually chestnut brown above, pale below, but variable; dark (melanistic) forms are common. In British Isles ear-tufts and tail gradually bleach until by mid-summer they are creamy white. Found in a wide variety of wooded habitats, except in Britain where they are mainly restricted to coniferous woodlands. They feed on tree seeds, foliage, shoots and fungi, as well as animals including eggs and nestlings of birds, and insects. Although widely distributed in Europe, east to China, Korea and Hokkaido, in the British Isles they have undergone longterm fluctuations within historic times. Extinct in S Scotland and Ireland by 18th century, and a century later close to extinction in the Scottish Highlands. Subsequently reintroduced into Scotland and Ireland, and flourished. The decline this century in England and Wales, where they are now close to extinction, is mainly due to habitat changes, but roles of disease, and competition from introduced Grey Squirrels *S. carolinensis* are not fully understood. Breed freely in captivity.

*** GIANT SQUIRRELS**
Ratufa spp.
HB 9.8–18.1ins; T 9.8–18.1ins; WT 3.3–6.6lb.
All four species are large, brightly coloured squirrels, some black above, pale yellow-fawn below, others marked with reds, browns and grey above. They are generally solitary or live in pairs, in dense forest where they feed on fruit, nuts and small animals, eggs etc. *R. macroura* **(543)** is found in S India and Sri Lanka; *R. indica* **(544)** in S and C India, except lowlands; *R. bicolor* **(545)** from Nepal east to Indo-China and south to Sumatra, Bali and Java and *R. affinis* **(546)** from the Malay peninsula to Sumatra and Borneo. Some, such as *R.bicolor*, have isolated populations on islands. Throughout their range, but particularly in Thailand and Malaysia, their tall forest habitat is increasingly under pressure. In Vietnam spraying of forests with defoliants during warfare has seriously affected their habitat. Frequently exhibited in zoos, and have been bred, but there are no self-sustaining populations. All are listed on CITES App.II, and have legal protection in India and Indonesia. They also occur in many national parks and reserves.

**** PYGMY FLYING SQUIRRELS**
Petaurillus spp.
HB 2.6–3.9ins; T ca.3.1ins.
The smallest of the flying squirrels, confined to Borneo and Peninsular Malaysia. *P. kinlochii* **(547)** is only known from Selangar in Malaysia; *P. hosei* **(548)** from a few scattered localities in Borneo, and *P. emiliae* **(549)** from a single pair collected in Sarawak in 1901.

*** FOUR-STRIPED GROUND SQUIRREL**
Lariscus hosei **(550)**
HB 6.8–10.2ins; T 4.3–5.6ins; WT 5.1–7.5oz.
Upperparts brownish, with 9 prominent stripes; one reddish-brown in the middle of the back and two buff and two blackish either side. Very little known, and confined to a limited area in North Borneo.

***** GIANT FLYING SQUIRREL**
Petaurista petaurista **(551)**
HB 14.6–16.9ins; T 14.4–18.5ins; WT 2.2–6.6lb.
Widespread from Kashmir, east to Vietnam and south to Indonesia, but locally threatened by forest destruction, notably in Vietnam.

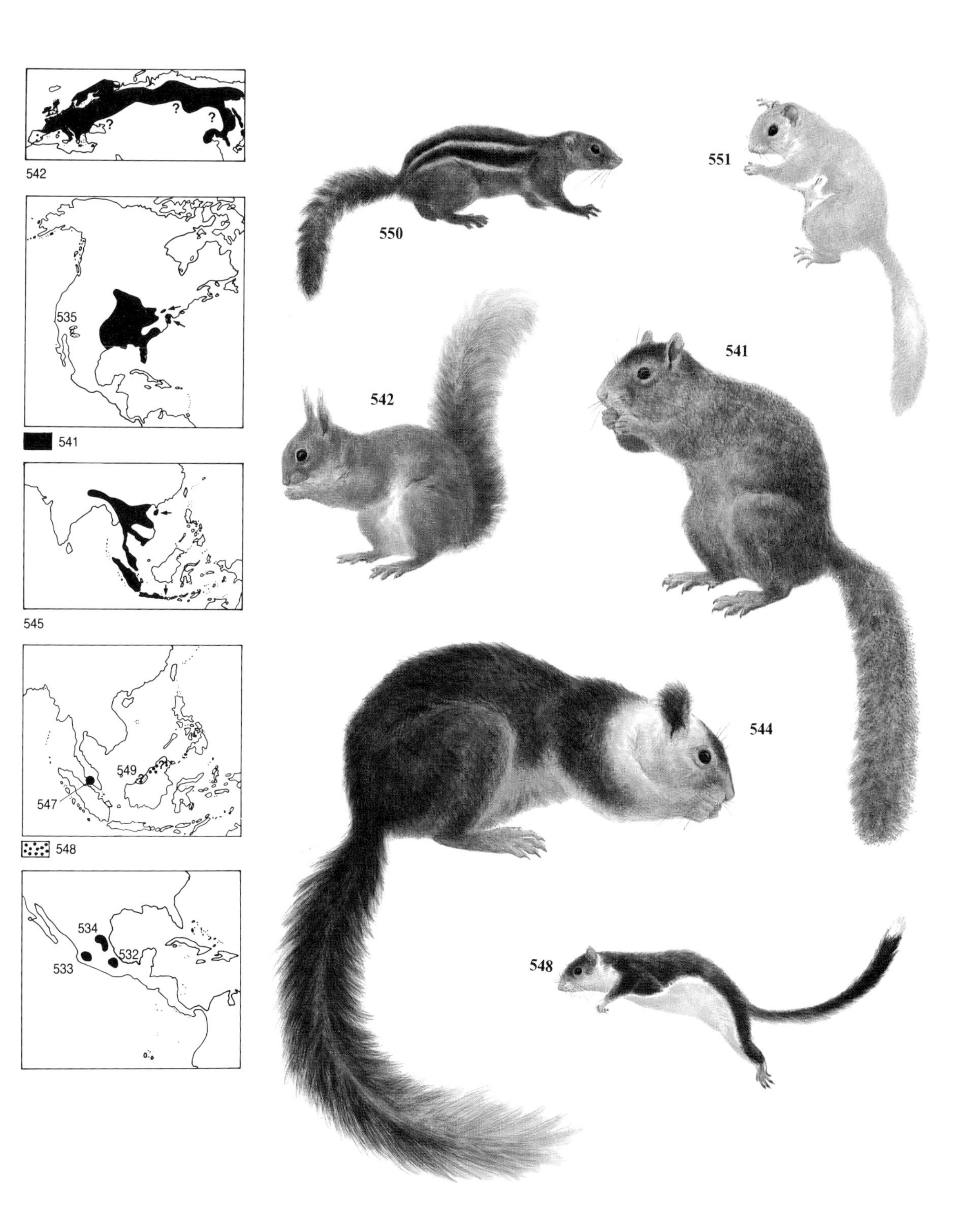

?
?
?
542
535
541
545
549
547
548
534
532
533
550
551
541
542
544
548

**** HAIRY-FOOTED FLYING SQUIRREL**
Belomys pearsoni **(552)**
HB 7.1–10.2ins; T 3.9–6.3ins.
The long, soft fur is glossy reddish-brown on the upperparts, and the underside whitish or reddish-whitish. Found in dense forests in the eastern Himalayas, but it is very poorly known, and may occur in different habitats elsewhere. Known from widely scattered localities in Nepal, Sikkim, Arunachal Pradesh, China, Taiwan, Burma, Thailand and Vietnam.

*** SIBERIAN FLYING SQUIRREL**
Pteromys volans **(553)**
HB 5.3–8.1ins; T 3.5–5.5ins; WT 3.3–5.9oz.
Colour varies from silvery to buffy-grey above, to whitish below. The tail is slightly flattened but not connected to the membranes, which connect the front and hind legs. Strictly nocturnal, they are found mainly in coniferous forests, though rarely seen even in areas where they are common. They are arboreal and feed on nuts, pine kernels, bark and buds, and some animal matter such as insects; nests are built close to the main trunk, or in holes. Their range extends across most of N Eurasia from Finland and the Baltic Sea to E Siberia, Korea and Sakhalin and Hokkaido. Apparently declining in the western part of European range, but little data exists.

**** SUNDA TREE SQUIRRELS**
Sundasciurus spp.
HB 5.1–11.4ins; T 2.8–11.4ins; WT 1.9–14.7oz.
A group of 14 species of greyish or brownish squirrels found in forests in SE Asia. Some species have adapted to cultivated and bush country, but others are found in forests; with the rapid destruction of these forests some of the lesser known species will disappear. ** *S. davensis* **(554)** from Mindanao, Philippines is only known from one locality; ** *S. hoogstraali* **(555)** from Busuanga Island, Philippines; ** *S. mollendorffi* **(556)** from Culion Island, in the Calamian Isles, Philippines; other species in the Philippines are poorly known.

****** MINDANAO ARROW-TAILED FLYING SQUIRREL**
Hylopetes mindanensis **(557)**
Known only from a single locality in Misamis Oriental, Mindanao, Philippines; only specimen destroyed in World War II. The closely related * *H. alboniger* **(558)** threatened in Vietnam; occurs north to Sichuan and west to Nepal.

*** DWARF FLYING SQUIRRELS**
Petinomys spp.
These little-known species of squirrel inhabit tropical forests (up to 3950ft) where they are mainly nocturnal. They have relatively small litters (1–3). Because their habitat is being destroyed several species could be at risk, in at least part of their range – which includes several islands. *P. bartelsi* **(559)** and *P. sagitta* **(560)** are confined to Java; *P. crinitus* **(561)** is only found on Basilan and Mindanao in the Philippines and *P. hageni* **(562)** occurs on Borneo (one specimen), Sumatra and some of the Mentawai islands. More widespread species *P. setosus* from Burma, Thailand, Malaysia, Sumatra and Borneo threatened by damage to habitat and fragmentation.

**** PANAMA MOUNTAIN SQUIRREL**
Syntheosciurus brochus **(563)**
HB 5.9–6.7ins; T 4.7–5.9ins.
Tawny brown above, and rusty below. Known only from 2 specimens collected near Cylindro, Panama, in 1902, and 6 near Cerro Pittier, Costa Rica, by 1984.

**** POAS MOUNTAIN SQUIRREL**
Syntheosciurus poasensis **(564)**
Similar in size to preceding species. Cinnamon-buff on back, with reddish fringe to tail. Known from one specimen, from Poas Volcano, Alajuela, Costa Rica. Another, possibly of this species, was collected in Panama in 1966.

** *Callosciurus* spp.
Several of these often spectacularly coloured squirrels from SE Asia have restricted ranges, and may be threatened. The Vietnamese population of * *C. finlaysoni* **(565)**, which is found from Burma, east into Indo-China, may have suffered from the effects of forest destruction by herbicides used in warfare.

*** BLACK AND RED BUSH SQUIRREL**
Paraxerus lucifer **(566)**
HB 7.9–12.4ins; T 6.3–10.4ins; WT up to 1.5lb.
Colour variable, but often brightly coloured, with orange-red upperparts, and greyish underparts. Three subspecies occur in isolated populations in montane forests in an area of over 386sq.mls: *P.l. lucifer* on the NW shore of Lake Malawi; *P.l. laetus* in mountains to W of Lake Malawi; and *P.l. byatti* on Kilimanjaro, the Usumbara, Ulunuru and Uzungwa Mountains. Isolated distribution, so in habitats under pressure some populations could be threatened, but status little-known.

**** SWYNNERTON'S BUSH SQUIRREL**
Paraxerus vexillarius **(567)**
HB 9.5ins; T 8.3ins.
Back greyish with orange head and feet and bright orange tip to tail, which is banded black and white. Known from 2 specimens from the Usumbara Mountains, from forests which have now greatly diminished. (NB: it might possibly be a hybrid *P. lucifer x P.palliatus.*) Other species, such as * *P. flavivittis* **(568)** include little-known subspecies, often isolated. ** Vincent's Squirrel *P.vincenti* **(569)** is only known from Namuli Mountain, C Mozambique.

**** BIG POCKET GOPHERS or TALTUZAS**
Orthogeomys spp.
Taltuzas occur in a wide variety of habitats up to 9850ft in altitude, including semi-desert and montane forests. They excavate shallow burrows and are largely nocturnal, feeding on vegetable matter, and are often pests of agriculture. Several species have very restricted ranges with isolated populations and may be threatened; these include *O. lanius* **(570)** which is only known from the SE of Mt Orizaba, in Veracruz, Mexico; and *O. cuniculus* **(571)** from Zanatepec, Oaxaca, Mexico.

*** POCKET GOPHER or TUZA**
Pappogeomys tylorhinus **(572)**
HB up to 14ins; T up to 4.2ins; WT up to 1.3lb.
As with closely related genera, the distribution and occurrence of these Pocket Gophers is poorly known; the Tuza only from a relatively small area in C Mexico, and the closely related *P. neglectus* **(573)** is from Cerro de la Calentura, Queretaro, Mexico. *P. merriami* **(574)** is also only known from a relatively small area in C Mexico. However, if rare in museums, such species can be locally abundant.

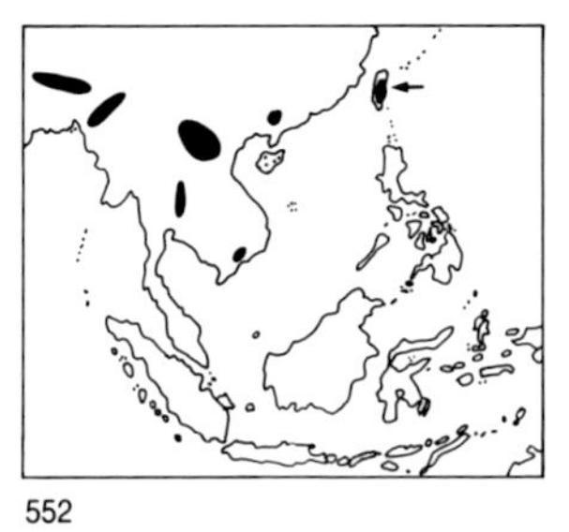

552

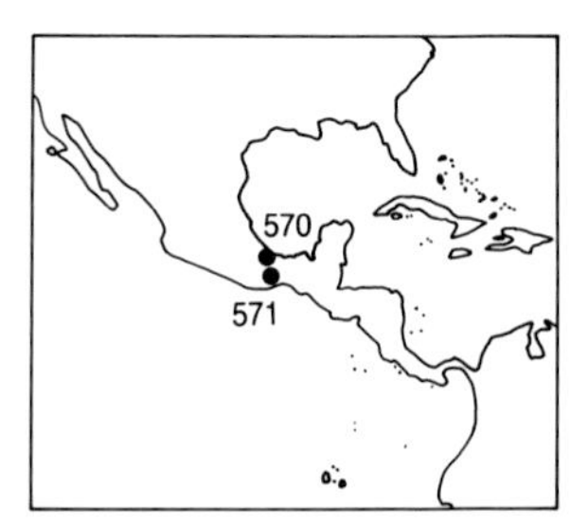

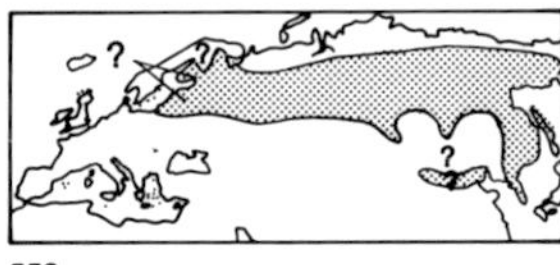

553

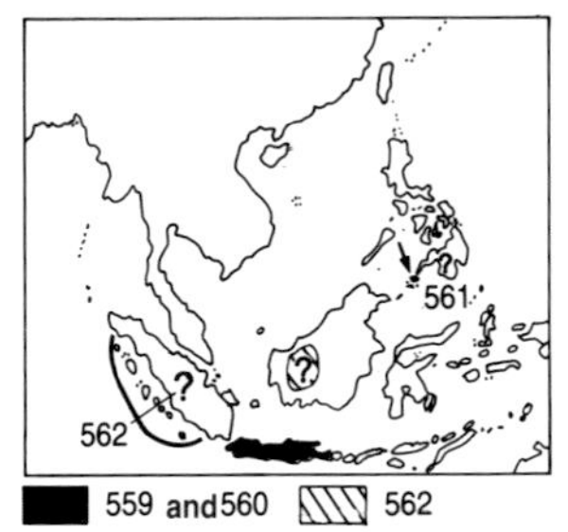

559 and 560 562

? 561

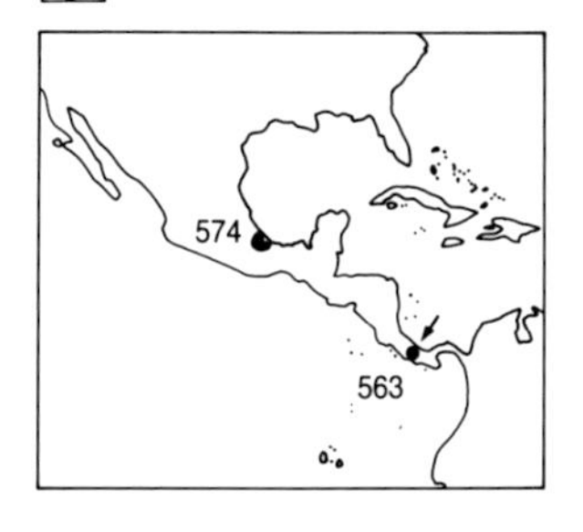

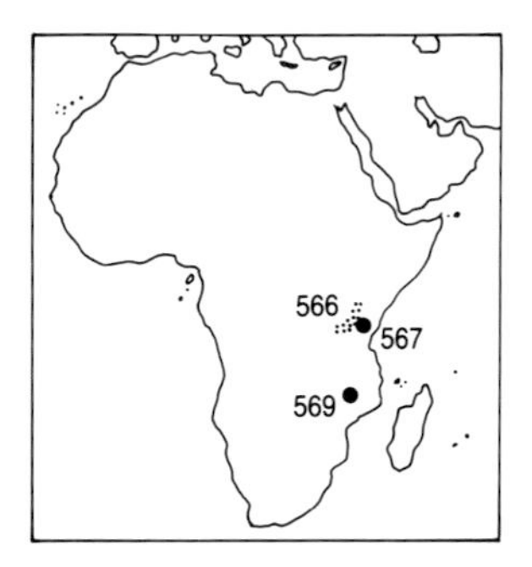

* MICHOACAN POCKET GOPHER

Zygogeomys trichopus (**575**)

Although reportedly abundant this species is only known from two areas in Michoacan, Mexico, at altitudes of 5900–11,800ft in the pine zone, and may possibly be vulnerable. Is reported to be of damage to crops.

*** SOUTH-EASTERN POCKET GOPHER

Geomys pinetis (**576**)

HB 6.1–7.9ins; T 2.8–3.7ins; WT up to ca. 12.2oz.

The short fur is brownish above, and paler below. The hands are modified for digging and insides of the external cheek pouches are furry. Found in sparsely wooded areas, particularly pinewoods, and fields. Breed throughout year, with up to 3 young in a litter. They usually live singly in burrows which they excavate themselves, feeding on underground roots, bulbs, rhizomes, tubers etc., and can do damage in agricultural areas. The coastal Georgia populations declined during the present century through habitat change; *G.p. fontanelus* is probably extinct, *G.p. goffi* may be extinct, and *G.p. cumberlandius* was thought extinct but recently rediscovered (1981); *G.p. colonus* survives in small numbers. These subspecies are sometimes regarded as full species. The ** **Tropical Pocket Gopher** *G. tropicalis* (**577**) is known from a limited area in SE Tamaulipas in Mexico and possibly declining due to habitat changes. Pocket gophers thrive in captivity, but there do not appear to be any self-sustaining populations of the rarer species.

KANGAROO RATS

Dipodomys spp.

HB 3.9–7.9ins; T 3.9–8.3ins; WT 1.2–6.3oz.

They are all characterised by long tails, and long hind legs. Of the ca.20 species, several species and subspecies are considered threatened. Some of these are listed below; however, there are a large number of subspecies, few of which have been studied in any detail. * **Chisel-toothed Kangaroo Rat** *D. microps* (**578**): in areas of Kansas where bush cover is being lost, this species is being displaced by **Merriam's Kangaroo Rat** *D. merriami*. Several subspecies have restricted ranges, and those isolated may be vulnerable. ** **Heermann's Kangaroo Rat** *D. heermanni* (**579**) has a subspecies *D.h.morroensis*, confined to south side of Morro Bay, California. Several other subspecies also have very restricted ranges, though not yet apparently threatened. The ** **Texas Kangaroo Rat** *D. elator* (**580**) has a small range around Clay county, Texas, and SW Oklahoma where loss of habitat due to agriculture, urbanisation, and destruction of cover have caused its decline. ** **Stephens' Kangaroo Rat** *D.stephensi* (**581**) has a restricted range in San Jacinto Valley of Riverside County, California. ** **Big-eared Kangaroo Rat** *D. elephantinus* (**582**) has a restricted range in San Benito and Monterey Counties, California. ** **Giant Kangaroo Rat** *D.ingens* (**583**) occurs in a small area on the west of San Joaquin Valley, California. A subspecies of the ** **Fresno Kangaroo Rat** *D. nitratoides exilis* (**584**) is known only from a small area in Fresno County, California, and the species is restricted to the San Joaquin Valley. * **Phillip's Kangaroo Rat** *D. phillipsii* (**585**) is endemic to Mexico, and some populations have very restricted ranges. * **Ord's Kangaroo Rat** *D. ordii* (**586**) has a number of subspecies, confined to islands off Texas; others from small areas inland. None reported as threatened yet, but restricted range makes them vulnerable to change, and may be losing habitat to agriculture in Mexico.

Other kangaroo rats with restricted ranges include: ** *D. insularis* (**587**) found only on San José Island, Baja California, Mexico; ** *D. margaritae* (**588**) found only on Santa Margarita Island, Baja California. Several of the closely related Forest Spiny Pocket Mice *Heteromys* spp. have little-known restricted ranges in Mexico and Central America. ** *H. nelsoni* (**589**) is only known from S Chiapas in Mexico, and ** *H. longicaudatus* (**590**) is only known from Tabasco, Mexico. ** **Anthony's Pocket Mouse** *Perognathus anthonyi* (**591**) has only been found on Cedros Island, W Baja California; other *Perognathus* spp. have restricted ranges, but there may prove to be fewer, rather variable species, when they have been further studied.

* ZENKER'S FLYING SQUIRREL

Idiurus zenkeri (**592**)

HB 2.8–3.1ins; T 3.5–4.3ins; WT 0.5–0.6oz.

Also known as the **Dwarf Anomalure**, **Flying Mouse** and **Pygmy Scaly Tail**. A tiny gliding rodent grey-brown above, whitish below. Found in primary forest in two separated areas, in W and C Africa, where it roosts in tree holes, often mixed with other flying squirrels or bats. A good climber, and can glide on membranes which stretch between their limbs, for 164ft, without losing height. The eastern population of Zenker's Flying Squirrel is (as far as known) confined to the Bwamba Forest, Uganda, which is rapidly being destroyed.

*** EURASIAN BEAVER

Castor fiber (**593**)

HB 29.5–39.4ins; T 11.8–15.8ins; Wt up to 88lb, usually smaller.

The largest European rodent, with a broad, flat, scaly tail. They always live in or close to water where they often build lodges, make dams, fell trees and clear canals; in some areas, such as the Rhône, France, they are much more secretive. Once found more in wooded areas north of the Mediterranean across the whole of Europe and N Asia. Although their range has been extensively reduced and fragmented within historic times, and they are now extinct over most of Europe, there have been numerous successful reintroduction programmes and they are now found in many parks and reserves. The *** **American Beaver** *C. canadensis* (**594**) is also much reduced in the southern parts of its range, but is now being extensively reintroduced, and has always remained widespread in the northern parts of its range.

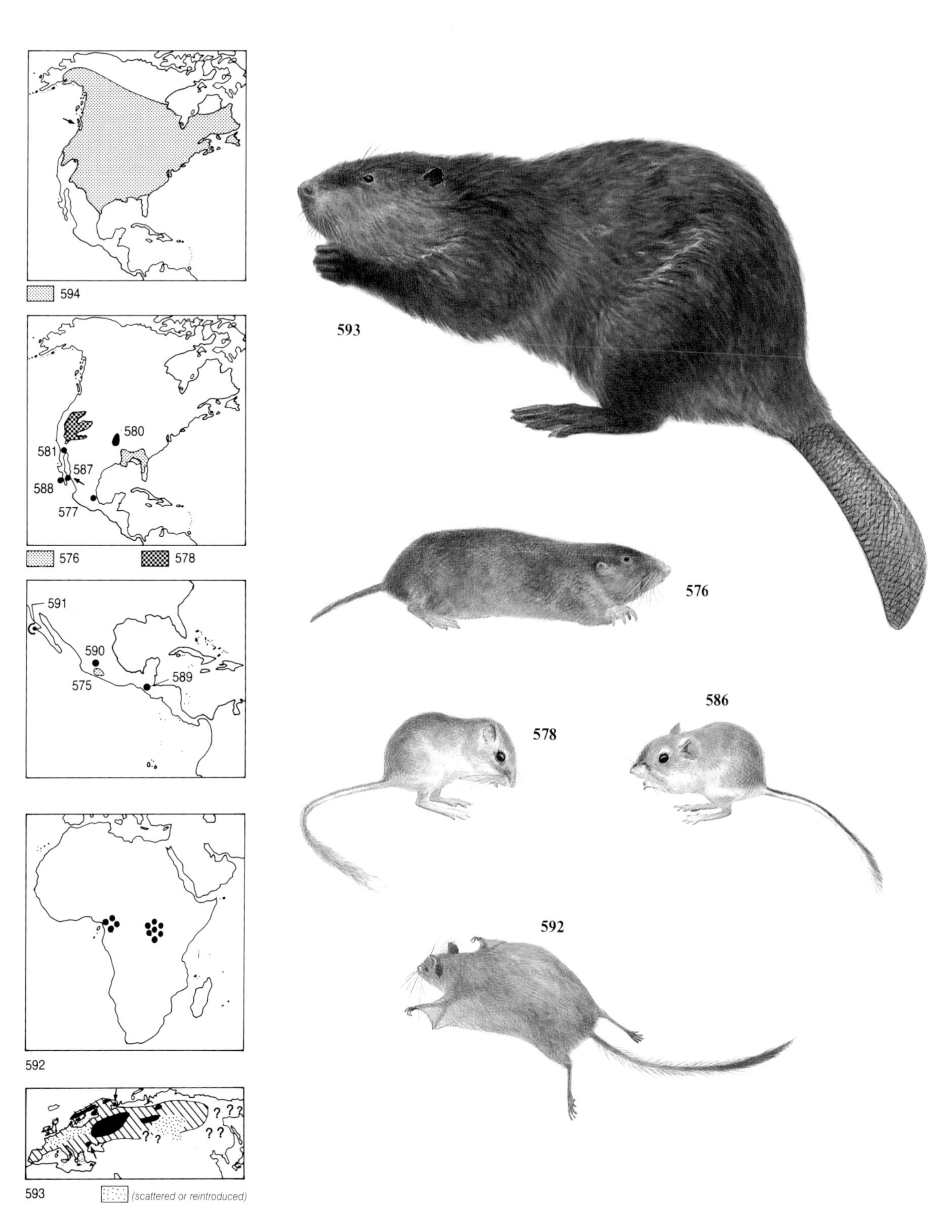
594
581
580
587
588
577
576
578
591
590
589
575
592
593
(scattered or reintroduced)
593
576
586
578
592

* DINAGAT ISLAND CLOUD RAT

Crateromys australis (**595**)

HB ca.10.4ins; T 11.1ins

Only known from a single specimen collected in 1975, and first described in 1985, from the island (259 sq.mls) of Dinagat, off the NE tip of Mindanao in the Philippines.

** ILIN ISLAND CLOUD RAT

Crateromys paulus (**596**)

HB ca.10ins; T ca.8.5ins.

The short coarse fur is dark brown above, creamy below, with a densely furred tail. First described in 1981, only known from one specimen from Ilin Island, West of Mindoro, Philippines. A closely related species, *C. schadenbergi* is apparently quite common in N Luzon, Philippines.

** CLIMBING RATS

Tylomys spp.

HB 6.7–9.8ins; T 9.8–11.4ins.

Several species have restricted ranges and may be threatened, as they are mainly found in forests. The **Chiapas Climbing Rat** *T.bullaris* (**597**) is known only from Tuxtla, Chiapas, Mexico; **Darien Climbing Rat** *T.fulviventer* (**598**) from Tacarcuna, Darien, E Panama, and **Tumbala Climbing Rat** *T.tumbalensis* (**599**) from Tumbala, Chiapas, Mexico.

*** SALTMARSH HARVEST MOUSE

Reithrodontomys raviventris (**600**)

HB 2.5–3ins; T 2.1–3.7ins; WT 0.3–0.5oz.

Reddish-brown above, pinker below. Confined to salt marshes in San Francisco Bay, USA, where they feed on seeds and vegetation. They construct globular nests, and also appropriate nests of Song Sparrows. Main threat to survival is spread of urban development, and drainage. They are protected. The closely related ** *R. sumichrasti* (**601**) has a small isolated population on the Panama/ Costa Rica border; ** *R.microdon* (**602**) also has isolated populations in Mexico and Guatemala. ** *R. rodriguezi* (**603**)is known from one locality in Costa Rica, and ** *R. spectabilis* (**604**) only from Cozumel Island, Mexico.

WHITE-FOOTED MICE

Peromyscus spp.

HB 2.8–6.7ins; T 1.6–7.9ins; WT 0.5–3.8oz.

Often among the most abundant small rodents, but some species and populations have very restricted ranges, and are threatened, mostly by habitat destruction for urban or industrial development. A subspecies of **Cotton Mouse** *** *P. gossypinus lapaticola* (**605**) is restricted to less than 0.6 sq.mls on Key Largo, Florida; *** *P. floridanus* (**606**) is restricted to dry scrub in Florida. Several *** 'Beach' Mice *P.polionotus* (**607**) have declined: the **Alabama Beach Mouse** *P. p. ammobates* was, in 1985, reduced to a single population of 875 in Fort Morgan State Park, and the **Choctawhatchee Beach Mouse** *P.p.allophrys* was reduced to about 500 by 1985; the **Perdido Key Beach Mouse** *P.p.trissyllepsis* is restricted to a single population in Alabama's Gulf State Park, Florida Point. The US has introduced measures to protect the mice and their habitat. The ** **Mexican White-footed Mouse** *P.aztecus* (**608**) is a little-known species with a restricted range in Mexico, Guatemala, Honduras and El Salvador, which although rare in museums is probably widespread. The ** **Jico Deer Mouse** *P.simulatus* (**609**) is only known from Veracruz, Mexico; ** *P.chinanteco* (**610**) from one locality in Oaxaca, Mexico. The ** **Yellow Deer Mouse** *P.flavidus* (**611**) is only known from 2 localities in Panama; ** *P.bullatus* (**612**) is only known from near Limon, Veracruz, Mexico. ** *P.caniceps* (**613**) is confined to Monserrate Island, Mexico, and ** *P.dickeyi* (**614**) is confined to Tortuga Island, Mexico. ** *P.grandis* (**615**) and ** *P.mayensis* (**616**) are only known from single localities in Guatemala. ** *P.pembertoni* (**617**) is found only on San Pedro Nolasco Island, Mexico. ** *P. pseudocrinitus* (**618**) on Coronados Island, Mexico; and ***P.slevini* (**619**) on Santa Catalina Island, Mexico. ***P.stephani* (**620**) is confined to San Esteban Island, Mexico; ** *P.winkelmanni* (**621**) from one locality in Michoacan, Mexico. Many *Peromyscus* spp. have very restricted ranges and/or are little-known and may be vulnerable. However, many will also almost certainly prove to be more widespread than current information suggests.

***** SILVER or KEY RICE RAT

Oryzomys argentatus (**622**)

HB 5.1–5.5ins; T 4.7–5.1ins; WT ca.3oz.

Discovered in 1973, and confined to freshwater marshes in Cudjoe Key, Florida, USA. Observed only on a few occasions in the saltmarshes and adjacent mangroves. Although primarily vegetarian, feeding on seeds and plant matter, they probably eat small animals as well. A litter of three was recorded from one of the few kept in captivity. The main threat to this species is destruction of freshwater marshes around the Florida Keys, now reduced to scattered remnants. There are a total of about 60 species of rice rats, several of which are endangered, while several have become extinct in the past 100 years or so. Others have very restricted ranges, including the remaining Galapagos species; ** *O.bauri* (**623**) from Santa Fé; ** *O.narboroughi* (**624**) and ** *O.fernandinae* (**625**) from Fernandina. ** *O.cozumelae* (**626**) is only known from Cozumel Island in Yucutan; ** *O.nelsoni* (**627**) from Maria Madre Island, W Mexico; ** *O.peninsulae* (**628**) from a restricted area in S Baja California; ** *O.aphrastus* (**629**) is a little-known species from Costa Rica and EC Panama. Many species are variable with numerous subspecies described, a number of which are poorly known or have very restricted ranges.

** GUIANAN WATER RAT

Nectomys parvipes (**630**)

Only known from one locality, near Cacao, on the Comte River in French Guiana. Closely related to *N.squamipes*, which may be a composite species comprising many separate, closely related species. Very little is known of the status of the populations involved.

SOUTH AMERICAN FIELD MICE

Akodon spp.

Vole-like mice, with dense, soft fur, occurring over a wide range, in a variety of habitats. In Argentina and adjacent countries, there are a number of populations on islands. Little is known of their status, but some could be vulnerable. Those with restricted ranges include ** *A. llanoi* (**631**), from Isla de los Estados, at the extreme tip of Tierra del Fuego and ** *A.kempi* (**632**) from islands in the Parana Estuary, between Argentina and Uruguay.

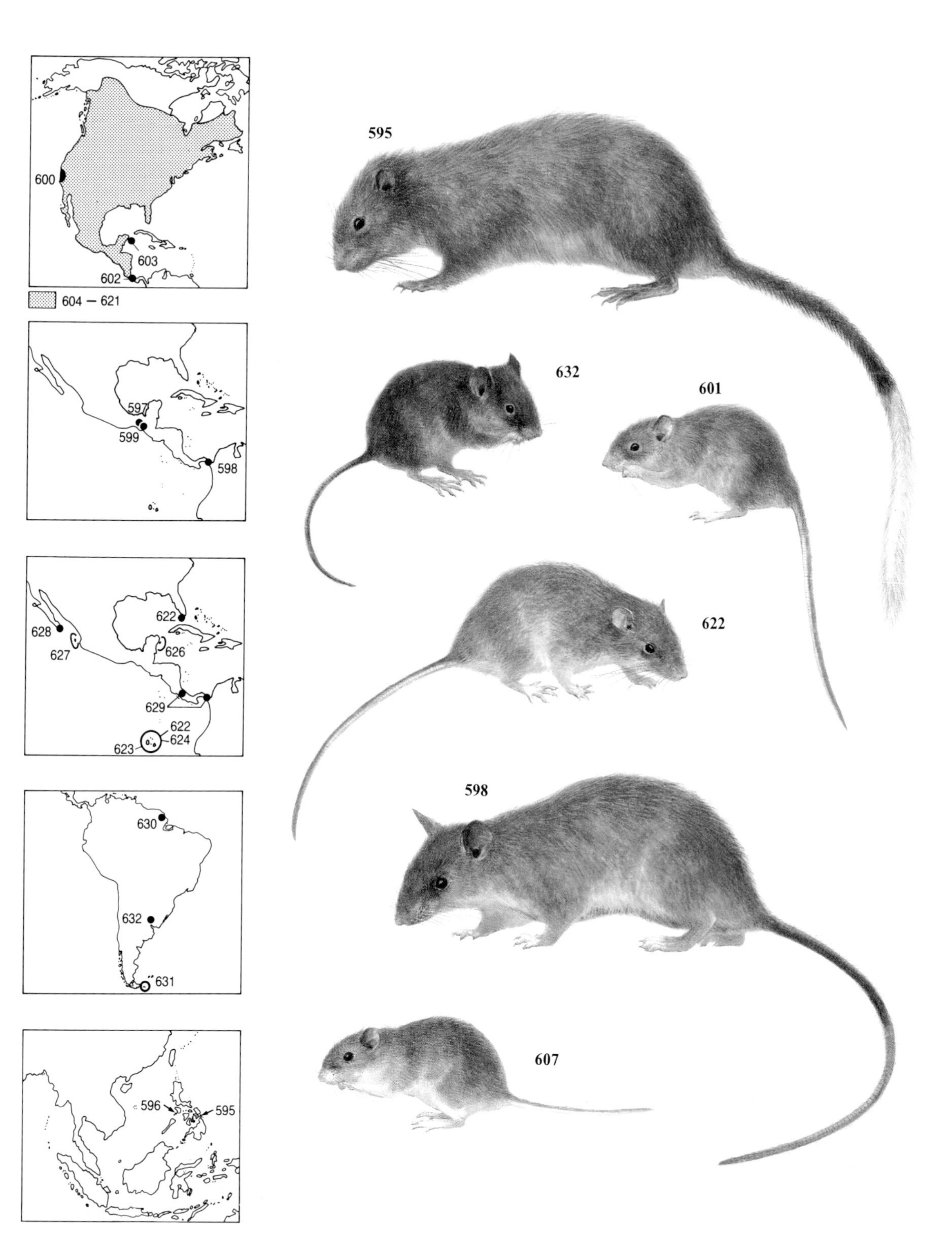

600
603
602
604 — 621
597
599
598
628
622
627
626
629
622
624
623
630
632
631
596
595
595
632
601
622
598
607

**** LEAF-EARED MOUSE**

Galenomys (=Phyllotis) garleppi **(633)**

HB up to 4.7ins; T up to 17.7ins.

Upperparts are buff with fine brown markings, and a yellower rump; the underside and legs are white. The only specimens known (5) were found at altitudes of 12,500–14,750ft in S Peru, W Bolivia and NE Chile. Little is known of its habits and status, but it is likely that its range is wider than currently estimated.

*** CHINCHILLA MOUSE**

Chinchillula sahamae **(634)**

HB 4.7–7.1ins; T 3.5–4.7ins.

Upperparts are greyish or buff with black streaking, the underparts pure white. The fur is thick and soft, tail furry, and ears large. Chinchilla Mice live among rocks and boulders at altitudes of 11,500–16,400ft in the Altiplano, on the borders of Chile, Peru, Bolivia and Argentina. They are vegetarians, feeding almost entirely at night. Their thick chinchilla-like fur is used for trimming, and making coats; one such coat may contain 150 individual skins. In some areas trapping for furs has reduced numbers considerably.

*** WOOD RATS**

Neotoma spp.

Similar in size to a house rat, wood rats have hairy tails and fine, soft fur, relatively large ears and a usually whitish belly and feet. Found in a wide variety of habitats, some species build complex houses of sticks, twigs and bones etc., which may reach more than 6ft 6ins high. Several of the 20 species have restricted ranges: ** *N. varia* **(635)**, ** *N. bryanti* **(636)**, ** *N. anthonyi* **(637)**, ** *N. martinensis* **(638)** and ** *N. bunkeri* **(639)** are restricted to islands off Baja California; ** *N. nelsoni* **(640)** is known only from the type locality in Vera Cruz, in the high plains of C Mexico. A large number of subspecies have been described for many species; very little is known about several species and it is possible that others have very restricted ranges and may be threatened. The Key Largo population of the * **Eastern Wood Rat** *Neotoma floridana smalli* **(641)** is threatened by habitat destruction; less than 60 sq.mls of suitable habitat remains.

**** MAGDALENA RAT**

Xenomys nelsoni **(642)**

HB 6.1–6.5ins; T 5.5–6.7ins.

Reddish-brown above, creamy white below, with a white patch over each eye, and below each ear. The large ears lack fur, and the tail is hairy. Similar in appearance to a small wood rat *Neotoma* spp., but little has been recorded of its habits. Has only been found in 3 localities in Mexico, up to altitudes of 1450ft, where it appears to be nocturnal, inhabiting tropical deciduous forest in coastal areas.

**** ECUADOR FISH-EATING RAT**

Anotomys leander **(643)**

HB 5ins; T 4.9ins.

Dark grey above, paler below, with a whitish patch around the ears. The only known specimens were collected on Mt Pichincha, near Quito, Ecuador; this specimen was collected at 11,800ft, in a fast flowing stream. Alleged specimens from Peru were probably misidentified. The diploid chromosome number is 92, the highest recorded for a mammal.

**** RORAIMA MOUSE**

Podoxymys roraimae **(644)**

HB 3.9ins; T 3.7ins.

Has long, soft, dark, slate coloured fur. Only known from a few specimens described in 1929, which were collected on Mt Roraima, which virtually straddles the border of Venezuela, Brazil, and Guiana. Almost nothing is recorded of its habits.

**** BRASILIA BURROWING MOUSE**

Juscelinomys candango **(645)**

Similar in general appearance to a common rat *Rattus* (757). First described as recently as 1965, from specimens collected in Brasilia Federal District, Brazil, this species may well prove to be more widespread than its present recorded range indicates.

**** VENEZUELAN FISH-EATING RAT**

Ichthyomys pittieri **(646)**

About the size of a common rat *Rattus* (757), but modified for aquatic life, with broad hind feet and flattened head. Not discovered until the 1960s, and only known from the Rancho Grande (Henri Pittier) NP, near the headwaters of the Rio Limon, in Aragua, Venezuela.

**** GOLDMAN'S WATER MOUSE**

Rheomys raptor **(647)**

Water mice live close to fast-running streams in tropical forests, where they feed on small crustaceans, insect larvae etc. They are extremely well adapted to an aquatic life with flap-like valves on the nostrils and large, compressed webbed hind feet. Goldman's Water Mouse is only known from Mt Pirri near the headwaters of the Rio Limon, at an altitude of 4600ft, where it was discovered in the early 1900s.

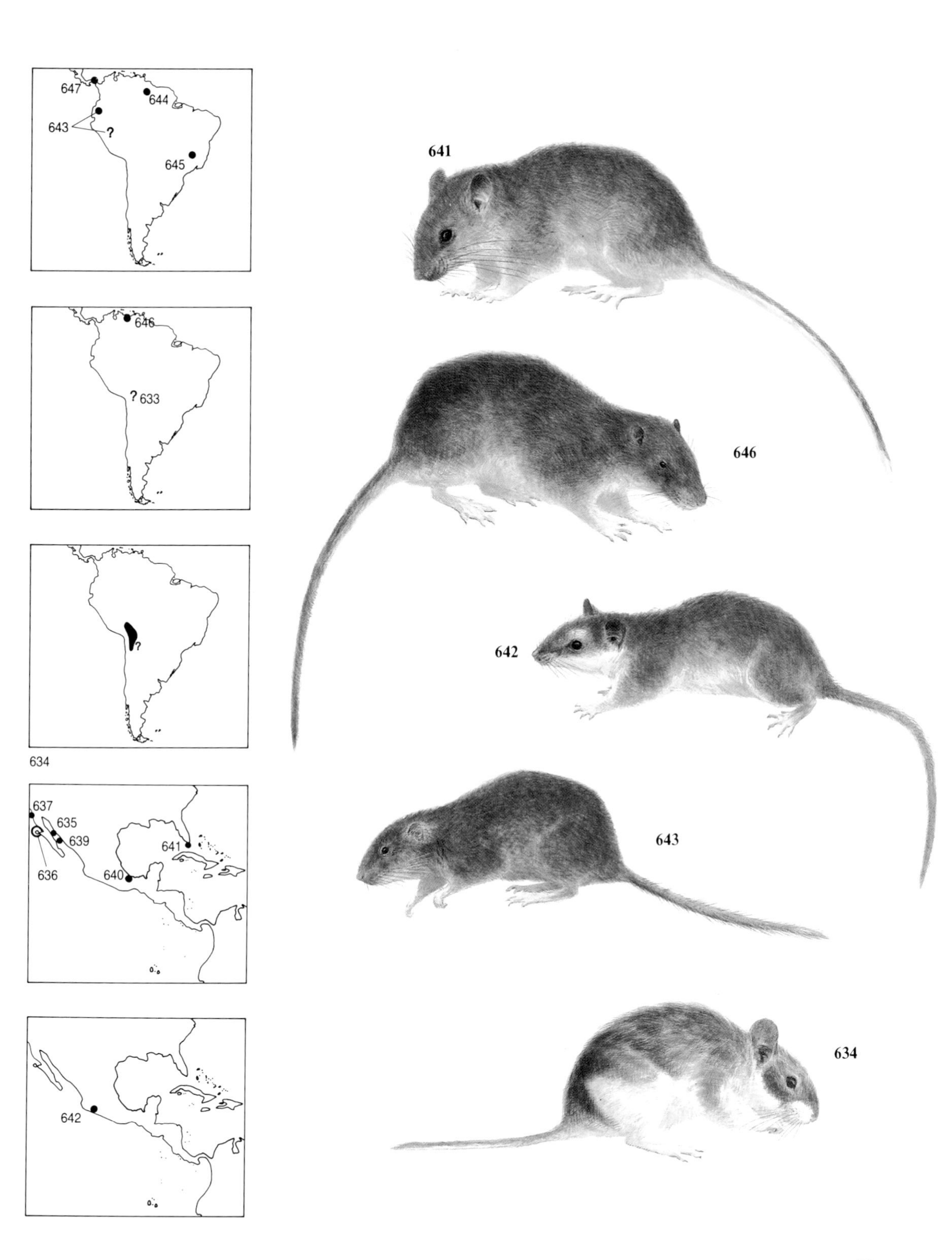
647
644
643
?
645
641
646
? 633
646
?
642
634
637
635
639
636
641
640
643
642
634

FISH EATING RATS

Daptomys spp.

HB 3.9–5.1ins; T 3.1–4.3ins; WT ca.1.6oz.

The three species are similar in size and appearance; they have thick fur, dark above, greyer below. All three species of *Daptomys* are known from very restricted areas, and virtually nothing is known of their status, or distribution. The ** **Venezuelan Fishing Rat** *D. venezuelae* (**648**) is known from only three specimens collected in Venezuela, at altitudes of up to 4600ft; the single specimen of the ** **Peruvian Fishing Rat** *D. peruviensis* (**649**) was collected at 950ft and described in 1974; and the ** **Guyanan Fishing Rat,** *D. oyapocki* (**650**) from a single specimen collected in southern French Guiana, first described in 1978, near the Oyapock River.

MOUSE-LIKE HAMSTERS

Calomyscus spp.

Although often grouped as one species ** *C. bailwardi* (**651**), the five species or populations are fairly distinct. They are found from S Turkestan to Baluchistan, Pakistan. They have been regarded as on the way to extinction, but some populations are abundant. ** *C. hotsoni* (**652**) is only known from Gwambuk Kaul, Pakistani Baluchistan.

*** COMMON HAMSTER

Cricetus cricetus (**653**)

HB 7.9–13.4ins; T 1.6–2.4ins; WT up to 2lb.

The thick fur is rather variable in colour but is usually brownish or yellowish above, and black on the belly, with some white on the flanks. Hamsters have well developed cheek pouches. The Common Hamster is found in steppes and also in cultivated areas, where it burrows extensively. The elaborate burrows contain nesting chambers, food storage and latrine chambers. They feed on a wide variety of cereals, seeds, fruit, roots, and also animal matter. During the winter months, hamsters hibernate. They usually produce 2 litters of 4–12 young each year, although in captivity they are much more prolific. Common Hamsters have in the past been serious pests to agriculture, and were also trapped for their fur. However, the mechanisation of agricultural technology has led to a decline in most of their European populations. Although widespread, their range is contracting and will probably continue to do so. The * **Golden Hamster** *Mesocricetus auratus* (**654**), is known only from a small area near Aleppo in northern Syria. All domestic Golden Hamsters descend from a single pregnant female captured in 1938, but little is known of its status or habits in the wild; it is closely related to the more widespread *M. brandti.* * *M. newtoni* (**655**) is confined to E Bulgaria and E Romania.

PYGMY GERBILS

Gerbillus spp.

Pygmy gerbils are small, long-tailed, with large hind feet and big eyes; the general colouring is sandy, greyish or brownish above, whitish below. They are mainly nocturnal and live in arid habitats. Although many species are widespread and common, some have very restricted ranges. The taxonomy is often confusing: some species being treated as subspecies by some authors and vice-versa. This confusion is, at least in part, due to a lack of specimens from many parts of their range, and some populations may prove to be more widespread when greater studied in detail. ** *G. mauritaniae* (**656**) is known only from a single (possibly aberrant) specimen from E Mauritania; ** *G. hoogstraali* (**657**) and ** *G. occiduus* (**658**) are each only known from single localities in SW Morocco. The closely related ** **Greater Short-Tailed Gerbil** *Dipodillus maghrebi* (**659**) is known only from a single locality in northern Morocco, and the ** **Tunisian Short-Tailed Gerbil** *D. zakariai* (**660**) is confined to the Kerkennah Islands off E Tunisia (Rhabi and Chergui). The ** **Somali Pygmy Gerbil** *Microdillus peeli* (**661**) is a little-known species which occurs on the edge of the desert in central Somalia and Ethiopia.

** DUNE HAIRY-FOOTED GERBIL

Gerbillurus tytonis (**662**)

HB 3.5–4.3ins; T 4.5–5.3ins; WT 0.8–1.1oz.

Rich reddish-brown above, pure white below. Has a very restricted range, confined to shifting dunes in a small area south of the Kuisela River in the Namib Desert, Namibia, but they are not in any immediate danger.

*** WHITE-FOOTED VOLE

Phenacomys (=Arborimus) albipes (**663**)

HB 3.5–4.3ins; T 2.2–2.9ins; WT 0.6–1oz.

Rich brownish above, greyer below, similar in general appearance to *Clethrionomys* and *Microtus* voles. One of the rarest and least-known of North American rodents, confined to alder woods along rivers and streams in coastal Oregon and NW California.

*** MALAGASY GIANT RAT

Hypogeomys antimena (**664**)

HB 11.8–13.8ins; T 8.2–9.8ins.

A large rat, generally greyish- or reddish-brown above, with white underparts and paws. The ears are large, and the back legs are well developed for jumping. They are nocturnal and confined to coastal forests in sandy areas of W Madagascar. They excavate long burrows, and have only a single offspring in a litter. Threatened by extensive cutting and burning of forests, and thought to have declined sharply.

* GREATER MOLE-RAT

Spalax microphthalmus (**665**)

HB up to 12ins; T vestigial.

A burrowing rodent with no visible eyes, still common in many parts of its range – it may even be a pest – but the population *S. m. arenarius* in the southern USSR, to the west of the Crimea, is considered threatened. The closely related * **Giant Mole-rat** *S. giganteus* (**666**) is also considered threatened in the USSR, and only occurs in three isolated populations near the Caspian Sea.

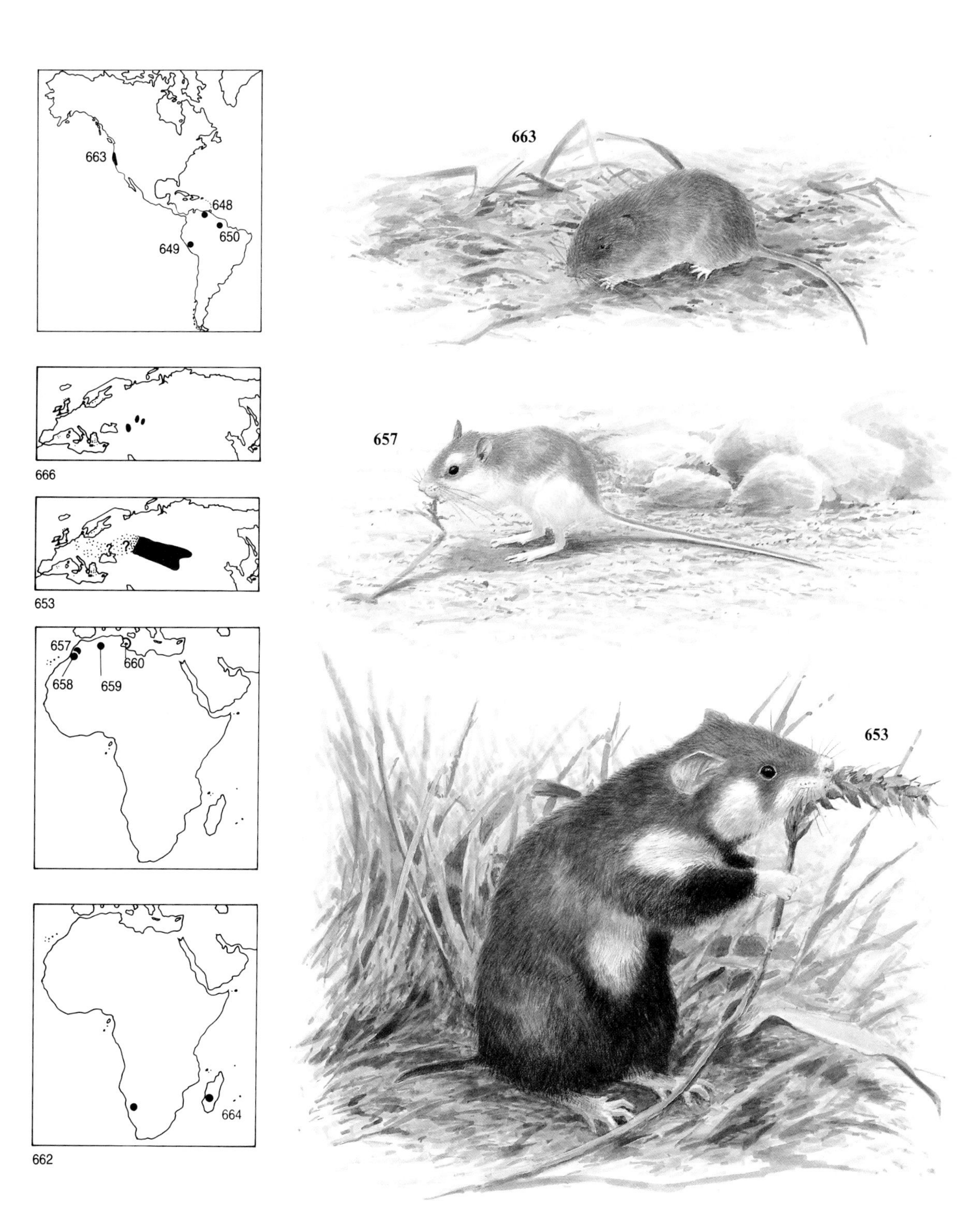
663
648
650
649
666
653
?
657
660
658
659
662
664
663
657
653

******* GHANA FAT MOUSE**
Steatomys jacksoni **(667)**
HB 4.7ins; T 2ins.
Dark, with a white spot at the base of the ears, and white hands and feet. It is the largest species of fat mouse, and was found in an area of forest that has now been destroyed, in the Ashanti District of Ghana. It was described in 1936, and is only known from a single specimen, although it may possibly occur in Togo. Fat mice accumulate fat in order to aestivate; because of this fat, they are often eaten by native peoples of Africa. The Ghana Fat Mouse, if not already extinct, is likely to be extremely rare, is threatened by habitat destruction, and vulnerable to hunting pressures.

******* GROOVE-TOOTHED FOREST MOUSE**
Leimacomys buettneri **(668)**
HB 4.6ins; T 1.5ins.
Also known as **Buettner's Togo Mouse**, the most noticeable feature of this relatively large mouse is its extremely short, naked, tapering tail. It is golden-brown above, grey-brown below. Since it was discovered in the 1890s, east of Yege in Togo, this mouse has not been found again, and is either extremely rare or already extinct.

**** CONGO TREE MOUSE**
Dendroprionomys rousseloti **(669)**
The Congo Tree Mouse is brownish above, white below, with velvety mole-like fur. Only known from 4 specimens from the vicinity of Brazzaville, Congo, described in 1966. Its status is unknown.

FIELD VOLES
Microtus spp.
There are numerous species and distinct populations of vole with isolated ranges, and their taxonomic status is far from certain, sometimes being considered full species, sometimes subspecies. Included are the ** **Beach Vole** *M. breweri* **(670)** from Muskeget Island, Mass., USA; ** *M. abbreviatus* **(671)** from Hall and St Matthew Isles, Alaska, USA; ** *M. canicaudus* **(672)** from Willamette Valley, Oregon, and adjacent Washington, USA; ** *M. mujanensis* **(673)**, found in 1978 in a single locality in USSR; *M. nesophilus* from Great and Little Gull Islands, New York, USA (now extinct); ** *M. oaxacensis* **(674)**, discovered in 1960, is confined to the mountains of NC Oaxaca, Mexico, and ** *M. umbrosus* **(675)** is known only from Mt Zempoaltepec, Oaxaca, Mexico. The subspecies *** *M. californicus scirpensis* **(676)** of the **California Vole** is restricted to a small area of marsh near Death Valley, and was rediscovered in 1979, after being believed extinct. *** *M. ochrogaster ludovicianus* **(677)** from the prairies of SW Louisiana and SE Texas is probably extinct, since it has not been seen for nearly 90 years. In Europe isolated populations of the * **Root Vole** *M. oeconomus* **(678)** are threatened by habitat loss and change.

NB: The **Pine Voles** *Pitymys* spp. are closely related to the field voles and often included in the genus *Microtus*. They are widespread in continental Europe. Their systematics are complicated, with many isolated populations. *P. bavaricus* from S Germany, first described in 1962, is probably extinct. Other populations may be threatened.

VARYING LEMMINGS
Dicrostonyx spp.
The taxonomy of this group is very poorly known, some zoologists recognising only 2 species, others 10 or more. Some of the populations/species have restricted ranges: ** *D. unalascensis* **(679)** is found on Umnak and Unalaska Islands, ** *D. vinogradovi* **(680)** only on Wrangel Island. As far as is known they are not threatened, but their isolation may make them vulnerable.

**** DOLLMAN'S TREE MOUSE**
Prionomys batesi **(681)**
HB 2.4ins; T 3.9ins.
The short, velvety fur is brownish above, buff below, and the eyes surrounded with dark rings. The tip of the tail is naked, suggesting it might be prehensile. It was described in 1910, from Bitye, near the Ja River, Cameroon, and since then has only been found there and around Bangui in the Central African Republic.

STEPPE LEMMINGS
Lagurus spp.
Small vole-like lemmings, the Asian species, * *L. luteus* **(682)** occurring from Kazakhstan to S Mongolia, and * *L. lagurus* **(683)**, from the steppes of the Ukraine east to Mongolia and Sinkiang, have declined in the west of their range; the range of *L. lagurus* formerly extended to Kiev, ca.186 miles further west than its present range, and *L. luteus* is possibly extinct in Kazakhstan. Their disappearance is linked to changes in their steppe habitat. They can be serious pests to agriculture.

*** SOUTHERN BOG LEMMING**
Synaptomys cooperi **(684)**
HB 4.6–6.1ins; T 0.5–0.9ins; WT 0.7–1.7oz.
Brownish above, silvery below with inconspicuous eyes and ears. They live in grassy meadows and forests from Manitoba, east to Newfoundland, south to Kansas, Arkansas, and west to North Carolina and Virginia. Populations fluctuate widely from year to year, and until recently the Dismal Swamp population in NE Carolina and SE Virginia was considered threatened or extinct, since it was not observed between the 1890s until its rediscovery in 1980.

**** AFRICAN CLIMBING MOUSE**
Dendromus kahuziensis **(685)**
Only known from a single specimen from Mt Kahuzi, E Zaire, described in 1969; it may prove to be more widespread.

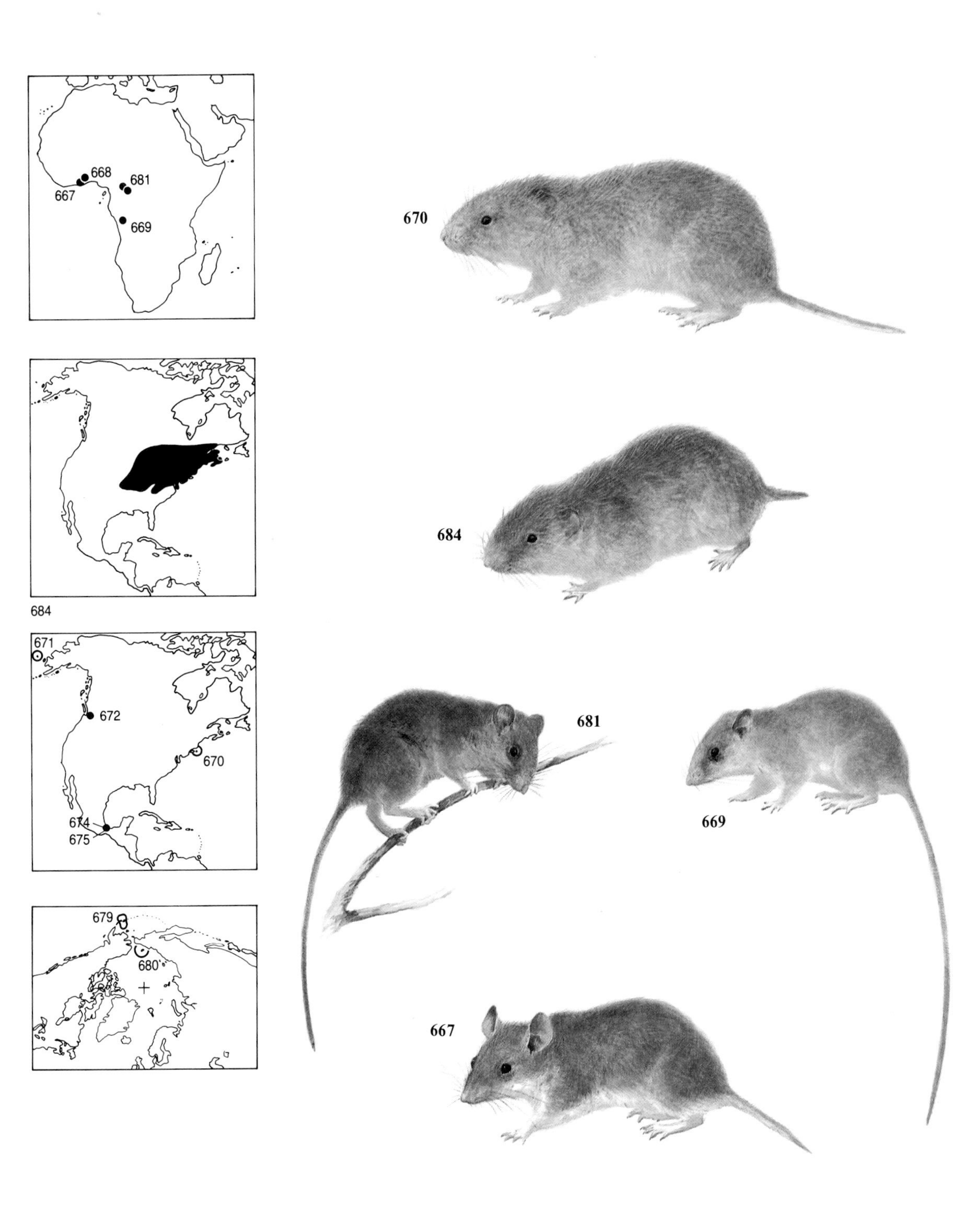
668
667
681
669
670
684
684
671
672
670
674
675
679
680
681
669
667

**** BRAZILIAN SPINY RICE RAT**
Abrawayaomys ruschii **(686)**
HB ca.7.9ins; T 3.3ins; WT ca.1.6oz.
Greyish-yellow above, whitish-yellow below, with darker head. Spines are mixed with fur. Only known from Castelo, Espirito Santo, Brazil, from where it was described in 1979; little is known of its distribution or habits.

**** ECUADORIAN SPINY MOUSE**
Scolomys melanops **(687)**
HB ca.3.5ins; T ca.2.8ins.
Generally brownish above and grey below. Much of the fur on the upperparts is modified into spines. Known from a few specimens collected in 1924 at Mera in E Ecuador, at an altitude of 3750ft.

**** MT PIRRI CLIMBING MOUSE**
Rhipidomys scandens **(688)**
Only known from a single specimen described in 1913, which was shot near the head of the Rio Limon, Mt Pirri, Darien, E Panama.

**** RIO RICE RAT**
Phaenomys ferrugineus **(689)**
HB ca.5.9ins; T ca.7.5ins.
Rusty red above, with dark tips to the hairs along the middle of the back; some whitish hairs behind the ears and the underside is whitish-yellow. They appear to be modified for climbing, but little is known of their habits. Only known from Rio de Janeiro, E Brazil.

*** RINTJA MOUSE**
Komodomys rintjanus **(690)**
HB 5.1–7.9ins; T 4.4–6.3ins.
The coarse, thick fur is sandy coloured above, darker in the middle, with whitish feet. Believed to be a ground-dwelling species, living in monsoon forests. Although not known to be immediately threatened or rare, it is confined to the islands of Rintja and Padar, near Komodo, Indonesia, and sub-fossils (ca. 3500 BC) have been found on Flores, indicating it was once more widespread.

**** SULAWESI SOFT-FURRED RAT**
Eropeplus canus **(691)**
HB 7.7–9.5ins; T 8.3–12.4ins.
The brownish-greyish fur is soft and thick, interspersed with longer guard hairs on the upperparts. It is similar to *Rattus callitrichus* of Sulawesi, but has several differences in the skull and teeth. Only known from 5 specimens, all collected above 4900ft, in C Sulawesi.

PHILIPPINE FOREST RATS
Batomys spp.
HB 7.7–8ins; T 4.9–7.3ins.
The 3 species are only known from a total of 9 specimens. ** *B.granti* **(692)** was described in 1895 from 5 specimens collected by native hunters using terriers on Mt Data, in N Luzon, in the Philippines. It is dark on the upperparts, brownish on the rump, with slaty underparts, and a thickly furred, dark tail; the eye is surrounded by a near-bald eye-ring. ** *B.dentatus* **(693)** is only known by a single specimen from Benguet in N Luzon, collected in 1911, and is paler and more buff-coloured than *B.granti.* ** *B.salomonseni* **(694)** was discovered in 1951 on Mt Katanglad, on the Island of Mindanao in the Philippines. *B.salomonseni* is generally darkish, but with some white on the paws. All have been found in dense cover, at altitudes of 5250–7850ft. A fourth, as yet undescribed species, occurs on Dinagat Island near Mindanao.

FLORES GIANT RATS
Papagomys spp.
There is only one living species, * *P.armandvillei* **(695)** which is confined to the island of Flores in Indonesia. It has also been found in remains ca.3500 years old. These contained also the remains of a closely related species *P.theodorverhoeveni*, which has never been found alive, and another extinct, closely related genus, *Spelaeomys.*

**** PREHENSILE-TAILED RAT**
Pogonomys mollipilosus **(696)**
HB 5.1–5.9ins; T 6.3–8.3ins; WT 1.8–2.5oz.
The long tail is prehensile at the tip, and has a bare patch on the upper surface. Grey-brown fur above, whitish below, with narrow black eye-ring. *Pogonomys* spp. are found mainly in New Guinea where they are generally fairly abundant in forests. This species is moderately abundant in Papua New Guinea and is probably the same species on Fergusson Island, but has only recently (since 1974) been recorded on the east coast of Cape York, Australia. The main threat is destruction of the remaining forests in Queensland.

LUZON RATS
Carpomys spp.
HB up to ca.7.9ins; T up to 8.3ins.
Only two species : ** *C.melanurus* **(697)** is the larger, and has a shiny black tail which is thickly furred for nearly a quarter of its length; ** *C.phaeurus* **(698)** is slightly smaller, with a proportionally shorter tail, furred for only a little of its length, which is dark brown or blackish, but not shiny. Both species are believed to be arboreal. *C. melanurus* is known from 4 specimens, described in 1895 from ca.7850ft on Mt Data, N Luzon, Philippines; *C. phaeurus*, also described in 1895, is known from 3 specimens from Mt Data and one from Mt Kapilingan, N Luzon, Philippines.

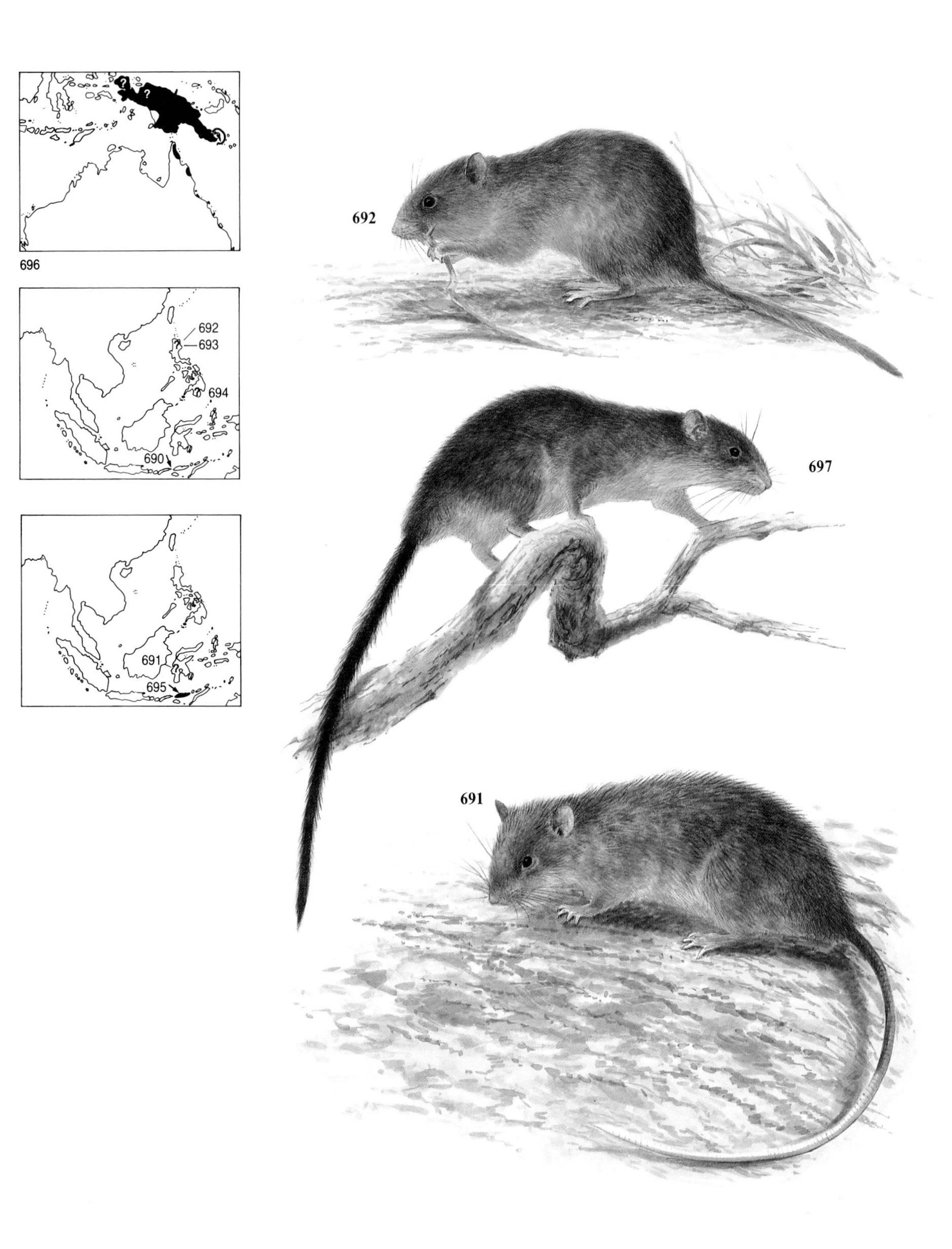
696
692
693
694
690
691
695
692
697
691

WOOD or FIELD MICE
Apodemus spp.
Some species of these Old World mice have well-marked island subspecies, notably *A.sylvaticus*, found on the islands around the British Isles, Yugoslavia etc; ** *A.sylvaticus krkensis* (**699**) from the tiny island of Krk in the Adriatic is sometimes regarded as a full species, but more often as a subspecies of *A.sylvaticus*. Some of these populations may be vulnerable to genetic swamping by introduction of closely related species or subspecies.

***** RYUKYU SPINY RAT**
Tokudaia osimensis (**700**)
HB 4.9–6.9ins; T 3.9–4.9ins.
Black and tawny above, whitish below tinged with orange. Relatively few specimens are known, and were collected in dense undergrowth of grasses and ferns on the islands of Amami o shima and N Okinawa in the Ryukyu Islands, Japan. Believed to be threatened by widespread destruction of its natural habitat.

AUSTRALIAN NATIVE MICE
Pseudomys (Thetomys) spp.
There are about 20 species of Australian native mice, most of which are rather like house mice in general appearance. Several species have very restricted ranges, and others are declining. They are fully protected in Australia.

******* SHARK BAY MOUSE**
Pseudomys praeconis (**701**)
HB 3.3–4.5ins; T 4.5–4.9ins; WT 1–1.7oz.
Confined to part of Bernier Island, but formerly occurred on adjacent mainland of Australia.

******* HASTINGS RIVER MOUSE**
Pseudomys oralis (**702**)
HB 5.1–6.7ins; T 4.3–5.9ins; WT 3.1–3.5oz.
Thought to be extinct since the 1840s until its rediscovery in 1969, in a scattered population around Warwick, Queensland, and more recently in Mt Boss State Forest, NSW. During the 1980s found to be more abundant than previously thought.

**** PILLIGA MOUSE**
Pseudomys pilligaensis (**703**)
HB 2.5–3.1ins; T 2.6–3.1ins; WT 0.3–0.4oz.
A recently discovered species, found only in a small area of Pilliga Scrub (3088sq.mls), northern NSW, in sandy areas dominated by Cypress Pine (*Callitris* spp.). Has been bred successfully in captivity.

***** HEATH RAT**
Pseudomys shortridgei (**704**)
HB 3.5–4.7ins; T 3.1–4.3ins; WT 1.9–3.1oz.
A large, native mouse first found in SW Western Australia, but not recorded there since 1906; it was discovered in 1961, in habitats regularly subjected to burning, in Portland Region and Grampian Mountains, Victoria.

***** ASH-GREY MOUSE**
Pseudomys albocinereus (**705**)
HB 2.5–3.7ins; T 3.3–4.1ins; WT 0.5–1.4oz.
Confined to SW Western Australia, where it is still abundant in coastal areas and in nature reserves, but increasingly rare in agricultural areas.

***** WESTERN MOUSE**
Pseudomys occidentalis (**706**)
HB 3.5–4.3ins; T 4.7–5.5ins; WT 1.9–3.1oz.
Although confined to shrublands in SW Western Australia, it once ranged into the Nullarbor Plain and around the coast into South Australia. Clearance of its habitat for wheat growing is one of the main reasons for its decline, although it was probably declining before the advent of agriculture.

***** DESERT MOUSE**
Pseudomys desertor (**707**)
HB 3.1–3.9ins; T 3.1–3.7ins; WT 0.4–1oz.
Despite its name it appears to inhabit moist areas, only venturing into deserts after rain. Its range has contracted considerably; the only population studied in detail became extinct after the introduction of cattle in the 1970s.

******* GOULD'S MOUSE**
Pseudomys gouldii (**708**)
Has not been recorded since 1857, despite occurring over a large part of NSW and South Australia, and also SW Western Australia. The ******* Alice Springs Mouse** *P. fieldi* (**709**) is also possibly extinct. The ***** Smokey Mouse** *P. fumeus* (**710**) is rare, with a restricted distribution in SE Australia, and not known from any protected area; the *** Pebble-mound Mouse** *P. chapmani* (**711**) is also considered in need of conservation attention.

**** SLENDER-TAILED CLOUD RAT**
Phloeomys elegans (**712**)
Known only from a single specimen from an unknown location in the Philippines. Probably belongs to one of the two other slightly better-known species.

***** GOLDEN-BACKED TREE-RAT**
Mesembriomys macrurus (**713**)
HB 7.3–9.6ins; T 11.4–14.2ins; WT 7–11.5oz.
Generally buff-brown fur with an orange-brown stripe down the centre of the back, and white tail-tip and feet. Mainly arboreal, descending to the ground to feed. Found in Western Australia and Northern Territory, in coastal woodlands and also in more dense forests, but throughout their range they are sparsely distributed and generally rather rare. They are fully protected.

***** BLACK-FOOTED TREE-RAT**
Mesembriomys gouldii (**714**)
HB 9.8–12.2ins; T 12.6–16.3ins; WT 15–30.4oz.
Similar to the Golden-backed Tree-rat (713) but with dark hands and feet, and a dark rump. Found from NE Western Australia to Cape York; also on Melville and Bathurst Islands; believed to have declined markedly in recent years, in many parts of their range.

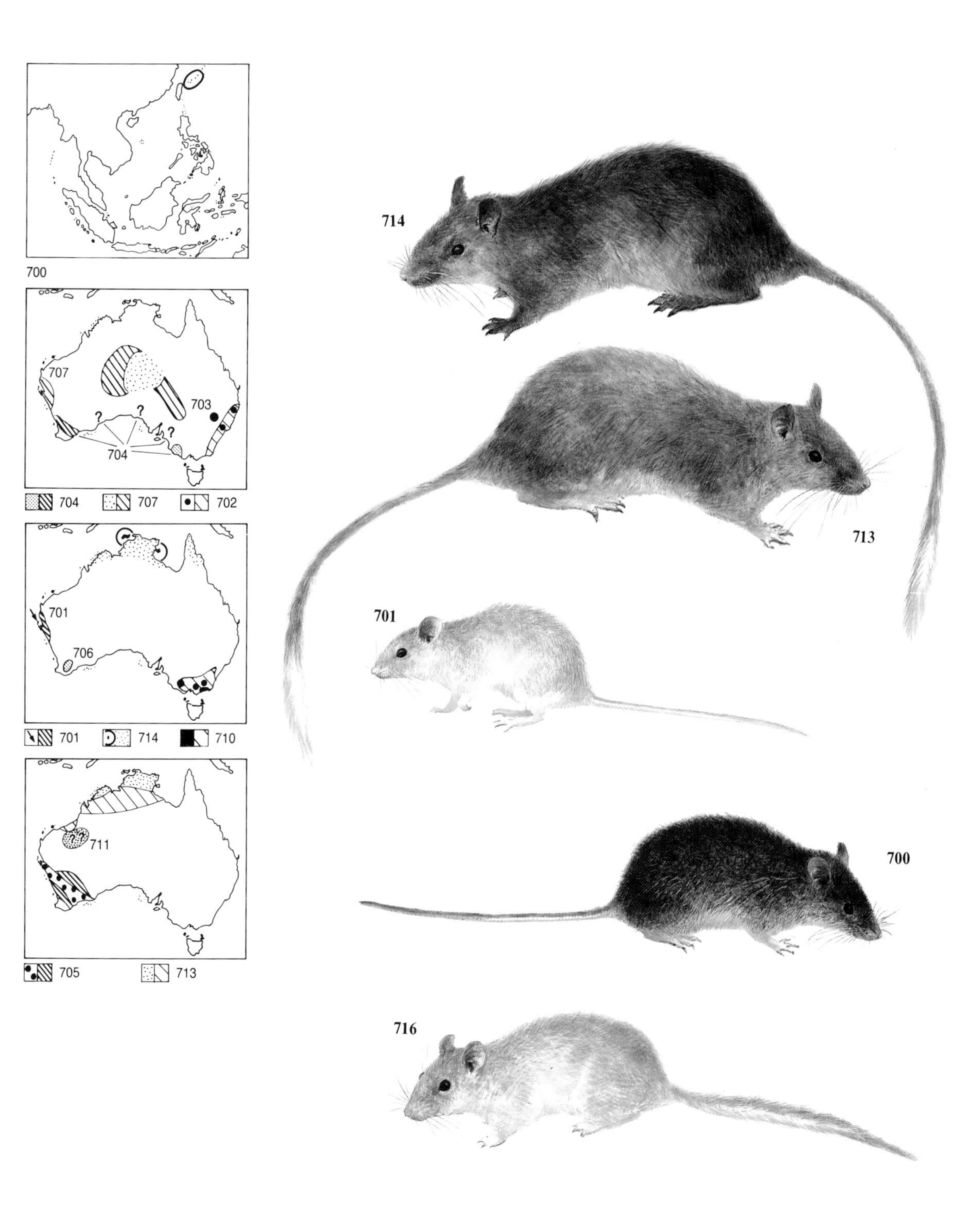
700
707
703
?
?
?
704
704
707
702
701
706
701
714
710
711
705
713
714
713
701
700
716

***** STICK-NEST RAT
Leporillus conditor **(715)**
HB up to 10.2ins; T up to 7.1ins; WT up to 1lb.
The soft, fine fur is generally grey-brown above and creamy-white below; the feet are marked with white. Notable for their habit of building large nests of sticks on the surface of the ground, up to 3ft high and 7ft in diameter. These protect them from the extreme temperatures of their desert habitat. Formerly widespread on the mainland of Australia, but now extinct except for a small population, *L. c. jonesi* which survives on Franklin Island (2sq.mls), which has been a Conservation Park since 1967. Although the population is reasonably stable (estimates vary from 1000 – 5000), they are highly vulnerable. Listed on CITES App.I and fully protected in Australia.

***** CENTRAL ROCK-RAT or MACDONNELL RAT
Zyzomys pedunculatus **(716)** (Illustrated on p. 115)
HB 4.3–5.5ins; T 4.3–5.5ins.
Soft-furred with a thick, hairy tail; when handled it readily sheds its tail leaving only a stump. Formerly found in the MacDonnel, James, Davenport, Granites and Napperby Ranges in Northern Territory. Although apparently easy to trap, only 5 recorded since its discovery in 1896 and none has been seen since 1960. Fully protected in Australia and listed on CITES App.I. The more widespread * **Woodward's Rock-Rat** *Z. woodwardi* **(717)** is rare and may be declining.

AUSTRALIAN HOPPING MICE
Notomys spp.
HB 3.6–7ins; T 4.9–8.9ins; WT 0.7–1.7oz.
The nine species are generally pale sandy brownish or greyish above, and white or greyish below. The fur is soft and they have a long tail and large ears. Found in sand dunes and other arid areas, as well as dry woodlands, where they are nocturnal, hiding in burrows by day. They feed on seeds, berries, leaves and other vegetable matter, and can live without any water. Although some species are widespread others are extremely localised or extinct.** **Northern Hopping Mouse** *N.aquilo* **(718)** is confined to Queensland and Groote Eylandt, Northern Territory. Listed on CITES App.I and protected in Australia.* **Dusky Hopping Mouse** *N.fuscus* **(719)** has a fairly wide range from Western Australia to SW Queensland, but is rare throughout. The **Short-tailed Hopping Mouse** *N.amplus*, the **Big-eared Hopping Mouse** *N.macrotis,* and the **Darling Downs Hopping Mouse,** *N.longicaudatus* are all presumed extinct, not having been seen since 1901. The * **Fawn Hopping-mouse** *N.cervinus* **(720)** from South Australia, SW Queensland and Northern Territory is also considered threatened. *** **Mitchell's Hopping Mouse** *N. mitchelli* **(721)** occurs in two main areas; that in the SW has declined markedly since the arrival of Europeans. All *Notomys* are listed on CITES App.II.

** ROSEVEAR'S STRIPED GRASS MOUSE
Lemniscomys roseveari **(722)**
Described in 1980 and only known from the Zambezi region of Zambia. The closely related ** *L.mittendorfi* **(723)** known only from a single locality in the Lake Oku region, Cameroon, may only be a distinct population of the widespread *L.striatus.*

** CHIRINDA ROCK RAT
Aethomys selindensis **(724)**
A small rat, only known from two specimens from Mt Chirinda (Silinda) on the border of Mozambique and Zimbabwe.

** SOFT FURRED RAT
Praomys hartwigi **(725)**
First described in 1969 and only known from near Lake Oku, Cameroon. May prove to be more widespread.

* LAKE VICTORIA GROOVE-TOOTHED RAT
Pelomys isseli **(726)**
Restricted to small islands on Lake Victoria, this species could be vulnerable to competition from the introduction of other species such as *P.hopkinsi* or even other genera such as *Arvicanthis.* ** *P.rex* **(727)** is only known from a single specimen from the Charanda Forest, Ethiopia.

** AFRICAN SWAMP RAT
Malacomys verschureni **(728)**
A small rat, known only from a single specimen described from Mamiki in NE Zaire; it may prove to more widespread.

** LAKELAND DOWNS MOUSE
Leggadina lakedownensis **(729)**
HB 2.4–2.8ins; T 1.6–1.8ins; WT 0.5–0.7oz.
One of the smallest Australian rodents. Generally greyish-brown above, white below. They are nocturnal, living in burrows in grasslands and woodlands. They were discovered in 1969 and found again in 1973. On both occasions the species occurred in plague proportions after flooding, but subsequently disappeared when normal dry conditions returned. They could prove to be isolated populations of the **Short-tailed Mouse** *Leggadina forresti* which is found over much of C Australia and W Western Australia. Although only 4 wild specimens are known, they have been bred in captivity.

* BROAD-TOOTHED RAT
Mastacomys fuscus **(730)**
HB 5.6–6.9ins; T 3.9–5.1ins; WT 3.4–5.1oz.
Rather like a large vole with a long tail; the long thick fur is deep brown above, paler below. Mostly solitary feeding on stems and leaves of sedges and grasses; they live in relatively moist habitats, including swamps and river valleys with tussock grass, and also montane woodlands and alpine heaths. They were formerly much more widespread in Australia than at present. But even in Tasmania where their range is fairly extensive, because of their specialised requirements they may be threatened by hydroelectric developments, cattle grazing, road construction and other changes of habitat. Found in several protected areas including Kosciusko State Park, New South Wales.

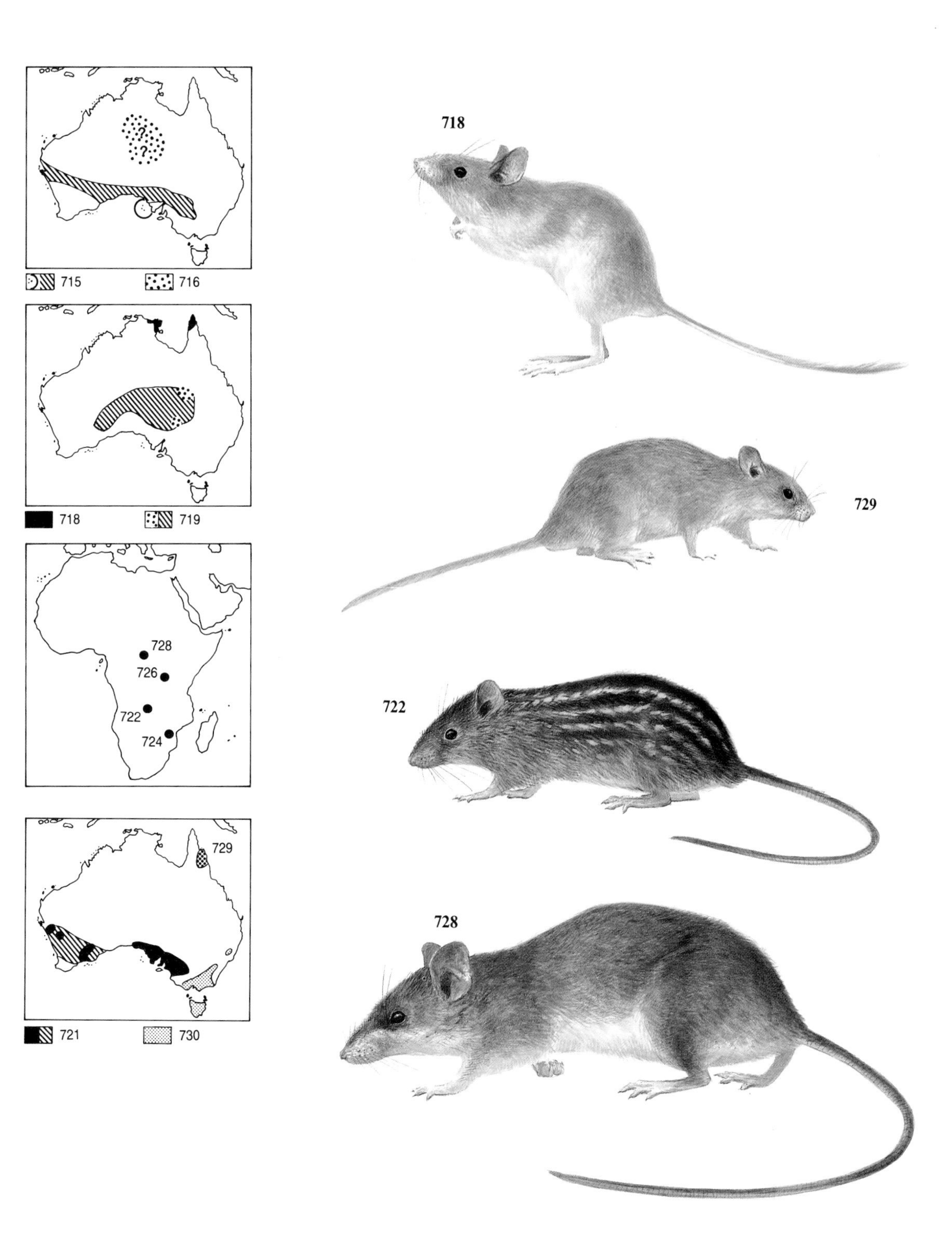

715
716
718
719
728
726
722
724
729
721
730
718
729
722
728

BANANA or MOUSE-TAILED RATS
Melomys spp.
HB 3.5–6.9ins; T 4.3–6.7ins; WT up to 7oz.
** *M.albidens* (**731**) is known only from Lake Habbema, near Mt Wilhelmina, Djajawidjaja, Irian Jaya, Indonesia, at an altitude of over 9850ft; ** *M. arcium* (**732**) is confined to Rossel Island (15.5 mls long) in the Louisiade Archipelago, Papua New Guinea; * *M. levipes* (**733**) is fairly widespread on New Guinea, but an isolated population on Thornton Peak, Queensland, Australia, may be at risk. Several species occur on islands, and the taxonomy of this genus is poorly understood; several new species are probably as yet unrecognised, including the Thornton Peak population. The * **Fawn-footed Melomys** *M.cervinipes* (**734**) from E Australia although widespread is locally threatened wherever its rainforest habitat is being destroyed. The closely related ** **Lowland Brush Mouse** *Pogonomelomys bruijni* (**735**) is only known from three localities in New Guinea, Salawatti Island and the Fly River.

SHREW-LIKE RATS
Rhynchomys spp.
HB 7.1–8.5ins; T 3.9–5.5ins.
Two species of shrew-like rodents, with pointed muzzles, small eyes and short, velvety fur, olive-grey above, whitish-grey below. ** *R.isarogensis* (**736**) has a slightly shorter tail than ** *R.soricoides* (**737**). Very little is known about them; 7 specimens of *R.soricoides* are known, all from Mt Data in N Luzon, and only one specimen of *R.isarogensis*, from Mt Isarog, in S E Luzon.

**** MOUNT APO RAT**
Tarsomys apoensis (**738**)
HB 5.3ins; T 4.7ins.
The long, coarse fur is brownish-grey above, yellowish-brown below. Yet another close relative of *Rattus* from the Philippines, known from only 5 specimens collected on Mt Apo, Mindanao, in montane forests.

***** FALSE WATER RAT**
Xeromys myoides (**739**)
HB 4.5–5.1ins; T 3.5–3.9ins; WT ca.1.9oz.
Dark grey above, paler below; although it has a streamlined shape it is not definitely aquatic. Only found in tropical Australia, on Melville Island, coastal Northern Territory, and Queensland. Apparently more common than once believed, but vulnerable to drainage and clearance of its habitat, as well as habitat change from introduced pigs and water buffalo; locally extinct. Listed as Endangered in Australia.

NEW GUINEA FALSE WATER RATS
Pseudohydromys spp.
HB 3.3–4.5ins; T 3.3–4.5ins; WT ca.0.7oz.
Rather shrew-like, with short dense fur, dark grey or brown above, paler below. Terrestrial, feeding on insects in montane forests at altitudes of 6900–11,800ft. Three species are known, all from relatively few specimens; ** *P. murinus* (**740**) from E Papua New Guinea; ** *P.occidentalis* (**741**) from W New Guinea, and an undescribed third species from the far north of New Guinea. As with many other little-known species from New Guinea, they will probably prove to be common and widespread when their habitat is fully investigated.

**** GROOVE-TOOTHED SHREW RAT or MOSS MOUSE**
Microhydromys richardsoni (**742**)
HB ca.3.3ins; T ca.3.5ins.
Greyish-brown above, paler below, with a mottled tail; the upper incisors are grooved. Found in forests at 1950–2950ft, where they are probably terrestrial and feed on insects. Only a few specimens are known from C and E New Guinea; they may prove to be more widespread and abundant with more detailed investigation of their habitat.

***** SOUTHERN BIRCH MOUSE**
Sicista subtilis (**743**)
HB 2.2–2.8ins; T 2.8–3.3ins.
A long-tailed mouse with a dark stripe down its back. Although widespread in the steppes of Asia, it has an increasingly fragmented range in Hungary and other parts of Europe, and has become extinct in many areas as the open grassy steppes have been ploughed for agriculture. The closely related * **Birch Mouse** *S.betulina* (**744**) also has scattered populations in the western parts of its range, and these may be threatened.

**** MANIPUR or CRUMP'S MOUSE**
Diomys crumpi (**745**)
HB 3.9–5.7ins; T 4.1–5.3ins.
Dark blackish-grey above, black on the rump, whitish below; the tail is black on top, white underneath. This species was described from a single skull in 1917 collected on Mt Paresnath, in Bihar, India, and then remained otherwise unknown until 1942, when some were collected in Manipur. In 1978 a few more were found in S Nepal. Its true status remains unknown, but it could prove to be more widespread.

DWARF JERBOAS
Salpingotus spp.
HB 1.4–2.2ins; T 3.3–5ins.
Sandy-coloured above, paler below. They are all desert-dwelling and feed mainly on seeds. ** *S.heptneri* (**746**) is only known from a single locality near the south of the Aral Sea; ** *S. thomasi* (**747**) from a single specimen from an unknown locality, possibly in Afghanistan; it may be conspecific with ** *S. michaelis* (**748**) which is known from two small areas in W Pakistan. ** *S. crassicauda* (**749**) is known from a few scattered localities in Kazakhstan, USSR, where it is is considered threatened, in S Mongolia, where it is more widespread, and Sin Kiang, China. ** *S. kozlovi* (**750**) is known from about 6 localities in Mongolia and N China.

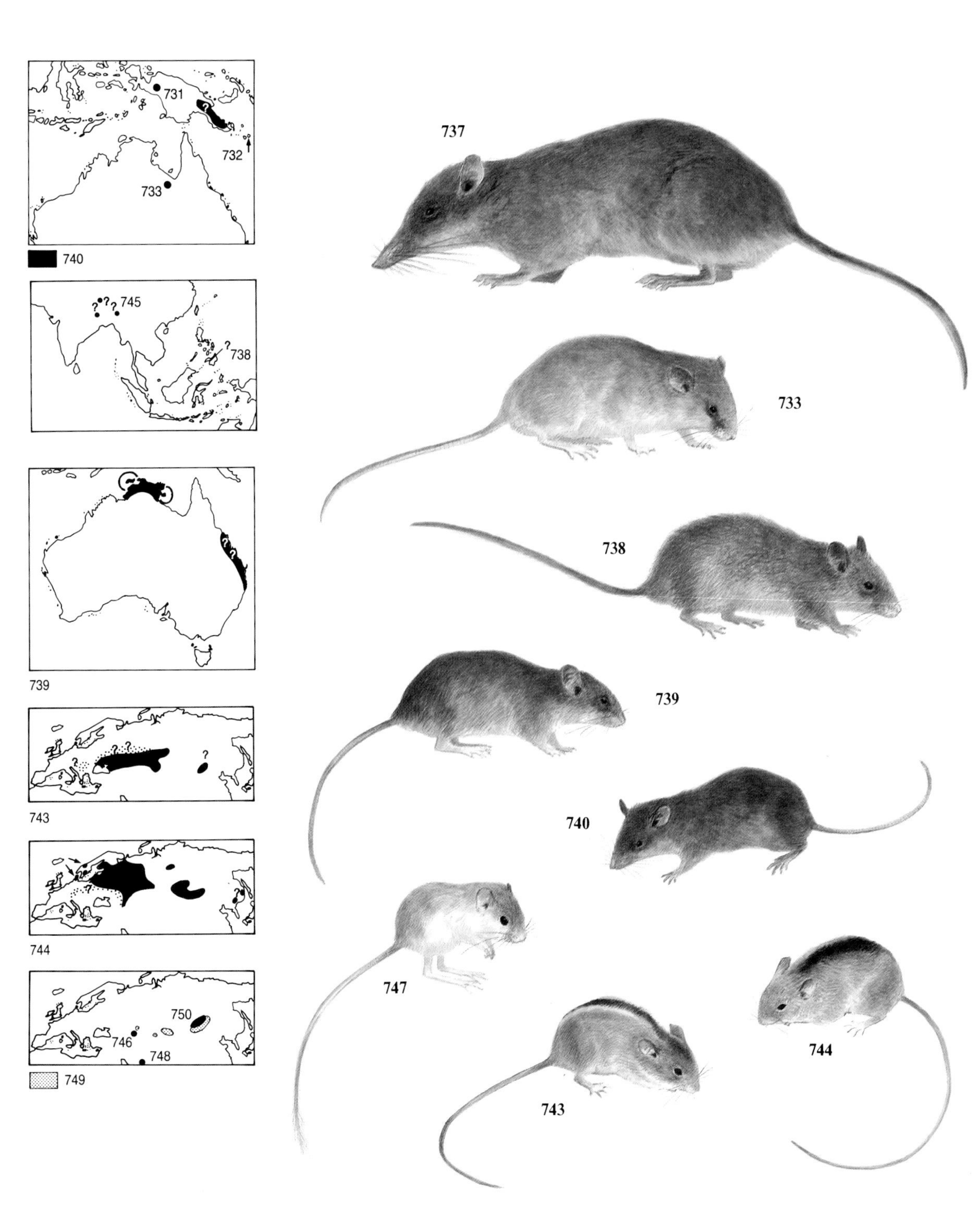
731
732
733
740
745
738
739
743
744
750
746
748
749
737
733
738
739
740
747
744
743

**** MINDORO RAT**
Anonymomys mindorensis (**751**)
HB ca.4.9ins; T ca.8.1ins.
The Mindoro Rat is bright tawny-brown or buff above, creamy below; the long tail is tipped with a tuft of fur. Only known from 3 specimens collected in the 1970s at Ilong (4500ft) in the Halcon Mountains, Mindoro, Philippines.

**** CEYLONESE RAT**
Srilankamys ohiensis (**752**)
HB ca.5.7ins; T ca.9.1ins.
Body and tail are dark greyish-brown above, and creamy below. Closely related to the rats *Rattus*, the Ceylonese Rat is restricted to montane forests in Uva Province, C Sri Lanka, where it is terrestrial.

**** JAVAN GREY TREE RAT**
Kadarsanomys sodyi (**753**)
HB 7.2–8.3ins; T 10.4–12ins.
Dark brown on the back, greyish on flanks with underparts white. This species was formerly regarded as a subspecies of *Lenothrix* (=*Rattus*) *canus*, but now considered to be sufficiently distinct to be placed in its own genus. Arboreal, occurring in bamboo forest at altitudes of 3300ft, where they make nests inside large bamboo stems. Only known from relatively few specimens, all collected from the SW slopes of Gunang Pangrango-Gede in W Java.

**** MARGARETA'S RAT**
Margaretamys parvus (**754**)
Only known from Gunung Nokilalaki, Sulawesi, Indonesia, at an altitude of 7400ft.

*** LIMESTONE RAT**
Niviventer hinpoon (**755**)
HB ca.5.5ins; T ca.5.5ins; WT ca.2.1oz.
Buff-grey above with spiny fur, tail bicoloured. Closely related to *Rattus*. Little is known of this species, which was discovered in 1973. It is restricted to a remote area of limestone cliffs on the Khorat Plateau in C Thailand, where it occurs in woodland and is considered to be rare and possibly endangered.

**** LONG-TAILED GIANT RAT**
Leopoldamys neilli (**756**)
Very similar to *Rattus*, with which they were once classified, but with a larger body and proportionately longer tail. Little is known of this species, which was first described in 1977, and is only known from a few localities in CW Thailand. Has been considered to be threatened.

RATS
* *Rattus* spp.
HB 3.1–11.8ins; T up to ca.11.8ins; WT up to 10.5oz.
The **Black Rat**, *R.rattus* (**757**) is a cosmopolitan pest species, which has locally been surplanted by the Brown Rat *R.norvegicus* in Europe, and has even been considered locally endangered. The **Canefield Rat** *R.sordidus* (**758**) from NE Australia, although abundant in the north of its range, is locally extinct in the south. The **Pale Field Rat** *R.tunneyi* (**759**) is still abundant in Australia, but its range has contracted markedly since the arrival of Europeans. Other species with restricted ranges may be threatened, should more aggressive species be introduced. However, most of the island rats appear to be successfully holding their own, since they tend to live in habitats not favoured by the commensals. These include *R.burrus* (**760**), *R.palmarum* (**761**) and *R.pulliventer* (**762**) from the Nicobars; *R.culionensis* (**763**) from Culion Island, Philippines; *R.elephinus* (**764**) from Taliabu, Sula Islands, Indonesia; *R.enganus* (**765**) from Engano Island, Sumatra; *R.latidens* (**766**) from Mt Data, Luzon, Philippines; *R.macleari* (probably extinct already) and *R.nativitatis* (**767**) from Christmas Island in the Indian Ocean; *R.omichlodes* (**768**) known only from one locality in Irian Jaya, Indonesia; *R.owiensis* (**769**) from Owi Island, off Papua New Guinea; *R.ranjiniae* (**770**) from Kerala, India; *R.rogersi* (**771**) and *R.stoicus* (**772**) from S Andaman Islands, India; *R. tyrannus* (**773**) from Ticao Island in the Philippines; and *R.remotus* (**774**), confined to Koh Samui and a few adjacent islands in the Gulf of Siam, Thailand. *R. (=Niviventer) cremoriventer langbianis* (**775**) is only known from a few isolated localities in Indo-China, Thailand, Burma and E India, and the type locality has been flooded by a dam. Several of these species may prove to be merely distinct populations, but doubtless further isolated species also remain to be discovered. There are several genera of rats closely related to *Rattus*, which are sometimes classified together, some of which are described here; some are known from single specimens or localities, others have ranges restricted to single islands, such as the ** **Seram Rat** *Nesoromys (=Rattus) ceramicus* (**776**), about which there is little or no information on precise status or distribution.

**** PHILIPPINE RATS**
Apomys spp.
Closely related to *Rattus* and often included in the same genus, *Apomys* are confined to the Philippines where they mostly occur in montane forests between 1950–8850ft. *A.sacobianus* (**777**) is only known from a single specimen collected in SW Luzon; *A.microdon* (**778**) from one specimen from Biga, in the Catanduanes Islands; and *A.littoralis* (**779**) from a single specimen from S Mindanao.

**** MEARNS' LUZON RAT**
Tryphomys adustus (**780**)
HB ca.6.9ins; T ca.5.9ins.
A medium-sized rat, closely related to *Rattus*; the fur has a rather 'scorched' appearance, which derives from its coarse texture, and the hairs bending towards the head; the tail is conspicuously scaled. Only known from a few specimens from the mountains of N Luzon, Philippines, status little known, although they may prove to be more widespread and locally abundant.

NOTE ON PHILIPPINE RODENTS
There are over 40 species of rats and mice in the Philippines, many of which are known from only a handful of specimens, and are described on this page and elsewhere in this book. Some of these come from areas of extensive habitat destruction, but others are perhaps considered rare because few scientists have visited the more remotes parts of the Philippines. Some will undoubtedly prove to be more widespread and abundant, but others are possibly already extinct. The Philippines are one of the most biologically diverse areas in the world – and one of the most threatened.

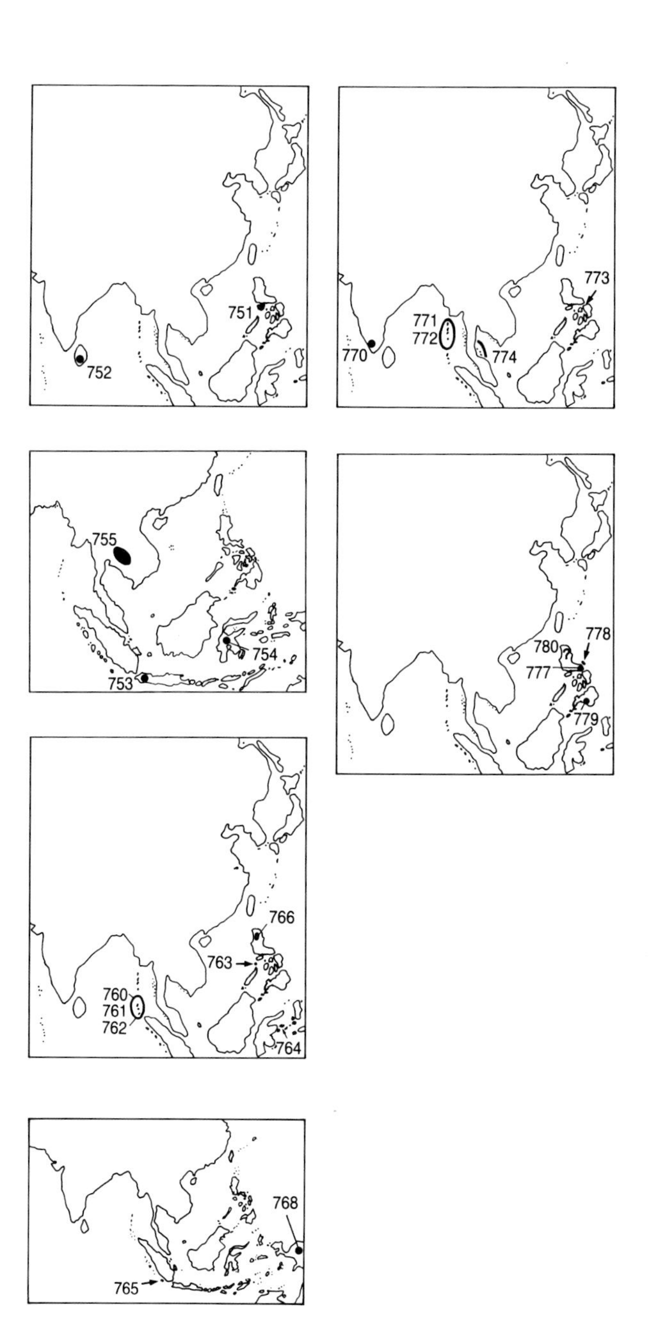

751
752
771
772
770
773
774
755
754
753
780
778
777
779
766
763
760
761
762
764
768
765

* HAZEL DORMOUSE

Muscardinus avellanarius (**781**)
HB 2.4–3.5ins; T 2.2–2.9ins; WT up to ca.1.7oz.
Orange-brown fur, thickly furred tail. Widespread over much of Europe and W Asia, numbers have dropped locally; in Britain it may have declined in the past 150 years, and in Scandinavia range has contracted and it is considered threatened. Changes to woodland habitat probably the main cause of decline, but collecting for pets in 19th century may have contributed. * **Garden Dormouse** *Eliomys quercinus* (**782**) is declining in Finland, but elsewhere holding its own.

** SECHUAN DORMOUSE

Chaetocauda sechuanensis (**783**)
Discovered in 1979, the two specimens known were found in the Wang-Lang Nature Reserve in mixed forest, in the subalpine region of Pinwn, N Sichuan, China.

** WOOLLY FOREST DORMOUSE

Dryomys laniger (**784**)
Discovered in 1968. Only known from SW Turkey at altitudes of up to 6550ft. Status unknown.

** PERSIAN DORMOUSE

Myomimus setzeri (**785**)
Described in 1976. Only known from Kurdistan, Iran.

*** DESERT DORMOUSE

Selevinia betpakdalensis (**786**)
HB 2.9–3.7ins; T 2.3–3ins; WT up to 0.8oz.
Greyish above, white below. When moulting, skin comes away in patches together with the fur, exposing dense new growth of fur below, which grows up to 0.4in./day. Patchy distribution in deserts to the west and north of Lake Balkhash, E Kazakhstan, USSR. Feeds on small invertebrates.

*** FOUR-TOED JERBOA

Allactaga tetradactyla (**787**) HB ca.4.3ins; T ca.6.7ins; WT ca.1.8oz.
Distinguished from other *Allactaga* jerboas by the absence of a fifth toe. Found in coastal salt marshes, desert areas, in N Libya and Egypt, where it digs short burrows in which it lives by day, emerging at night to feed on seeds, vegetable matter, and some insects. May be threatened by desert reclamation projects; extinct in certain valleys near Alexandria. Closely related Five-toed species ** *A.firouzi* (**788**) and ** *A.hotsoni* (**789**) known only from Isfahan and Kerman, respectively, in Iran.

*** FIVE-TOED PYGMY JERBOA

Cardiocranius paradoxus (**790**) HB 2–3ins; T 2.8–3.1ins.
Greyish-buff above, whitish below; the tail is constricted at the base, but swells before tapering towards the tip. Found in stony deserts, where they are mostly nocturnal, feeding on seeds. Known from north of Lake Balkhash, E Shan Mountains of N China; in USSR considered threatened. Recently more recorded proving not so rare.

HOUSE MICE

Mus spp.
The taxonomy and systematics of this genus is complex and constantly changing. Although *Mus musculus* and its allies among most serious pest species, some populations and species have very restricted ranges. ** *M. baoulei* (**791**), described in 1980, is only known from one locality in Ivory Coast. Some populations, such as *M.poschiavinus* from the Swiss Alps, are indistinguishable from *M.musculus*, except by chromosome examination, and may have very restricted ranges.

** STRIPE-BACKED MOUSE

Muriculus imberbis (**792**) HB 2.8–3.7ins; T 1.8–2.4ins.
Similar to house mice, *Mus* spp., with a dark stripe down the centre of the back. Inhabits grassy slopes at altitudes of 6250–11,150ft either side of the Rift Valley, in Ethiopia. Not seen in recent years; its range may have been reduced over the last half century due to the spread of agriculture.

** MINDANAO RAT

Limnomys sibuanus (**793**) HB ca.4.9ins; T ca.5.9ins.
A tawny-coloured rat with long dense fur, and long black guard hairs scattered on the back and rump; the underside is cream. Only known from 4 specimens collected on Mt Apo, Mindanao, Philippines, in 1904 and 1906 at altitudes of 6550–9200ft, in montane forests.

** SULAWESI LESSER SHREW RAT

Melasmothrix naso (**794**) HB ca.4.9ins; T ca.3.5ins.
A rat with dense, velvety fur; the snout is elongated and the ears are short. Only known from a single specimen trapped in Rano Rano, C Sulawesi in 1918. Status unknown.

** TATE'S RAT

Tateomys rhinogradoides (**795**) HB ca.5.4ins; T ca.6.6ins.
The fur is short and velvety, brown above, paler on the sides and underside. Similar to the Sulawesi Lesser Shrew Rat, but has a larger head with smaller eyes and several differences in the skull. The only specimen known was collected in 1930 at an altitude of 7200ft, in cloud forest in C Sulawesi.

** WHITE-TAILED NEW GUINEA RAT

Xenuromys barbatus (**796**) HB 10.8–12.2ins; T 8.7–11ins.
These large rats are reddish-brown or greyish above, buff or whitish below, with white on the tail. The first one was found in 1900, and two others found since. Only known from the central third of New Guinea.

PHILIPPINE SWAMP RATS

Crunomys spp. HB ca.3.9ins; T ca.2ins.
The blackish-brown fur is mixed with flattened spines, and the belly is slightly paler than the back. Two closely related species are known: ** *C.fallax* (**797**), a single specimen collected in the 1890s from Isabella Province, N Luzon, and ** *C.melanius* (**798**) collected in 1906 on Mt Apo, Mindanao; another specimen was collected in 1923.

** RED-SIDED HYDROMYINE

Paraleptomys rufilatus (**799**) HB ca.5.1ins; T ca.5.3ins.
Only known from 2 specimens, described in 1945, collected at 4750ft on Mt Dafonsero, in the Cyclops Mountains, NE Irian Jaya, Indonesia.

** LUZON SHREW-RAT

Celaenomys silaceus (**800**) HB ca.7.7ins; T ca.4.3ins.
Rather shrew-like, with velvety fur, grey above, paler below. Only 6 specimens are known, mostly from dense montane forest, 6900–7850ft around Mt Data in N Luzon.

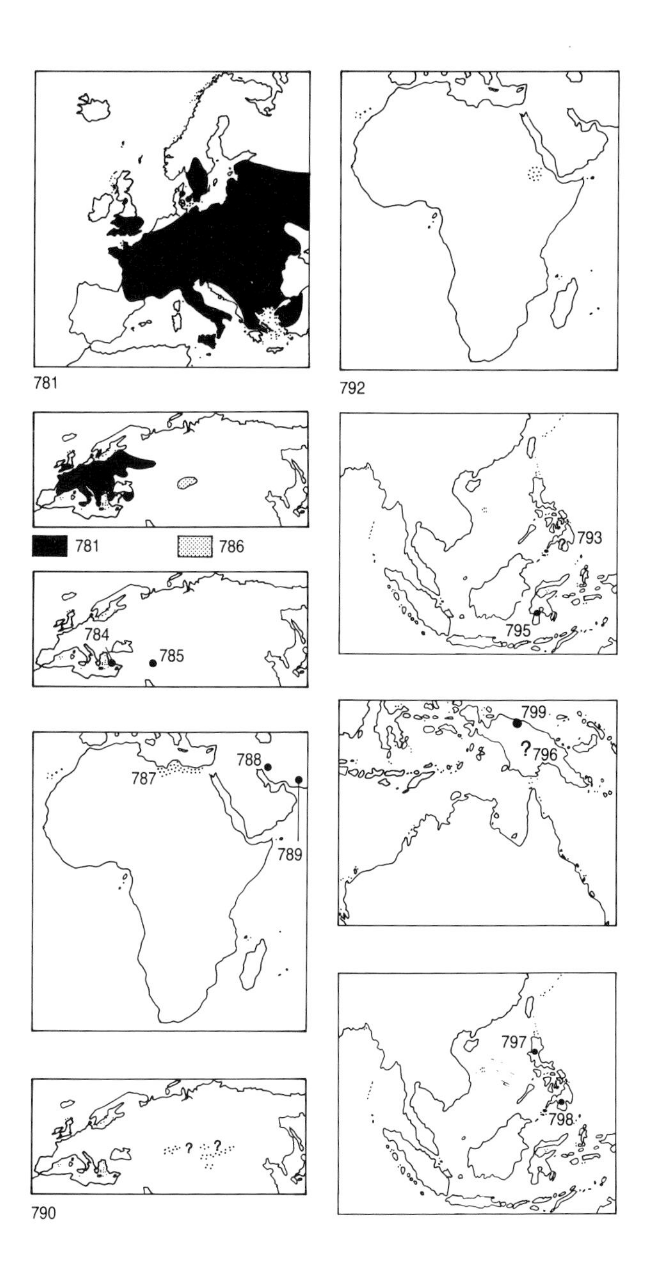
781
792
793
795
781
786
784
785
788
787
789
799
796
797
798
790

*** CRESTED PORCUPINE**
Hystrix cristata **(801)**
HB 19.7–27.6ins; T 2–4.7ins; WT 22–37.5lb.
Easily recognised by the long quills on much of the back and distinct crest of stiff bristles on the head and shoulders. They feed on a wide variety of vegetable matter and are often pests of crops. Found from North Africa, south to N Zaire and Tanzania. Their occurrence in Europe is thought to be as a result of introductions by man. Declining in Italy, and many other parts of the Mediterranean, and although portrayed in Egypt on carvings from the 5th and 6th Dynasties, are on the brink of extinction there, if not already extinct. They have declined in other parts of Africa, including Uganda, where they are hunted for food and their quills. The closely related * **Hodgson's Porcupine** *H. hodgsoni* **(802)** from NE India, east to S China and Indo-China, is regarded as threatened in India.

***** THIN-SPINED PORCUPINE**
Chaetomys subspinosus **(803)**
HB 16.9–19.7ins; T 10–11.8ins; WT ca.4.4lb.
The fur on the upper side is bristle-like, but there are spines on the foreparts. The tail is naked towards the tip. Found in forest edge habitats, usually associated with *Bombax* trees, and are nocturnal. Confined to SE Brazil in the states of Bahia and Espirito Santo, where their range is considerably reduced through deforestation. Protected in Brazil and occur in several reserves.

HUTIAS
Several genera of hutias formerly occurred in the Caribbean, but most are known only from skeletons found in caves. All appear to have become extinct after the arrival of human settlers. Some were probably exterminated by Indian colonists (Caribs), and it is thought that *Isolobodon portoricensis* was domesticated on Puerto Rico and introduced into other islands. It became extinct after the arrival of Europeans. (It has been suggested that small populations may survive on Hispaniola or Puerto Rico.) A giant Barn Owl *Tyto ostolaga* which preyed on the hutias also became extinct. Four genera, comprising seven species, are known only from skeletal remains, and two genera survive. Some species are known only from few specimens and study may show them to be variations of one of the better-known species.

HUTIAS
Capromys spp.
HB 8.7–19.7ins; T 1.4–1.6ins; WT up to 15.4lb.
The fur colour is variable, including brown, red, grey and black; there is a soft underfur with long coarse guardhairs. They are found in forested areas and also rocky habitats, where all except the Jamaican and Bahamian are arboreal. Once widespread in the Caribbean, several species are now extinct and others depleted. Although hutias are partially protected in Cuba, and the cays and islets declared a wildlife protection area, visiting fishermen apparently still kill them. The precise status of most of them is unknown, and some may be extinct.

******* CABRERA'S HUTIA**
Capromys angelcabrerai **(804)**
Known only from a few specimens collected in the mid 1970s on the islands of the Cayos da Ana Maria, off SC Cuba.

******* LARGE-EARED HUTIA**
Capromys auritus **(805)**
Known only from Cayo Fragoso off NC Cuba, first described in 1970.

**** GARRIDO'S HUTIA**
Capromys garridoi **(806)**
Described from a single specimen collected in 1967, on Cayos Maja, off SC Cuba.

***** BUSHY-TAILED HUTIA**
Capromys melanurus **(807)**
Scattered populations exist in E Cuba. Deforestation has reduced their habitat, and they are hunted. Thought to occur in a few reserves.

******* DWARF HUTIA**
Capromys nanus **(808)**
Formerly occurred throughout Cuba, but now confined to the Zapata swamp in S Cuba, where it is threatened by the spread of agriculture and predation by introduced mongooses.

******* LITTLE GROUND HUTIA**
Capromys sanfelipensis **(809)**
First described in 1970 from Cayo Juan Garcia off SW Cuba. In 1980 an expedition failed to find any signs of them.

***** CUBAN HUTIA**
Capromys pilorides **(810)**
Perhaps the commonest of the surviving hutias, but even it has declined in both numbers and range.

***** JAMAICAN HUTIA**
Capromys brownii **(811)**
C.b. brownii occurs on Jamaica in very much reduced numbers. A captive breeding programme has been established by Jersey Zoo, and they are totally protected in Jamaica. *C.b. thoracatus* formerly occurred on Little Swan Island between Jamaica and Honduras, but almost certainly became extinct after the introduction of cats, and a hurricane in 1955.

***** BAHAMIAN HUTIA**
Capromys ingrahami **(812)**
Now extinct over most of its range, which until the arrival of Europeans extended over most of the Bahamas. They survived on East Plana Cay, where perhaps 12,000 now exist. Have been introduced into a small cay in the Exuma NP.

HISPANIOLAN HUTIAS
Plagiodontia spp.
HB ca.12.6ins; T ca.5.9ins; WT ca.2.6lb.
Seven species have occurred on the island of Hispaniola in the past, five of which are extinct. The surviving species *** *P. aedium* **(813)** and *** *P. hylaeum* **(814)** are sometimes considered as a single species. Their fur is short and dense, greyish or brownish above, buff below; the tail is scaly and naked. Found in forests, where they are usually nocturnal. Threatened by hunting, deforestation, and predation by introduced mongooses. Between 1958 and 1978 the forest cover in Haiti was estimated to have fallen from around 80% to 9%. They occur in some protected areas and have legal protection in the Dominican Republic.

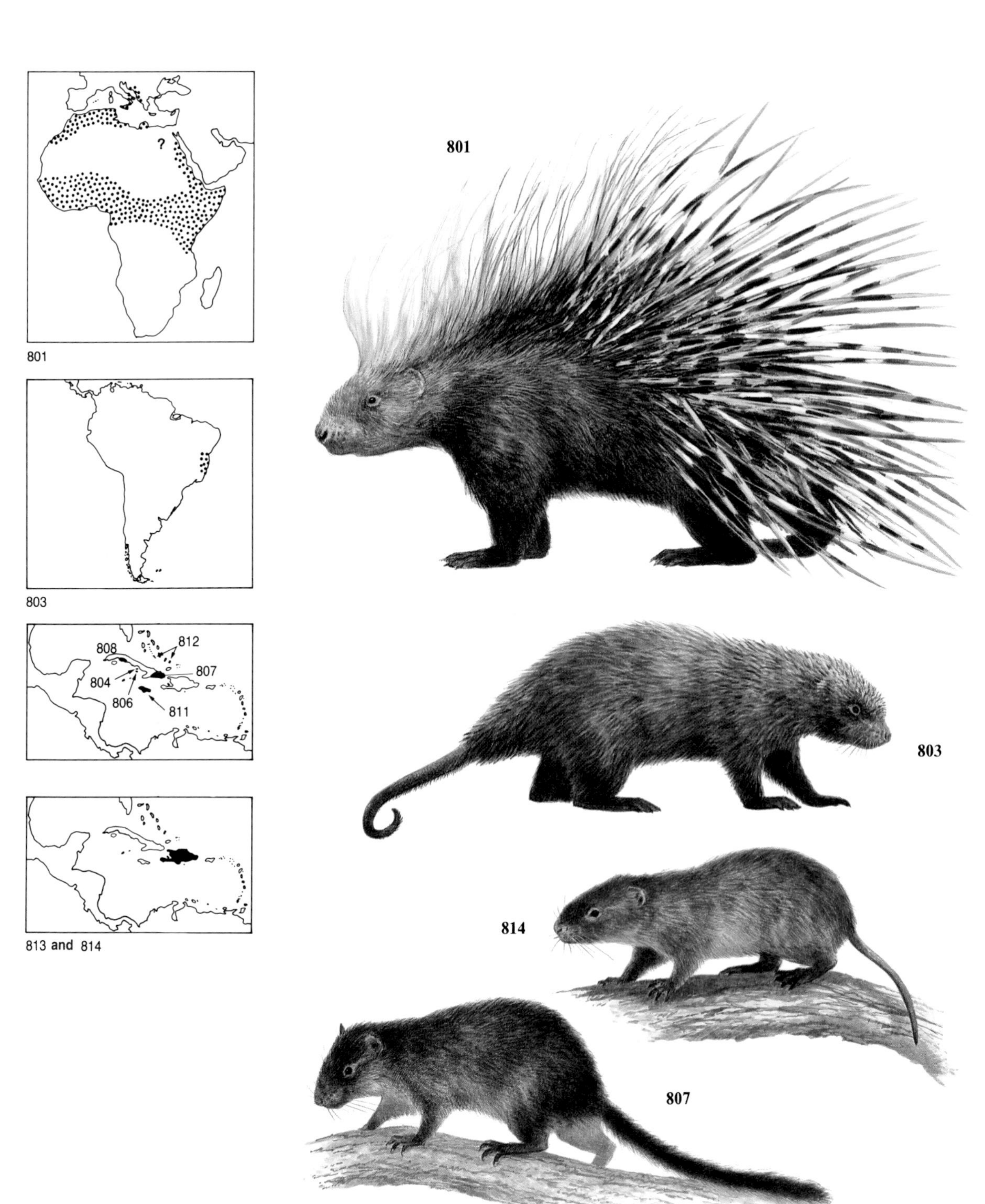
?
801
803
808
812
807
804
806
811
813 and 814
801
803
814
807

* UPPER AMAZON PORCUPINE

Echinoprocta rufescens (**815**)

HB ca.19.7ins.

Generally pale brown to blackish with a darker tail and feet. The spines are most dense on the rump, gradually thinning towards the forequarters. A little-known species, with a restricted range, confined to Colombia. Apparently locally abundant at altitudes of 2600–3950ft.

*** NORTH AMERICAN PORCUPINE

Erithizon dorsatum (**816**)

HB 25.4–36.6ins; T 5.7–11.8ins; WT 7.7–39.7lb.

Black or brownish in east of range, yellowish in west with spiny quills on rump and tail, up to 3ins long. Their preferred habitat is mixed forest where they are mainly nocturnal, hiding by day in hollow trees and logs. Also found in many other habitats, including more open tundra and pasture. They feed on a wide variety of vegetation including Skunk Cabbage, leaves, shoots, and in winter twigs and bark. Although widespread in North America, they are largely absent from the eastern States south of Minnesota, N Wisconsin, Michigan, and Pennsylvania. In the 18th and 19th centuries they were present in parts of Maryland and Virginia, but have since been exterminated. They are trapped as pests, since they cause damage by gnawing trees and wood, and they are also hunted as human food. However, they remain abundant and widespread over most of their range.

* PATAGONIAN MARA

Dolichotis patagonum (**817**)

HB 27.2–29.5ins; T absent; WT 19.8–35.3lb.

Rather hare-like, with long hindlimbs, and capable of fast running. Greyish-brown above, whitish below. Found in open grassy habitats, where they are largely diurnal, feeding on a wide range of vegetation. In some areas they are declining, and are now rare in Buenos Aires province, Argentina; this is largely due to competition with the European Brown Hare, which is spreading, and destruction of habitat. Exhibited in many zoos, breeding freely; a colony exists in N France under semi-natural conditions.

*** PACARANA

Dinomys branickii (**818**)

HB 28.8–31.1ins; T ca.7.9ins; WT 22–33.1lb.

Rather like a large guinea-pig with a short tail. Upperparts are dark brown or blackish with two broad white stripes along the back, and two rows of spots on the flanks. Found in montane forests from Colombia to W. Bolivia, at altitudes of 800–6550ft. They feed on fruit, leaves and herbage. Rare throughout their range, but reports of their extinction are believed to be unfounded. They are hunted for food, and also suffer from loss of habitat. They occur in the Manu NP, Peru. There are small numbers in captivity, but although bred regularly, do not appear to be self-sustaining.

PACAS

Agouti (=Cuniculus) spp.

HB 23.6–31.5ins; T 0.8–1.2ins; WT up to 22lb.

There are two species: *** *A. taczanowskii* (**819**) from the Andes of NW Venezuela, Colombia, Ecuador, and *** *A. paca* (**820**) which occurs from Mexico to Paraguay; the latter has rather coarse fur without underfur. They are mainly nocturnal, hiding by day in burrows which they excavate themselves, in caves, or the burrows of other animals. Often found close to water – they swim well – and they use regular paths in the forest. They feed on a wide variety of vegetable matter, including fallen fruit, roots, leaves and seeds. In some areas they are persecuted, because of the damage they cause to crops; they are also intensively hunted for their flesh which often commands exceptionally high prices. This, together with destruction of their forest habitat, has resulted in their extermination in many areas. They occur in many reserves and national parks within their range, and both species are being bred regularly in zoos.

AGOUTIS

Dasyprocta spp.

Slightly smaller than the pacas, agoutis have coarse shiny fur, generally rather variable in colour from blackish to orange-brown. Several species have declined, mostly through habitat destruction, and also from intensive hunting pressure; * *D. prymnolopha* (**821**) from E Brazil is hunted and much of its habitat has disappeared. **D. leporina* (**822**) was introduced into many of the Caribbean islands before the arrival of European settlers, and had been established long enough for several island subspecies to develop; most of these became extinct by the early 20th century. The status of several species, particularly those with restricted ranges, is very poorly documented; these include:** *D. ruatanica* (**823**) from the island of Roatan, off N Honduras; ** *D. coibae* (**824**) from the island of Coiba, off SW Panama. Isolated populations may occur on other small islands around Central and South America.

* CAPYBARA

Hydrochaeris hydrochaeris (**825**)

HB up to 51ins; T vestigial; SH up to 19.7ins; WT up to 176.5lb.

The largest living rodent, found in herds close to water over much of South America, east of the Andes from E Panama, south to N Argentina. They feed on a wide variety of grasses as well as aquatic vegetation. Although still widespread, they have undergone drastic declines in many areas. They are hunted extensively for meat and hides, and the fat is used in medicine. In recent years they have been ranched, but are now extinct over much of their former range. They occur in many reserves and national parks, where suitable habitat occurs. Their main threat is the draining of their habitat for cattle pasture.

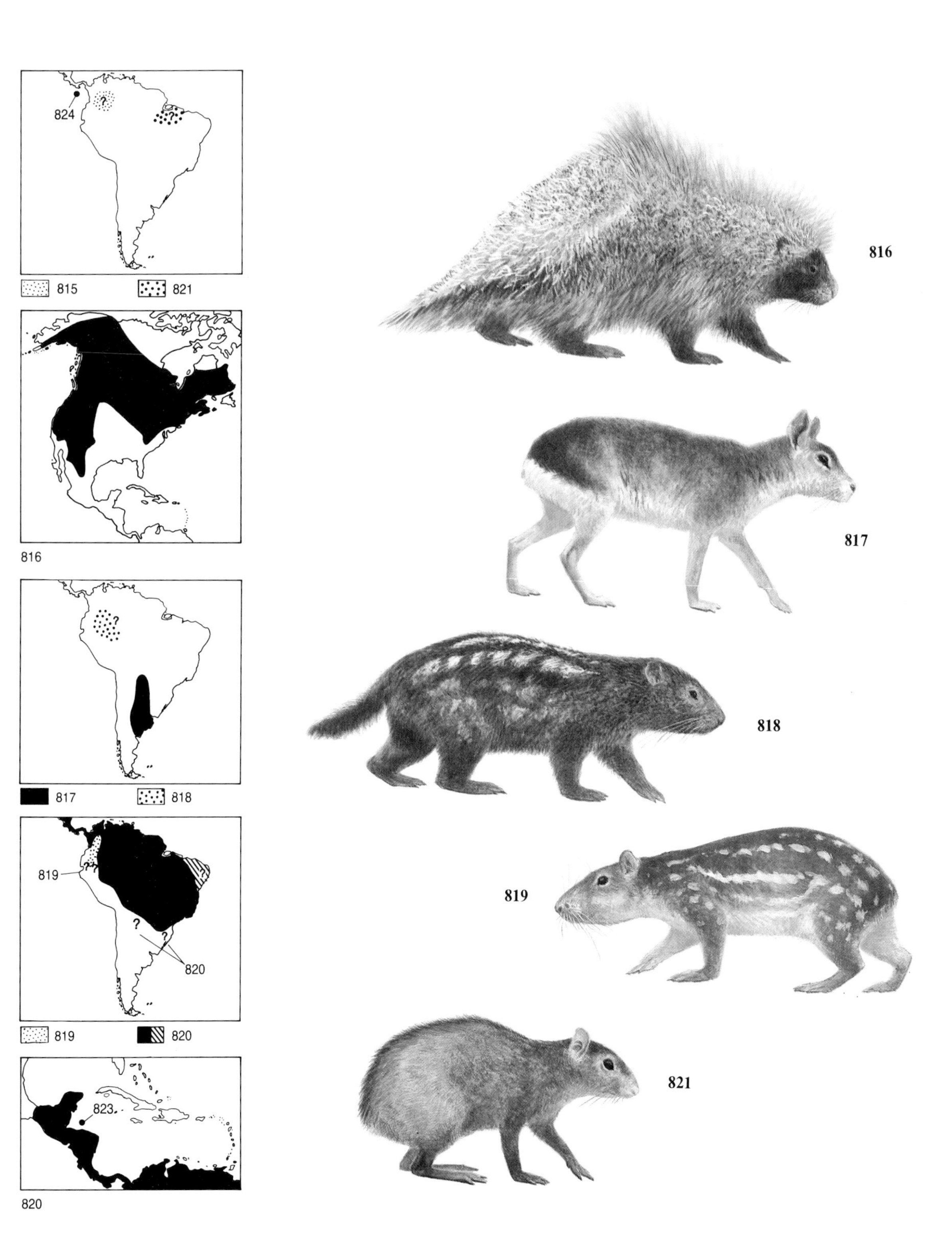
824
?
?
815
821
816
?
817
818
819
?
?
?
820
819
820
823
820
816
817
818
819
821

*** PLAINS VISCACHA

Lagostomus maximus (**826**)

HB 18.5–26ins; T 5.9–7.9ins; WT 4.4–17.6lb.

Male is nearly twice the weight of female. Colouring varies with habitat, from light sandy brown to dark greyish above, whitish below. The underfur is dense and soft, with coarse dark guard hairs.The nose is thickly furred with a complicated fold, to prevent sand and dust entering the nostrils. Found in barren pampas habitats of Argentina at altitudes of up to 800ft, they live in colonies of 15–30, occasionally up to 50, in extensive burrows which they excavate themselves. The burrows have up to 40 entrances which may be 20ins deep. They are normally regarded as pests, on the grounds that they compete with domestic stock for food, and their burrows cause accidents to cattle, horses, and humans. Consequently they have been subjected to extensive and systematic extermination campaigns often involving poisoning, and have also been extensively hunted for their fur, and flesh. They were introduced into Uruguay, but subsequently died out. Kept in a few zoos, and appear to breed freely.

* MOUNTAIN VISCACHA

Lagidium spp.

HB up to 17.7ins; T up to 15.8ins; WT up to 6.6lb.

Superficially like a long-tailed rabbit, their soft, dense fur varies in colour with altitude; at lower elevations it is dark grey, becoming chocolate brown at high altitudes. They live in arid rocky habitats from 9850–16,400ft, where they are diurnal and active throughout the year, feeding on almost any vegetation available. There are three species: *L. peruanum* (**827**), endemic to Peru; *L. viscacia* (**828**) which is found in the south of Peru, W and S Bolivia, N and C Chile and W Argentina, and *L. wolffsohni* (**829**) which occurs in S Chile and SW Argentina. They are hunted for their meat and fur, and in Peru and parts of Chile numbers have declined. Found in the Lauca NP, Chile, and a few other protected areas.

*** CHINCHILLAS

Chinchilla spp.

HB 8.9–15ins; T 2.9–5.9ins; WT up to 1.8lb.

Female is larger than male. The fur is soft and extremely dense, with up to 60 hairs growing from each follicle. There are two very closely related species: the **Short-tailed Chinchilla** *C. brevicaudata* (**830**), from the Andes of Peru, Bolivia, and NW Argentina; and the **Long-tailed Chinchilla** *C. lanigera (=laniger)* (**831**) from the Andes of N Chile; the Long-tailed Chinchilla is the smaller of the two (under 1lb, whereas the Short-tailed is up to 1.9lb) with a longer tail and longer limbs. The former are found at altitudes of 4900–8200ft, while the latter occupy the higher altitudes, up to 16,400ft, in barren, arid and rocky habitats where they are rarely seen by day – though they occasionally emerge to sun themselves. They live in colonies, which once numbered up to 100 or more, in rock clefts and crevices; they feed on almost any available vegetation. Their original distribution is not known, since they have been hunted for fur since pre-Columbian times. The Incas valued the fur for robes, and after European colonisation the demand increased greatly. They were once observed in their thousands, and at the end of the 19th century around 500,000 skins were being exported annually from Chile alone. Despite the introduction of protection in 1910, poaching for skins continued and as recently as 1981 coats made from wild trapped skins were allegedly being sold in Japan. The precise status of both species is not known in any detail, largely because of the inaccessibility of their remaining range. Currently habitat destruction is probably a greater threat to the Long-tailed Chinchilla, particularly at lower altitudes. The Long-tailed Chinchilla was first bred in captivity at the end of the 19th century, but it was not until the 1920s that commercial breeding commenced. However, they have never been particularly easy to breed on a large scale, and the more valuable Short-tailed Chinchilla has proved even more difficult to breed; it is likely that commercial breeding activities stimulated the demand for illegally caught live wild chinchillas, and also for skins from wild animals. Domesticated chinchillas are common in zoos, wild Long-tailed Chinchillas are only occasionally exhibited. The Short-tailed Chinchilla may occur in the Lauca NP, Chile. They are listed on CITES App. I.

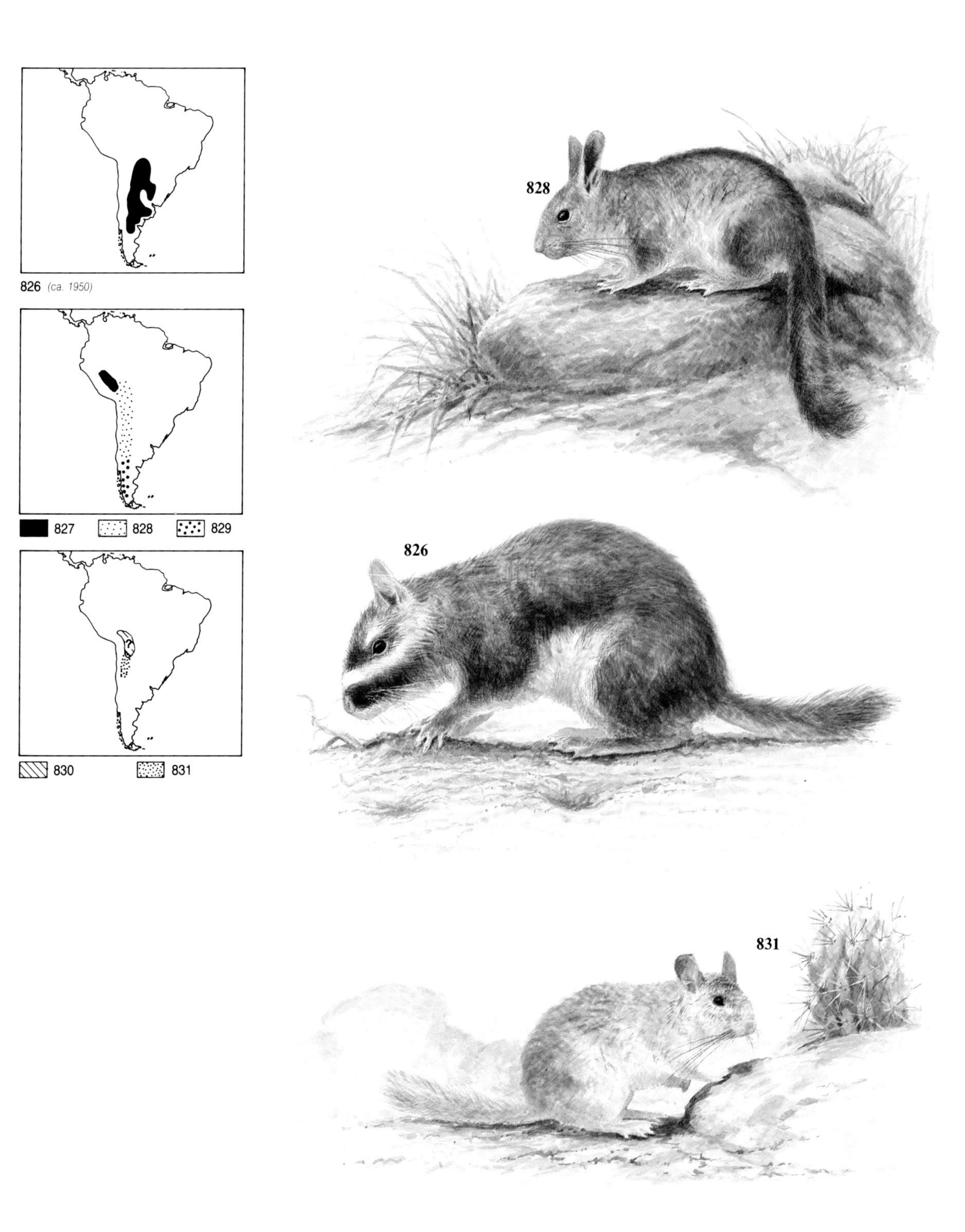
826 (ca. 1950)
827
828
829
830
831
828
826
831

*** CORSAC FOX

Vulpes corsac (**832**)

HB 19.7–23.6ins; T 9.8–13.8ins.

The thick, soft fur is reddish-grey or brown above, and whitish or yellowish below. In general apearance it is like a long-legged Red Fox *Vulpes vulpes*, with large pointed ears. Corsac Foxes live in arid steppes and desert areas where they are nocturnal, living in burrows, normally made by other animals. They feed mainly on small rodents, and other small birds, pikas and insects as well as fruit and other vegetable food. Sometimes gregarious, they live in small groups and hunt in small packs. The litter contains 2–11 young. Their thick fur has been in demand by the fur industry and they have been trapped commercially, which, together with the extensive spread of agriculture into the steppes in the 19th century, has greatly reduced and fragmented their range.

*** BLANFORD'S FOX

Vulpes cana (**833**)

HB up to 19.7ins; T 13–16ins.

A small, soft-furred, bushy-tailed fox, grey blotched with black above, and white below, with a dark tip to the tail; often there is a dark line down the back and the legs are dark. Found in open montane country, it is said to eat more vegetable matter than other foxes, often raiding grapevines, melons and other fruits. It is hunted for its pelt and in some parts of its range, notably USSR, it is very rare. An isolated population has recently been discovered in Israel. Listed on CITES App.II, and in recent years very little international trade has been noted.

*** SWIFT FOX

Vulpes velox (**834**)

HB 14.8–20.7ins; T 8.9–13.8ins; WT 4–6.6lb.

A small fox (male slightly larger than female) with long, thick fur, greyish-brown above, orange-brown on the flanks, legs and underside of tail, and a whitish belly. Has wider eyes and smaller ears than the closely related Kit Fox (835). Found in open prairies and similar short grass habitats, where it is largely nocturnal, living in burrows by day, feeding on cottontails, rodents, and other small animals. This species normally lives in pairs, and the litter is usually 3–6 young. Its natural range has largely been destroyed by ploughing, and during the 19th century it was extensively poisoned during campaigns to eliminate coyotes and wolves. Although it is spreading its range in some areas, it is still very much reduced in range, and the **Northern Swift Fox** *V. v. hebes* is extremely rare, and is listed on CITES App. II. The species has been kept and bred in captivity and has lived for nearly 13 years.The Southern Swift Fox *V.v. velox* is still reasonably widespread.

*** KIT FOX

Vulpes macrotis (**835**)

HB 14.8–19.7ins; T 8.9–12.7ins; WT up to 4.8lb.

Generally greyish with a yellowish tinge above, orange-buff sides and white belly. Very similar to the Swift Fox (834), but has larger ears and a longer tail. Found in open prairie and semi-desert habitats; densities of 500 adults per sq.ml. have been recorded in optimum habitat, or 1 per 4sq.mls in poor habitat. Like the Swift Fox, its numbers have been reduced by habitat destruction, and poisoning. The Californian populations *V. m. devia* disappeared by 1910, and *V. m. mutica* is now threatened by habitat loss.

*** FENNEC FOX

Fennecus zerda (**836**)

HB 13.8–16.1ins; T 6.7–11.8ins; WT 2.2–3.3lb.

A very pale, creamy-coloured fox, with very large (up to 6ins long) ears. The feet have hairy soles adapted to soft sand. Found in desert and semi-desert regions of N Africa, and the Arabian Peninsula; they feed on small animals such as lizards and locusts and are probably capable of surviving without free water. They live in groups of up to 10, led by a dominant male, and there are 2–5 young in a litter. Fennec Foxes have been intensively hunted in many parts of their range, and in the more populated parts of North Africa their range is much reduced and fragmented; in Arabia they are extremely rare. They breed freely in captivity, and are often seen in zoos.

** ISLAND GREY FOX

Urocyon (=*Vulpes*) *littoralis* (**837**)

HB 18.9–19.7ins; T 4.3–11.4ins; WT 5.5–15.4lb.

The fur is rather coarse, and generally greyish above, tipped with black and orange-buff on the sides. They are rather arboreal, and will even make their dens in hollow trees, up to nearly 33ft off the ground. The Island Grey Fox is closely related to the more widespread Grey Fox *U. cinereoargenteus* which occurs over much of N and C America. The Island Grey Fox is confined to San Miguel, Santa Rosa, Santa Cruz, Santa Catalina, San Nicolas and San Clemente Islands off SW California, where, because of its restricted range, is fully protected by state law. Six subspecies have been described. Grey Foxes are hunted commercially with over 225,000 being sold in one season, but the Island Grey Fox does not appear to be exploited for fur. *U. littoralis* may prove to be isolated populations of the Grey Fox.

* ARCTIC FOX

Alopex lagopus (**838**)

The Arctic Fox is widespread with a circumpolar distribution, and in many areas almost a pest species, and commercially important for its fur, which is white in winter, and bluish-grey or brown in summer. However, in a few areas it has undergone local declines, particularly in the south of its range. However, it is generally common, and is bred extensively in captivity.

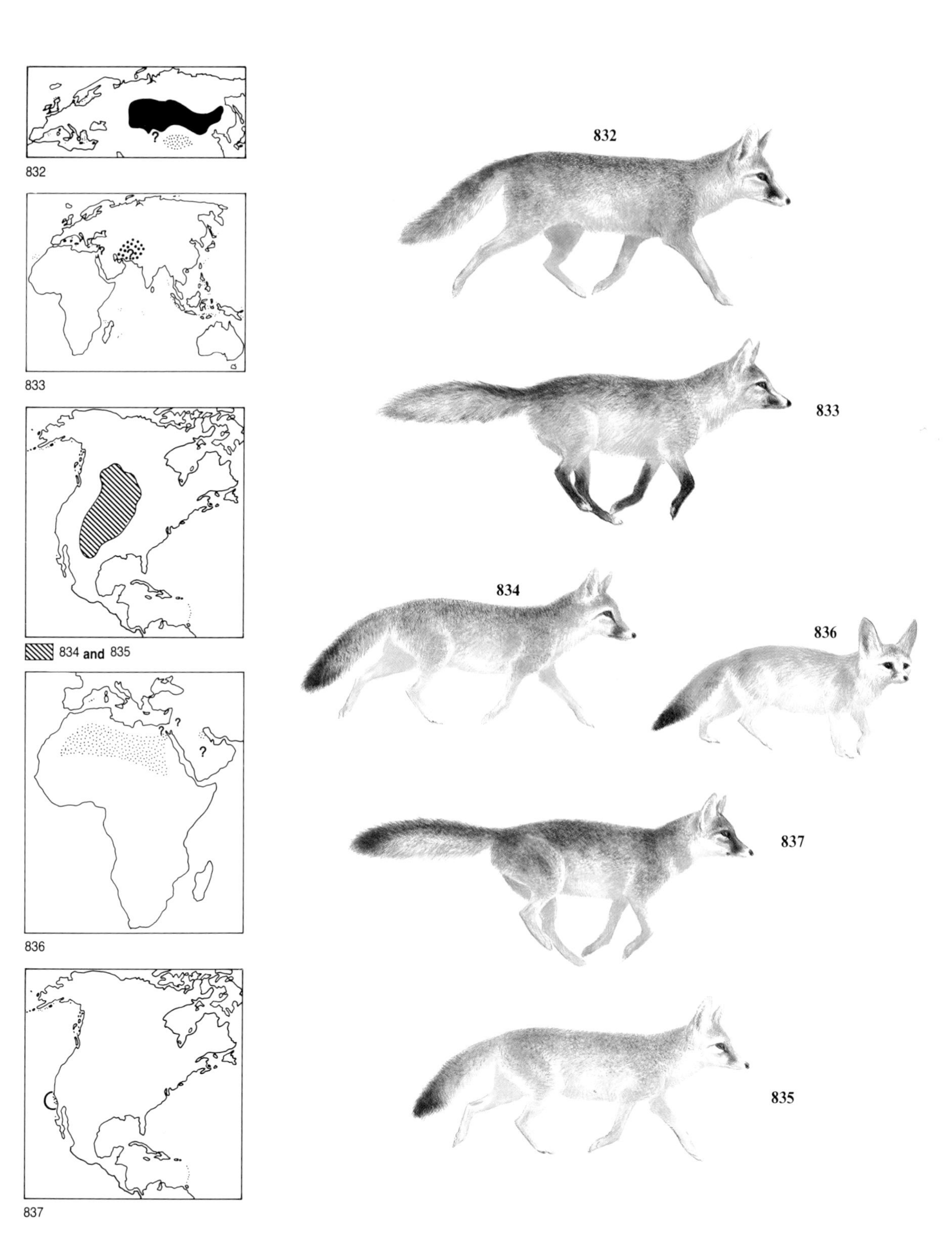
832
?
833
834 and 835
?
?
?
836
837
832
833
834
836
837
835

*** MANED WOLF

Chrysocyon brachyurus **(839)**

HB 47.3–51.2ins; T 11–16.9ins; SH 29.5–31.5ins; WT 44.1–66.1lb.

General colouring is orange-red above, white below with darker legs, and a paler tail. Fur is long and soft, with a slight mane on neck and shoulders. They live in grasslands and savannah, as well as swamps, where their long legs enable them to see above tall vegetation. Prey includes mammals, birds, reptiles, insects and some fruit and other vegetable matter. Outside the breeding season they are usually solitary, and the normal litter is 2–4 young. Their range is considerably reduced, both by the spread of agriculture, and direct persecution. In spite of an overall decline they are believed to have spread westwards since the end of the 19th century. Protected by law in Argentina, Brazil, and some other parts of their range, and found in several Brazilian parks and reserves, including the Serra da Canastra NP. Exhibited in many of the larger zoos and now being bred regularly in captivity. Both in the wild and in captivity there is a fairly high incidence of *Cystinuria*, an hereditary disease, causing mortality.

*** GREY WOLF

Canis lupus **(840)**

HB up to 63ins; T up to 22.1ins; WT up to 176.4lb.

Considerable variation in size, the smallest found in the south of its range. Fur colour is variable, generally greyish, but almost totally white and blackish individuals occur. Similar to the Red Wolf (841). Packs generally consist of a pair and their offspring, but there is much variation of behaviour. They feed mostly on mammals, including those larger than themselves, such as deer, Bison, and Elk (Moose). Formerly widespread throughout the northern hemisphere in a wide variety of habitats, the present range is considerably depleted. They have been persecuted because of their plundering of domestic livestock and alleged attacks on humans; in fact proven attacks of pure-bred, healthy wolves on humans are extremely rare. More recently they have been extensively trapped for furs. They are kept in many zoos and breed freely, and a number of the rarer subspecies are being bred; by 1985 there were ca.30 Mexican Wolves in captivity. Listed on CITES App.II, with the Indian sub-continent's populations on App.I; protected throughout most of their range, though still persecuted. Occur in many reserves, including most of the larger parks of North America; Coto Donana in Spain and Abruzzo in Italy.

***** RED WOLF

Canis rufus **(841)**

HB 39.4–51.2ins; T 11.8–16.5ins; SH 26–31.1ins; WT 44.1–88.2lb.

Generally greyish with buff or reddish tinge to upperparts, overlaid with black, and whitish or pale buff below; tail tipped with black. Very similar in general appearance to Grey Wolf, but has relatively longer legs, larger ears and shorter fur. Red Wolves were found in a wide variety of habitats including forests, prairies and wetlands, and they make dens in hollow trees, rock clefts, etc. They are mostly nocturnal, feeding on mammals up to the size of small deer, pigs, coypu, and will also take carrion. They normally have a litter of 4–7 (max 12). Originally they were widespread, in the southern States of USA, but with the arrival of Europeans and the changes to many habitats, combined with persecution, they declined rapidly. The closely related but more adaptable Coyote *C.latrans* was able to expand its range and started to interbreed with the remaining Red Wolves; by the 1970s there were very few pure Red Wolves left, and captive breeding programmes commenced in 1973. The first reintroduction was made in 1977, when they were re-established on Bull Island in Cape Romain National Wildlife Refuge, S Carolina. By 1980 there were over 50 in captivity, but some may have had Coyote ancestry.

*** SMALL-EARED DOG or ZORRO

Atelocynus (=Dusicyon) microtis **(842)**

HB 28.4–39.4ins; T 9.8–13.8ins; SH ca.14ins; WT 19.8–22lb.

Has relatively small ears (less than 2ins); upperparts grey or blackish, belly reddish-brown, and tail mostly black. Found in tropical forests from sea level to 350ft, but little is known of its habits, and nothing of breeding. Occurs in the Amazon, Orinoco, and Parana River basins, but throughout range very sparsely distributed. Undoubtedly, habitat destruction will fragment its range. Protected in Brazil and Peru, and occurs in a few reserves and national parks within range. Only occasionally kept in captivity.

SOUTH AMERICAN FOXES

Pseudalopex spp.

HB 23.6–47.3ins; T 11.8–19.7ins; WT 8.8–28.7lb.

Dense, soft fur, generally sandy or greyish, or reddish depending on species, with a bushy, black-tipped tail. They are generally nocturnal, feeding on small animals such as mice, lizards, frogs, insects and some fruit; will also take animals up to the size of lambs and hares. They have been extensively hunted, both because of their depredations on livestock, and also for skins. The *** **Small Grey Fox** ***P. griseus*** **(843)** has the finest fur and is particularly threatened in many parts of its range. In the 1980s ca.270,000 fox pelts a year were handled in Argentina alone. In parts of their range, the South American foxes receive protection (ranging from total protection, to closed seasons) and they occur in many national parks and reserves. The Small Grey and the * **South American Red Fox** ***P. culpaeus*** **(844)** are listed on CITES App.II. The closely related Falkland Islands Wolf *Dusicyon australis* was exterminated in the 19th century.

***** SIMIEN JACKAL

Canis simensis **(845)**

HB ca.39.5ins; T ca.9.8ins; SH 23.6ins.

The Simien Jackal (also referred to as a wolf or fox) is reddish-ginger above, white below, with a distinctive white collar. It has a rather narrow snout. Lives mostly in open grassland habitats, where it feeds mainly on small rodents, at altitudes of 9850–13,100ft. Although they are mostly active by day, they also hunt by moonlight, and in areas where they have been persecuted they are usually nocturnal. In the 19th century, they occurred over most of Ethiopia, but have declined with the spread of agriculture and depredation of habitat that lead to reduction in rodent numbers. Now extremely rare, and by 1978 only 500 were estimated to survive in four populations. The northern race *C.s. simensis* occurs in the Simien Mountains, and NW Shoa, and the southern race, *C.s. citernii* in the Bale Mountains and the Arussi Mountains. They occur within the Simien and (proposed) Bale National Parks, and are protected by law. None in captivity, though it is likely they could be bred.

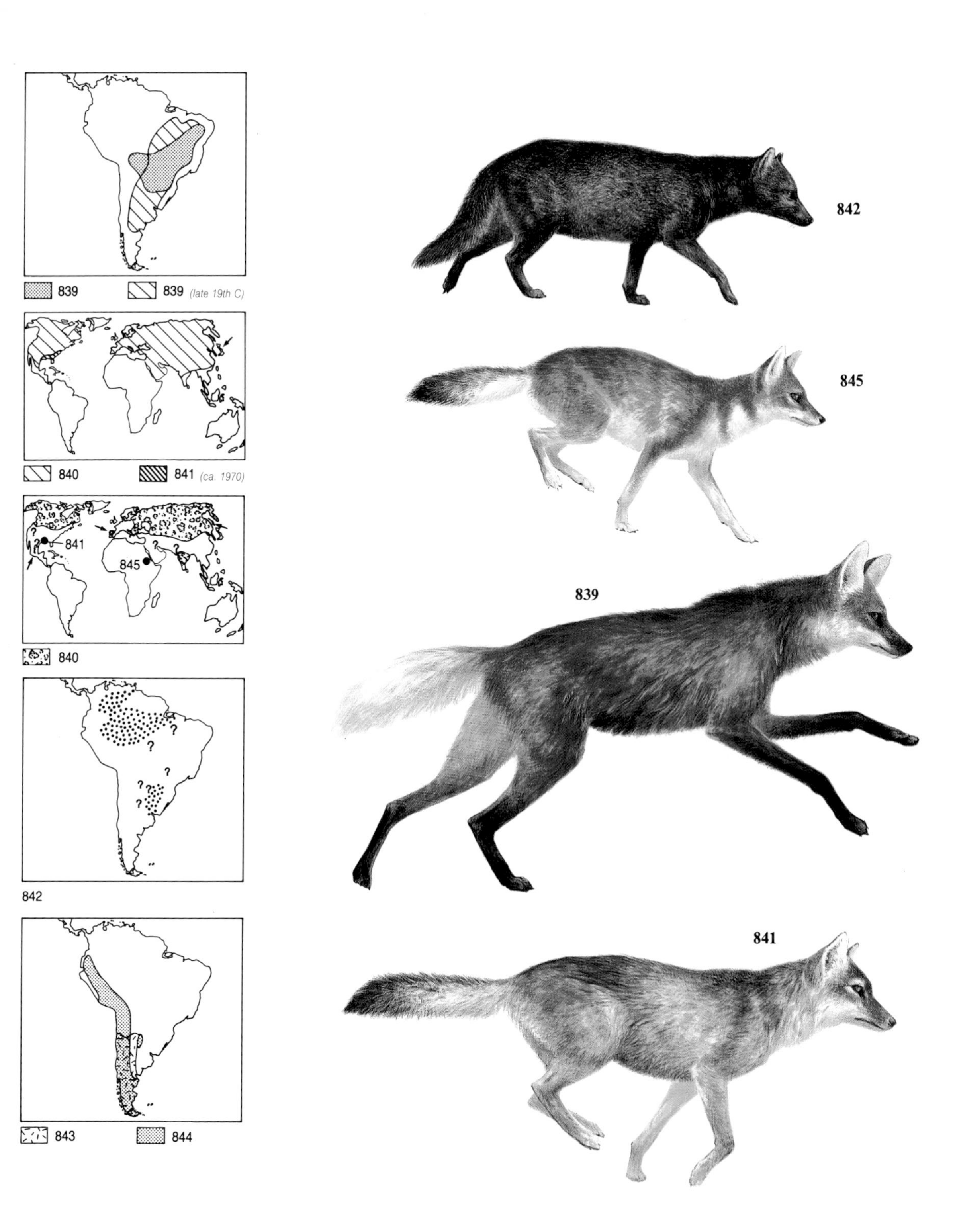
839
839 (late 19th C)
840
841 (ca. 1970)
841
845
840
842
843
844
842
845
839
841

******* HUNTING (PAINTED) DOG**
Lycaon pictus (**846**)
HB 29.9–43.3ins; T 11.8–15.8ins; SH 24–30.7ins; WT 37.5–79.4lb.
The colouring is extremely variable, with no two animals identical; the short fur is often so sparse that the bare black skin is visible. They normally occur in savannah and other open habitats, where they live in groups of up to 20 or more, and have a very complex and variable social structure. (Before the destruction of the huge herds of Springbok and other game in South Africa, they occurred in packs of hundreds.) Except when pups are present, and the whole pack stays relatively close to the den, they wander extensively; a pack will have a range of 1500 sq.mls. They hunt mostly in the morning and evening and chase their prey at speeds of up to 41 m.p.h.; they can run at 31 m.p.h. for over 3 miles. They often hunt in packs to kill larger prey such as zebra and antelope, but also hunt alone for rodents and other small animals. There has been a massive decline in the numbers of Hunting Dogs in historical times. Formerly found in suitable habitats over almost all of Africa, their numbers are now considerably reduced throughout, and many populations are quite isolated. It was estimated that by the late 1970s Hunting Dogs numbered less than 7000. Their decline has in part been due to direct persecution by man, but disease may also have played a part; they are known to have been affected by distemper which was introduced into E Africa in 1906. Small numbers are kept in captivity and a few are bred most years. They have legal protection in many parts of their range, and occur in many national parks, notably Serengeti and Kruger.

***** DHOLE or RED DOG**
Cuon alpinus (**847**)
HB 34.7–43.3ins; T 15.8–19.7ins; SH 16.5–21.7ins; WT 22–46.3lb.
Males are generally larger than females. Rather variable in colour, but generally rufous above, pale below with a black-tipped tail; in the more northern parts of their range and at higher altitudes they have much redder, silkier, longer fur in winter. Occur in a wide range of mainly forest habitats, from the Altai and Pamirs, east to Siberia and south to peninsular India, Malaya, Sumatra, and Java. They feed on antelope, wild boar, wild sheep, and also smaller prey such as rodents, carrion and insects. Will hunt larger prey in packs of up to 40 (usually smaller). The normal litter is 4–6 and the young are fed by all the pack. They have been persecuted throughout most of their range, and have also declined through habitat loss and reduction of available prey. Protected in India, USSR and some other parts of their range, and listed on CITES App.II. Occasionally kept in zoos, and have bred there.

***** BUSH DOG**
Speothos venaticus (**848**)
HB 22.6–29.5ins; T 4.9–5.9ins; SH 11.8ins; WT 11–15.4lb.
Darkish brown over most of body including the underparts, but paler on the head and neck. Rather short and stocky, with a short tail. They are found in forest and savannah, where they are usually diurnal, hiding in a den by night. Often found near water, they are apparently rather aquatic, diving and swimming well. They prey on rather large rodents such as Agouti and Paca and live in small packs of up to 10 with litters of 3–6 young. Bush Dogs have probably always been fairly rare, but in recent years they have declined in many areas and are threatened by habitat destruction. In the more northern parts of their range (Panama and Colombia) they are particularly threatened. Protected in several countries within their range and occur in several Brazilian national parks and possibly Manu NP, Peru. There are very few in captivity, but Frankfurt Zoo has been particularly successful in breeding them. Listed on CITES App. I.

******* GIANT PANDA**
Ailuropoda melanoleuca (**849**)
HB 47–59ins; T ca.44.3ins; WT 165–353lb.
The thick woolly fur has probably the most distinctive and best-known markings of any mammal. The front paws are modified for holding bamboo, upon which they feed almost exclusively. Confined to montane forests in the central Chinese provinces of Szechuan, Kansu, Shensi, and possibly Tsinghai and Yunnan at altitudes of 8850–12,800ft, but may descend to much lower altitudes during winter. There are normally 1–3 young in a litter, but only one survives. Giant Pandas have been declining for thousands of years, partly due to hunting by humans and partly due to changes in habitat, resulting from longterm climatic changes. By the early 1980s less than 1000 survived in the wild, separated into three isolated groups. In the 1970s many died when there was a major die-off of bamboo. Poaching for skins has occurred on a small scale, and in the 1960s and 1970s small numbers were also captured for zoos; although a few have been bred in captivity, the success rate is extremely low and they are not self-sustaining. They receive strict protection in China and are listed on CITES App.I.

***** LESSER or RED PANDA or CAT-BEAR**
Ailurus fulgens (**850**)
HB 20.1–25ins; T 11–19.1ins; WT 6.6–13.2lb.
Easily distinguished by bright chestnut colouring, ringed tail, and rounded head with large pointed ears and short muzzle. Found in temperate forests of Himalayas above 5000ft; they are nocturnal, spending the day curled up asleep in the topmost branches of a tree. They descend at dusk to feed, mostly on the ground, on roots, grubs, bamboos, grasses, fruits and some small animals. One to four (usually 1 or 2) young are born in a den in hollow tree. Red Pandas are little-known in the wild, and have a fairly sparse distribution. The thick, soft pelt has been involved in trade, and they are very popular zoo animals; in the past they were traded in small numbers for the pet trade, as they are easy to tame. Protected in India and Nepal, and listed on CITES App.II.

848
846
847
849
850
846 (historic times)
847
848
849
850

*** BROWN OR GRIZZLY BEAR

Ursus arctos (**851**)

HB 5ft 7ins–6ft; T 2.4–8.3ins; SH 35–59ins; WT 155–1720lb.

The largest living carnivore, although its size varies considerably. Some of the smallest occur in S Europe, the largest in the Kodiak Islands, Alaska. Because of the enormous variation in appearance within the species, many subspecies have been described, which have even been treated as full species in the past; however, they are generally agreed to represent variation within a single species. Formerly one of the most widespread of all mammals, occurring in a wide variety of habitats, from deserts and tundra to temperate and tropical forests and coasts. They feed on a wide variety of plant and animal matter, including fruit, roots, honey, fish and carrion; they may raid crops and kill livestock, and can do substantial damage to bee hives. They rarely kill humans, but deaths have occurred in many parts of their range. Wherever they occur in close proximity to man and his livestock, they have tended to decline. They became extinct in the British Isles around the 12th century, and are now confined to montane areas in W Europe in extremely small numbers. In N and E Europe populations are isolated and fragmented. In N Africa they became extinct in the 19th century and by the 1960s were extinct in most of the Arabian Peninsula, Levant and Near East. Widespread and still locally abundant in the USSR. In North America the populations in the south of their range are among the most threatened, but they are increasingly fragmented and isolated throughout their range. They occur in numerous national parks and reserves throughout their range; in fact they are increasingly confined to protected areas. It is estimated that there are ca.1000 in the western USA (excluding Alaska), of which ca.200 are in Glacier NP, and over 100 in Yellowstone NP. In the mid 1970s the world population was estimated at ca.100,000, of which ca.70,000 were in the USSR. They breed freely in captivity, are protected in many parts of their range; some populations are listed on CITES App.II, with several subspecies on App.I.

*** SPECTACLED BEAR

Tremarctos ornatus (**852**)

HB 47–71ins; T ca.0.8ins; SH 27.5–31.5ins; WT 132–309lb.

Uniform black, or dark brown except around the eyes and on the neck. Found in a wide variety of habitats in the montane regions of western South America at altitudes of up to 11,800ft. They prefer forested areas, but also occur in more open habitats. They are usually nocturnal, hiding in rock clefts and hollow trees by day, emerging in twilight to feed on a wide variety of fruit, leaves and other plant matter, such as bromeliads and bamboo, as well as small animals such as insect grubs; they also raid arable crops. The young (litters of 1–3) weigh only about 11.2oz at birth. Although very much reduced in numbers, particularly in areas close to human settlements, the Spectacled Bear is probably reasonably safe in some of the more remote parts of its range. Has been hunted both for meat and as a trophy, and because of its depredations on crops. Protected in Peru, Bolivia and some other parts of its range, and occurs in some of the larger national parks and reserves within its range. Kept in many zoos and an increasing proportion are captive-bred. Listed on CITES App.I.

*** ASIATIC BLACK BEAR

Ursus (=Selenarctos) thibetanus (**853**)

HB 47–71ins; T 2.6–3.9ins; WT up to 381.5lb.

Longish black (occasionally tinged with red) fur with a distinct white 'V' on the chest. Habits are similar to other bears, and they have a wide distribution in the mountains and upland areas of Asia, up to an altitude of 11,800ft in summer, descending in winter. Wherever they come into close contact with humans they are declining, but they are still relatively abundant in the more remote parts of their range. They are on the edge of extinction in Bangladesh. Listed on CITES App.I.

*** MALAYAN SUN BEAR

Ursus (=Helarctos) malayanus (**854**)

HB 39–55ins; T 1.2–2.8ins; WT 59.5–143.3lb.

The smallest bear, with short black fur and a white 'U' shaped marking on the chest. Generally nocturnal, they live in dense forests, using their powerful claws to climb trees. They feed on a wide variety of plant and animal matter, but insects such as termites, bees and earthworms form a large part of their diet. Their range extends from NE India and S China, south to Indo-China, Sumatra, and Borneo. It is declining rapidly in many areas through loss of its forest habitat, and is on the verge of extinction in Bangladesh. Nominally protected in most parts of its range and listed on CITES App.I.

* AMERICAN BLACK BEAR

Ursus americanus (**855**)

HB 51–75ins; T 2.8–7.1ins; SH ca.40ins; WT 202.9–588.7lb.

Smaller than the Grizzly. Nearly black in the eastern part of its range, but black to reddish-brown with white on chest in the west. Some populations on Alaskan islands are nearly white. Found mainly in forests and swamps up to 6900ft. They have been extensively hunted, and although still abundant and widespread in north of range, are confined to national parks or more scattered in south and east. They are subject to controlled hunting. The skins were (and still are) used in military uniforms for 'busbies' or 'bearskins'.

The * **Polar Bear** *Thalarctos maritimus* (**856**) is protected throughout its circumpolar range and in recent years numbers have increased; it cannot be considered threatened, although should protection be removed it would probably decline rapidly. The * **Sloth Bear** *Melursus ursinus* (**857**) from India and Sri Lanka is not known to be in any immediate danger, though undoubtedly declining through loss of forest habitat.

851
852
853
854
855
?
854
852
857
855
851

RACCOONS

Procyon spp.

HB up to 23.6ins; T up to 15.8ins; SH up to 11.8ins; WT up to 26.5lb.

Although the mainland Raccoon *P.lotor* is common and widespread in most parts of North America, and has been introduced into Europe, where it is spreading, the Central American species have much more restricted ranges. ***** *P.gloveralleni* (**858**) is only known from Barbados, and may already be extinct; ** *P.maynardi* (**859**) from New Providence Island in the Bahamas; ** *P.minor* (**860**) from Guadeloupe Island in the Lesser Antilles; ** *P.pygmaeus* (**861**) from Cozumel Island, Mexico; ** *P.insularis* (**862**) from Maria Madre Island (*P.i.insularis*) and Maria Magdalena Island (*P.i.vicinus*). Because of their restricted range, these raccoons are vulnerable to hunting pressures and competition from *P.lotor*, should it be accidentally or deliberately introduced.

** CHIRIQUI OLINGO

Bassaricyon pauli (**863**)

HB up to 18.7ins; T up to 18.9ins; WT up to 3.3lb.

Olingos are arboreal forest-dwellers, where they are mainly nocturnal, feeding on small animals and fruit. The Chiriqui Olingo is only known from Cerro Pando in W Panama and ** **Harris's Olingo** *B.lasius* (**864**) from Cartago in Costa Rica.They are closely related to the more widespread *B.gabbii*, and may prove to belong to a single species.

** COZUMEL COATI

Nasua nelsoni (**865**)

Closely related to the mainland Coati, it is confined to the island of Cozumel, Qintana Roo, Mexico.

***** BLACK-FOOTED FERRET

Mustela nigripes (**866**)

HB 15–19.7ins; T 4.5–5.9ins; WT 1.7–2.4lb.

One of the world's rarest mammals. Outside the breeding season they are normally solitary; the litter of 1–5 young is born in the underground nest. They are normally associated with Prairie Dog 'towns', and were probably once widespread and abundant on the Great Plains. However with the spread of agriculture and cattle ranching, and destruction of Prairie Dogs, they declined rapidly, until by the 1950s they were considered extinct. A colony discovered in 1964 in S Dakota became extinct in the wild in 1974, and the captive colony taken from these animals also died out. In 1981 another colony was discovered in Wyoming. It is strictly protected and research programmes involving captive propagation have been developed; its survival depends on the survival of Prairie Dog towns. Payments are made to farmers to protect Prairie Dogs, and some of the areas where they may still occur are protected as reserves. Listed on CITES App I.

* **European Polecat** *M.putorius* (**867**) is superficially rather similar in appearance to the Black-footed Ferret; it underwent declines in Britain and the more densely populated parts of W Europe, during the 19th and early 20th centuries, but is now recolonising much of its former range. An adaptable species, decline was mostly due to persecution by gamekeepers.

** WATER WEASEL

Mustela felipei (**868**)

HB 8.5ins; T 4.4ins.

Dark brown above, light brown below. The most notable feature of this species is the webbing between the toes, suggesting it is aquatic. The only two specimens known were found close to water at 5750–8900ft, in the Cordillera Central, Colombia. It was not described until 1978. ** **Black-striped Weasel** *Mustela strigidorsa* (**869**) is a fairly large weasel with a distinctive white stripe down the centre of the back and a yellowish-white stripe down the belly. Only known from 8 specimens, since first being described in 1853. These are scattered from Nepal eastwards through Burma, Yunnan (China), to Thailand and Laos. It has been found in evergreen forest at altitudes of 3950–7200ft. Status unknown.

* MARBLED POLECAT

Vormela peregusna (**870**)

HB 11.4–15ins; T 5.9–8.3ins; WT 12.9–25.2oz.

Distinctive marbled patterning on the back, it has a display reminiscent of a skunk in which the fur is erected, the tail arched over the back, and a foul-smelling secretion from the anal gland ejected. Mainly occurs in open steppe country where it excavates its own burrow. Found from E Europe to China, its open steppe habitat has been extensively ploughed in the past 150 years; in areas close to human habitation it is often trapped because of its depredations on domestic poultry, and for its fur, though of relatively little commercial value. Only occasionally seen in zoos and there do not appear to be any self-sustaining populations in captivity, although they breed fairly readily.

*** EUROPEAN MINK

Mustela lutreola (**871**)

HB 11–16.9ins; T 4.7–7.5ins; WT up to 1.6lb.

Similar to the more familiar American Mink *M.vison*, which escaped from ranches and is now established in the wild in many parts of Europe; distinguished by white on upper and lower lip (American has white only on lower lip). Found in aquatic habitats, they feed on mammals, birds, frogs and fish. Found in two widely separated populations, which probably divided within historic times. The range extends from E Europe and Finland east to W Siberia. Occurs also in W France, and NW Spain. Its range in N Europe was formerly more widespread, and in many parts of its E European range it has declined. Habitat changes are probably responsible in part, but in recent decades the spread of the American Mink is probably more significant, as is water pollution; *M.lutreola* is still trapped for its fur in the parts of its range where it is still reasonably abundant.

* SIBERIAN WEASEL

Mustela sibirica (**872**)

HB 9.8–15.4ins; T 5.1–8.3ins; WT 12.6–28.7oz.

Fur colour very variable, but generally yellowish-brown above, paler below. They are forest-dwellers, widely distributed from European Russia eastwards to Siberia, Japan, and south to Taiwan, Thailand and Java. They are important furbearers with large numbers trapped, particularly in USSR. The populations from Java and Sumatra, *M.s. lutreolina*, often treated as a full species, are only known from 11 specimens all collected at altitudes of above 3000ft, and are likely to be threatened. The ***Malaysian Weasel** *M. nudipes* (**873**) from Peninsular Malaya, Sumatra and Borneo, includes a distinctive subspecies, *M.n. hamakeri*, known from Java from a single specimen; its status is unknown.

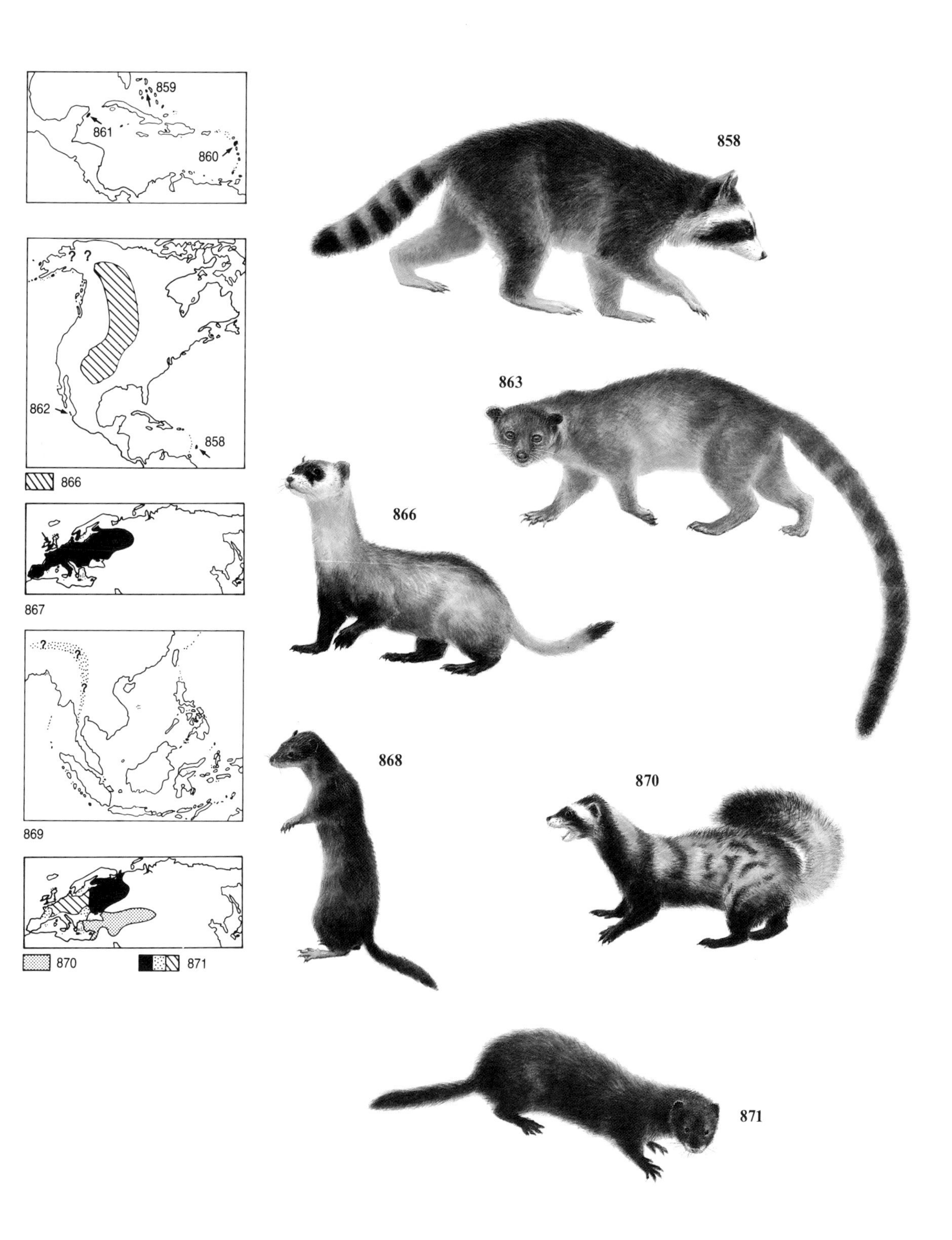
859
861
860
?
?
862
858
866
867
?
?
?
869
870
871
858
863
866
868
870
871

*** WOLVERINE or GLUTTON

Gulo gulo (**874**)

HB 25.6–43.3ins; T 6.7–10.2ins; WT 15.4–70.6lb.

Male larger than female. Long thick fur, generally dark brown with a lighter patch along the sides and over the rump. Found in taiga and tundra zones of the northern hemisphere, but used to be found in many other habitats. They feed on carrion, small mammals, berries, and in winter are able to hunt larger mammals such as deer and reindeer, when the Wolverine can run faster than its prey across soft snow. Have been hunted for their fur, which is used for trimming and lining, and extensively persecuted for alleged depredations on reindeer and other livestock. Will follow trappers and raid their traps. Their range once extended as far south as S Scandinavia and N Germany in Europe, and south to central California in North America. Now protected in most parts of their range, which is extending in some parts of the USA and Canada.

*** PATAGONIAN HOG-NOSED SKUNK

Conepatus humboldti (**875**)

HB up to 19.7ins; T up to 15.8ins; WT up to 9.9lb.

The typically skunk-like black and white pattern is rather variable, but the white line or spot on the face found in *Spilogale* or *Mephitis* skunks is always lacking. Mainly nocturnal, feeding on insects and other small animals, and small amounts of fruit. Occasionally hunted for their skins, but the main threat comes from changes in habitat. Occur in the Perito Moreno NP, Argentina, and probably in other parks and reserves. (The Argentinian populations, *C.h. castaneus* are sometimes treated as a full species.) The Patagonian Hog-nosed Skunk is listed on CITES App.II.

The * **Patagonian Weasel** *Lyncodon patagonicus* (**876**) is a little-known species found in the pampas of Argentina and S Chile.

* EURASIAN BADGER

Meles meles (**877**)

HB 22.1–35.5ins; T 4.6–7.9ins; WT 22–35.5lb.

Greyish with distinctive black and white face. The Eurasian Badger is very adaptable, living in a wide variety of habitats; despite centuries of persecution it survives throughout most of its range, but often in reduced numbers. Although popularly thought to be threatened, this is usually only a local problem. Extensively trapped and hunted in many parts of its range, both as an alleged pest and for its hair, which is used in brush manufacture, particularly men's shaving brushes. In the UK the species has been exterminated locally as part of a government programme to control Bovine TB. Protected in many countries, they also occur in a large number of national parks and reserves.

* JAPANESE MARTEN

Martes melampus (**878**)

HB 18.5–21.5ins; T 6.7–8.7ins; WT up to 3.3lb.

Similar in size to the Pine Marten (882) but its colouring is more sandy-brown, with a white neck patch. Very little has been recorded of its habits, which are presumed to be similar to other martens. Occurs on Honshu, Kyushu, Shikoku, Tsushima and has been introduced into Sado, Japan, and Korea. Little precise information on status, but has declined due to overhunting for its fur, and has also been poisoned by agricultural pesticides.

* YELLOW-THROATED MARTEN

Martes flavigula (**879**)

HB 17.7–25.6ins; T 14.6–17.7ins; WT 4.4–6.6lb.

The short, harsh fur is rather variable in colour, but usually dark brown or blackish above, pale brown below, with a bright yellow patch on the neck and chest. Live in a wide variety of forest habitats up to an altitude of 9000ft. They hunt by day and night, preying on a variety of animals up to the size of small deer, and can eat nectar and fruit. They sometimes hunt in family groups; the normal litter is 2–3. Found from the Himalayas and S Siberia south to the Malay Peninsula, Sumatra, Java and Borneo. Although the fur is of little commercial value, they are trapped for local use and also because they predate poultry. The subspecies in Taiwan *M.f. chrysospila* is thought to be threatened, and the status of *M.f. robinsoni* from mountains in Java needs investigation. The closely related * **Nilgiri Marten** *M. gwatkinsi* (**880**), sometimes considered to be conspecific with *M. flavigula*, is deep brown, with reddish foreparts, and found only in hill forests in the Nilgiri Hills of S India above 3000ft. The * **Sable** *M. zibellina* (**881**) originally from Siberia, west to eastern Scandinavia, was once reduced to a few scattered remnant populations, but a conservation and reintroduction programme in the USSR has enabled it to spread back over much of its former range, and it is now being trapped for fur once more. The * **Pine Marten** *M. martes* (**882**) originally found over much of the northern hemisphere, and some other martens, have often declined in areas close to human habitation, largely due to direct persecution. When protective measures are instigated they usually recolonise suitable habitat, albeit rather slowly.

* RATEL or HONEY BADGER

Mellivora capensis (**883**)

HB 23.6–30.3ins; T 5.9–11.8ins; SH 9.8–11.8ins; WT 15.4–28.7lb.

Greyish or pale yellowish above, contrasting markedly with black or dark brown underparts; occasionally uniformly dark. The hair is coarse and rather sparse, and the skin exceptionally tough, and very loose; no external ears. Considered exceptionally dangerous, capable of dealing with any other predator of comparable size or larger. They live in a wide variety of habitats, including forests, open rocky country, and wetlands. They feed on mainly small animals up to the size of young antelope, and grubs, fruit etc; particularly fond of honey, and in Africa have developed an association with the bird Black-throated Honeyguide *Indicator indicator*, which leads them to bees' nests. Its large range is over much of Africa, the Near and Middle East, to India and southern USSR. In many parts, particularly close to human habitation, it has become rare. It is hunted, poisoned and trapped because of its depredations on commercial bee hives. Protected in USSR, India, and many parts of Africa. Often exhibited in zoos, has been bred, though not self-sustaining in captivity.

FERRET BADGERS

Melogale spp.

HB 13–16.9ins; T 5.7–9.1ins; WT 2.2–6.6lb.

The most characteristic feature is the contrasting facial markings of black and white or yellowish. ** *M. everetti* (**884**) is only known from Mt Kinabalu, Borneo, at 3500–9850ft; the Javan populations of * *M. personata* (**885**) may be a distinct species and need further study; and the Hainan and Taiwan populations of * *M. moschata* (**886**) may be threatened.

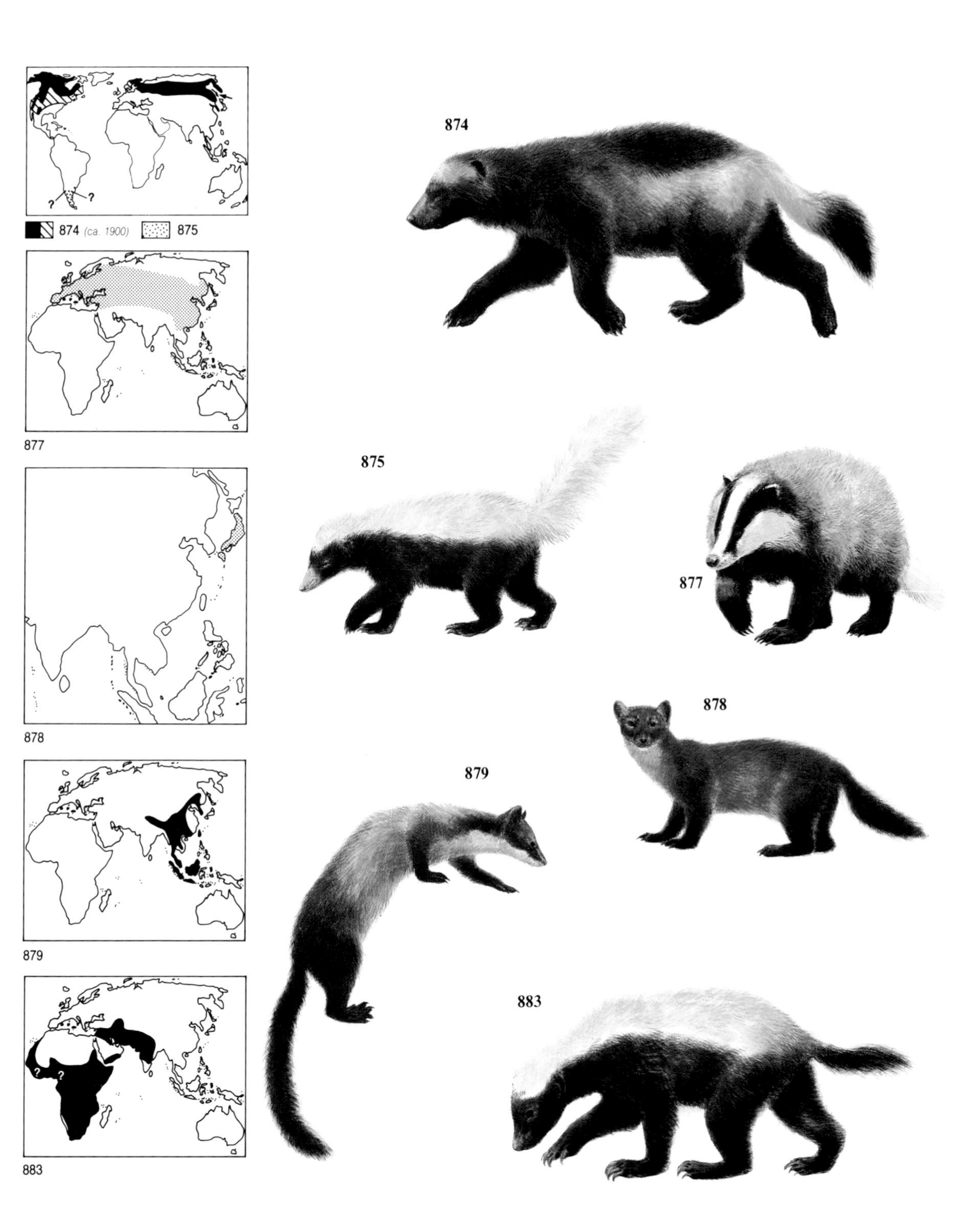
874 (ca. 1900)
875
?
?
877
878
879
883
874
875
877
878
879
883

*** EURASIAN OTTER

Lutra lutra (**887**)

HB 23.6–31.5ins; T 13.8–17.7ins; WT 8.8–28.7lb.

Brownish above, pale below with a long, tapering tail. Very agile swimmers, also travelling extensively overland. They occur in a wide variety of aquatic habitats, incuding coasts, rivers, and lakes, where they feed on a wide variety of animals up to the size of small birds, and water voles, but mostly fish such as eels and perch. They were formerly widespread and common over most of Europe and N Asia, and south to Java and S India. In the past century they have undergone massive declines in most of lowland Europe. In the more remote parts of the British Isles, Scandinavia, USSR, and elsewhere they remain widespread and often fairly numerous. About 10 subspecies are generally recognised. They are fully protected in most of their European range, and receive some protection in most other countries. The * **Hairy-nosed Otter** *L.sumatrana* (**888**) is still widespread and abundant over most of its range, but in some areas, such as Vietnam, is giving cause for concern.

* SPOT-NECKED OTTER

Lutra maculicollis (**889**)

Has a wide distribution over much of Africa south of the Sahara, is declining in many areas. The introduction of exotic fish into many African lakes has had an adverse effect; despite providing additional food for the otters, the nets used to catch fish destroy many of them.

*** NORTH AMERICAN OTTER

Lutra canadensis (**890**)

HB 35.1–51.2ins; T 11.8–20ins; WT 11–30lb.

Very similar to the Eurasian Otter, and most readily distinguished by its larger rhinarium (hairless part of the nose). Originally found in a wide variety of aquatic habitats from Alaska south to the southern USA, their range has contracted dramatically during the present century. Their decline is due to habitat destruction, water pollution, and unregulated trapping. In some states, such as Colorado, reintroduction programmes have been started.

*** SMOOTH-COATED OTTER

Lutra perspicillata (**891**)

HB 25.6–29.5ins; T 15.8–17.7ins; WT 15.4–24.2lb.

Similar to the Eurasian Otter, but with the tail tip flattened. Although widely distributed and often abundant in many parts of its range, isolated populations may be threatened. The population confined to the marshes of the Tigris, Iraq, *L.p.maxwelli* (named after Gavin Maxwell) is ca.1240 miles from any others. A single specimen was collected in Borneo last century; the status of the species in many parts of SE Asia remains unclear.

*** SOUTH AMERICAN RIVER OTTER

Lutra longicaudis (**892**)

HB ca.31.5ins; T ca.19.7ins; WT up to 30.9lb.

Occur in a wide variety of aquatic habitats from S Mexico to N Argentina. It is closely related to and may be conspecific with the North American Otter (890). Declines have been noted in several areas; in Argentina there has been a marked decrease since the 1930s and it is locally extinct in Uruguay. They are protected in many parts of their range and occur in a large number of national parks and protected areas.

*** SOUTHERN RIVER OTTER

Lutra provocax (**893**)

Confined to estuaries, lakes and rivers in C and S Chile and W Argentina. Virtually nothing is known of their status, but they are believed to have declined from overhunting for their fur. They are protected in Chile, and they occur in several protected areas in both countries, including the O'Higgins NP, Chile.They are listed on App.I of CITES.

*** MARINE OTTER or CHINGUNGO

Lutra felina (**894**)

The Marine Otter is mainly coastal in its distribution, occasionally going up rivers to hunt for freshwater prawns; it also feeds on a wide variety of other crustaceans, and to a lesser extent on fish. It formerly occurred along almost the entire coast of Peru and Chile from about 6°S to Cape Horn. Its range and population density is now considerably reduced. The main reasons for its decline include hunting for its fur, and persecution by fishermen for alleged damage to fisheries. It is protected in Peru and Chile, but the law is difficult to enforce in many of the more remote parts of its range. It occurs in several parks and reserves including Paracas NP, Peru, and on Chiloe Island, Chile. It is listed on App.I of CITES.

*** ZAIRE CLAWLESS or SWAMP OTTER

Aonyx congica (**895**)

HB 31.1–38.2ins; T 16.1–22.1ins; WT 30–75lb.

The least adapted of all otters for aquatic life, with short soft fur, poorly developed webbing on the hindfeet, and unwebbed forefeet. They are sometimes considered conspecific with the more widespread * **African Clawless Otter** *A.capensis* (**896**). Confined to rivers, lakes and swamps in the rainforests of Central Africa. In areas close to dense human settlement both species become rare; *A.capensis* is reportedly extinct in Niger. Although not generally exploited for international trade, they are often hunted for local use.

All otters not listed on CITES App.I are listed on CITES App II.

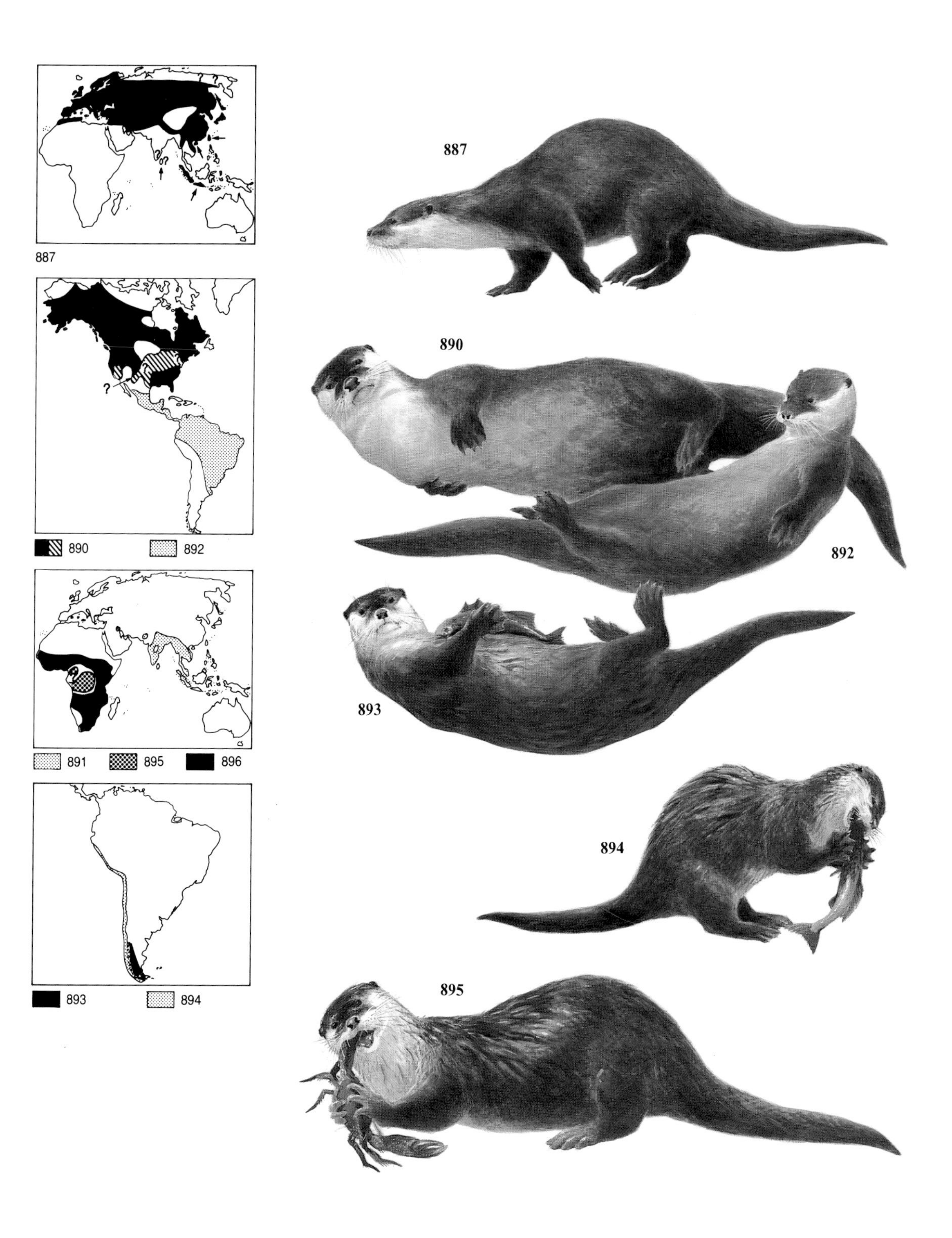
887
?
?
??
887
?
890
892
890
892
891
895
896
893
894
893
894
895

***** GIANT OTTER**

Pteronura brasiliensis **(897)**

HB 33.9–55.2ins; T 13–39.4ins; WT 48.5–75lb.

The short, velvety fur is brown when dry and blackish when wet; there are numerous creamy white blotches on the underparts. The feet are large with extensive webbing, and the muscular tail is flattened. They live in slow-moving rivers, lakes and marshes, often within forests, where they are diurnal, feeding mainly on fish and crabs. Groups usually consist of up to 20 (usually 4–8), and litters of 1–5 cubs are born in a den in a river bank burrow. Formerly found from Colombia and the Guianas south to Argentina and west to the foothills of the Andes, it is now reduced to scattered remnants within that area. The main reason for this decline has been hunting for the fur trade. Between 1950 and 1965 it is estimated that more than 7500 hides were exported from Brazil alone. In the 1970s protection was introduced in many parts of its range, but illegal trade continued. The Giant Otter occurs in some national parks and reserves, and is listed on CITES App.I.

***** SEA OTTER**

Enhydra lutris **(898)**

HB 39.4–47.3ins; T 9.8–14.6ins; WT 33.1–99.9lb.

Male is larger than female. The head is large and blunt, and the legs and tail rather short; the hindfeet are webbed and flipper-like; the only carnivore with only four incisors in lower jaw. Unlike most marine mammals, does not have blubber for insulation, but relies on dense, long, soft fur. Lives in coastal waters; rarely goes more than half a mile from land. The pup (1 or 2, but only 1 normally survives) is born at sea, but does not dive until about a month old. Sea Otters feed on fish, crabs and molluscs, which they eat while lying on their backs; they are one of the few tool-using mammals, using rocks to break open molluscs. They once occurred from Hokkaido, Sakhalin and Kamchatka eastwards across the Bering Straits and south along the coast of North America to Baja California. From the 17th century until 1911, they were hunted for pelts which were probably the most valuable of all mammals'. In 1911 the world population was estimated at 1000–2000. Under protection its numbers have slowly increased, helped by reintroduction programmes. The **Southern Sea Otter** *E.l.nereis* was believed to have become extinct in 1920, but ca.100 were discovered in 1938, and this population has grown to nearly 2000. The Sea Otter occurs in many reserves and protected areas and has legal protection in all its range states. *E.l.nereis* is listed on CITES App.I; other populations on App.II.

***** ORIENTAL or LARGE SPOTTED CIVET**

Viverra megaspila **(899)**

HB 28.4–33.5ins; T 11.8–14.6ins; WT 17.6–19.8lb.

Generally grey with small black spots, and 5 white rings on the tail. Feeds on small animals, eggs and some vegetable matter, and is found in a variety of mainly wooded and forested habitats, often near to human habitations.The civet used in perfumery is derived from *Viverra* spp. In the eastern part of its range it is often fairly common, but the Indian population *V.m.civettina* is extremely rare, if not already extinct; last seen 1974. It has been persecuted for its raids on poultry, and lost much of its habitat to agriculture. Protected in India. The more widespread and abundant *** Large Indian Civet** *V.zibetha* **(900)** is apparently rapidly declining in Bangladesh. The *** African Civet** *Civettictis civetta* **(901)** is exploited for its scent and is locally declining. Fifty years ago nearly 2.8 tons were produced a year; each animal produces 7oz of the secretions a year. Status unknown or poorly documented.

*** OTTER CIVET**

Cynogale bennettii **(902)**

HB 22.6–26.8ins; T 5.1–8.1ins; WT 6.6–11lb.

The Otter Civet is rather otter-like, with thick, short fur, and numerous exceptionally long whiskers. The nostrils and ears can be closed, and the feet are webbed. Always found close to water, but also climb well. Feed mostly on crustaceans, molluscs, fish, birds, small mammals and some fruit. Two or three young are born each year but little is known of their breeding or other habits. Occurs in the Malay Peninsula, Sumatra and Borneo. A single specimen found in Northern Vietnam is sometimes considered to be a separate species, ** *C.lowei* **(903)**. Although widespread it is rare throughout most of its range and is listed on CITES App.II.

**** AQUATIC GENET or CIVET**

Osbornictis piscivora **(904)**

HB ca.17.5ins; T ca.13.4ins; WT ca.11lb.

The long, dense fur is reddish or chestnut, blackish on the tail. The sides of the face are whitish, and there are two white spots between the eyes. Known from very few specimens, all collected in dense forest at 1650–4900ft close to streams. Its range includes the Semliki Forest, NE Zaire and the shores of Lake Victoria. Virtually nothing is known of its habits, biology or status.

*** AFRICAN LINSANG**

Poiana richardsoni **(905)**

HB up to 15ins; T up to 15.8ins; WT up to 1.5lb.

Generally sandy-coloured, with dark blotches above and dark banded tail. Occurs in forests from Sierra Leone east to N Zaire. It is poorly known and, particularly in the west of its range, its status needs investigation.

***** FOSSA**

Cryptoprocta ferox **(906)**

HB 23.6–31.5ins; T 23.6–31.5ins; WT 15.4–26.5lb.

The largest native carnivore in Madagascar, with short, thick, reddish-brown or blackish fur. Found in woodland and more open country, they are good climbers and feed on lemurs and other small mammals, birds, amphibians, reptiles and some insects. They occasionally raid poultry; they have been extensively hunted and are now rare in most parts of their former range, particularly near to human habitation. There are small numbers in captivity, and they have occasionally been bred. Listed on CITES App.II.

*** SMALL SPOTTED GENET**

Genetta genetta **(907)**

HB 18.5–23.6ins; T 15.8–18.9ins; WT 4.4–5.3lb.

One of the most widespread of the genets, found over much of Africa, SW Europe, and the Middle East. Within this range its abundance is very variable, and it is often hunted for its fur, and also because of its raids on poultry. Its distribution is patchy in southern Africa and the published accounts are confusing as to its precise distribution. In Europe, where it was probably introduced, although sometimes considered rare, it appears to be spreading its range.

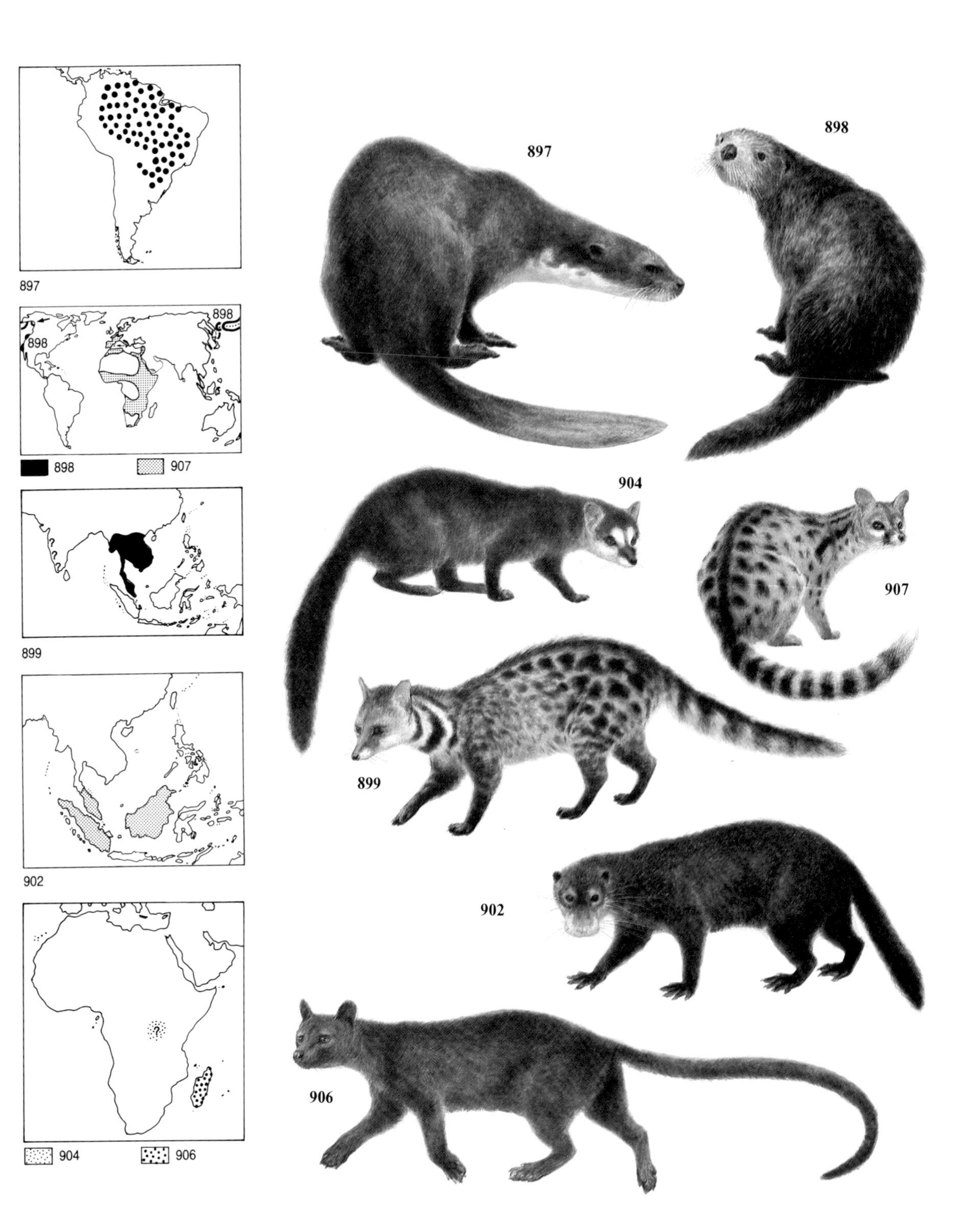
897
898
898
898
898
907
899
902
904
906
?
897
898
904
907
899
902
906

**** OWSTON'S PALM CIVET**
Chrotogale owstoni **(908)**
HB 19.7–25ins; T15–18.9ins.
It is marked with broad, dark bands on the back and base of the tail and spotted on the sides and limbs; the neck has longitudinal stripes. A unique feature is the broad close-set incisor teeth. Very few specimens have been found, from Laos, Tonkin, N Vietnam and Yunnan. Nothing is known of its status nor of the possible threats to its survival.

*** BANDED PALM CIVET**
Hemigalus derbyanus **(909)**
HB 16.1–20.3ins; T 10–15.1ins; WT 2.2–6.6lb.
The back and base of tail are banded with broad dark bands against a whitish or buff background. The hairs on the neck region point forwards. Only found in tall forests from S Burma, south through the Malay Peninsula to Borneo, Sumatra and possibly the Mentawai Islands. Although widespread, it is rare throughout its range and is likely to be declining in areas where forest is being cleared.

**** HOSE'S CIVET**
Hemigalus hosei **(910)**
HB 18.5–21.3ins; T 11.8–13.2ins.
Dark brown above, whitish below, with feet partly webbed. Little is known of habits, but webbed feet suggest it may be associated with streams. Only a few observations from mountainous regions of NW Borneo.

*** MALAGASY CIVET**
Fossa fossa **(911)**
HB 15.8–17.7ins; T 8.3–9.1ins; WT up to 4.4lb.
Generally greyish with a reddish tinge, and rows of dark spots on the back and thighs, which sometimes merge to form lines. The tail is banded with brown. Found in forests, hiding in hollow trees and rock clefts; normally nocturnal. Frogs, crustaceans, fish and other small animals, as well as some fruit, form the main part of their diet. They live in pairs and indicate their territory with calls. A single young is born. The Malagasy Civet is becoming rare through habitat loss, as well as overhunting, and is now restricted to the east and northwest forested areas of Madagascar. Occasionally kept in captivity and listed on CITES App.II.

***** FALANOUC**
Eupleres goudotii **(912)**
HB 17.7–25.6ins; T 8.7–9.8ins; WT 4.4–10.1lb.
The soft, dense reddish-brown fur is paler below. Found in lowland forests of Madagascar, where they hide by day in rock clefts, hollow trees and burrows. They feed, mostly at dusk or night, on earthworms, and other invertebrates and amphibians. They live in small groups or alone, and the usual litter is 1–2 young. Forest habitat is declining and the 2 subspecies, *E.g.goudotii* from the east and the larger *E.g.major* from the north are both declining. Listed on CITES App.II. Two other mongooses from Madagascar may also be threatened, though little is known of their status: **** Broad-striped Mongoose** *Galidictis fasciata* **(913)** from the forests of E Madagascar, and the **** Brown-tailed Mongoose** *Salanoia concolor* **(914)** from the NE of the central plateau of Madagascar.

**** HOSE'S MONGOOSE**
Herpestes hosei **(915)**
A rather reddish-brown, short bushy-tailed Mongoose, known from one specimen from Baram, Sarawak, Borneo, collected in 1893. The eastern subspecies of the **** Snouted Mongoose** *Herpestes naso microdon* **(916)** has only been found in the forests of E Zaire; nothing is known of its status. The status of * *Herpestes smithi* **(917)** from India and Sri Lanka is poorly known; since it is believed to be confined to forest habitats, it is likely to be threatened, particularly *H.s.thysanurus*, which occurs in lowland forests. Other Asian mongooses that may be threatened include: * *H.vitticollis inornatus* **(918)** from SW India; * *H.brachyurus* **(919)** from the Malay Peninsula, south to Borneo, Sumatra, Calamian Islands and Palawan.

**** LIBERIAN MONGOOSE**
Liberiictis kuhni **(920)**
HB 16.5ins; T 7.9ins; WT 5lb.
Predominantly dark brown, with a dark stripe on the neck, bordered by paler stripes. They have rather long claws and a long snout with small teeth, which suggests that they are mainly insectivorous. One of the few seen alive was found in a burrow close to a termite nest. Has been found in closed forest in the Upper Cess River, NE Liberia. The status of the Liberian Mongoose is uncertain; at the time of its discovery, as recently as 1958, it was eaten by human hunters, and since then very few have been encountered.

***** SPOTTED LINSANG**
Prionodon pardicolor **(921)**
HB 13.8–14.6ins; T 12.2–13.4ins; WT 8.8–11lb.
Unlike other linsangs, the spots are always separated and do not merge. They occur in montane and hill forests where they are often arboreal, feeding on birds; they also eat mammals and insects. Little is known of their precise distribution and status, but they appear to be declining through loss of habitat in India, and are noted as very rare in Thailand. The status of several populations on islands need investigation.

*** SULAWESI PALM CIVET**
Macrogalidia musschenbroeki **(922)**
HB ca.39.5ins; T ca.23.6ins.
The upperparts are brownish, with brown spots or bands; whitish underparts with a reddish breast. They have been recorded from montane forests as well as scrub and lowland forest, where they feed on small rodents and fruit; they have also been killed while raiding poultry. Although once considered threatened or even extinct, they are now believed to occur in many parts of Sulawesi, and are relatively abundant in some areas.

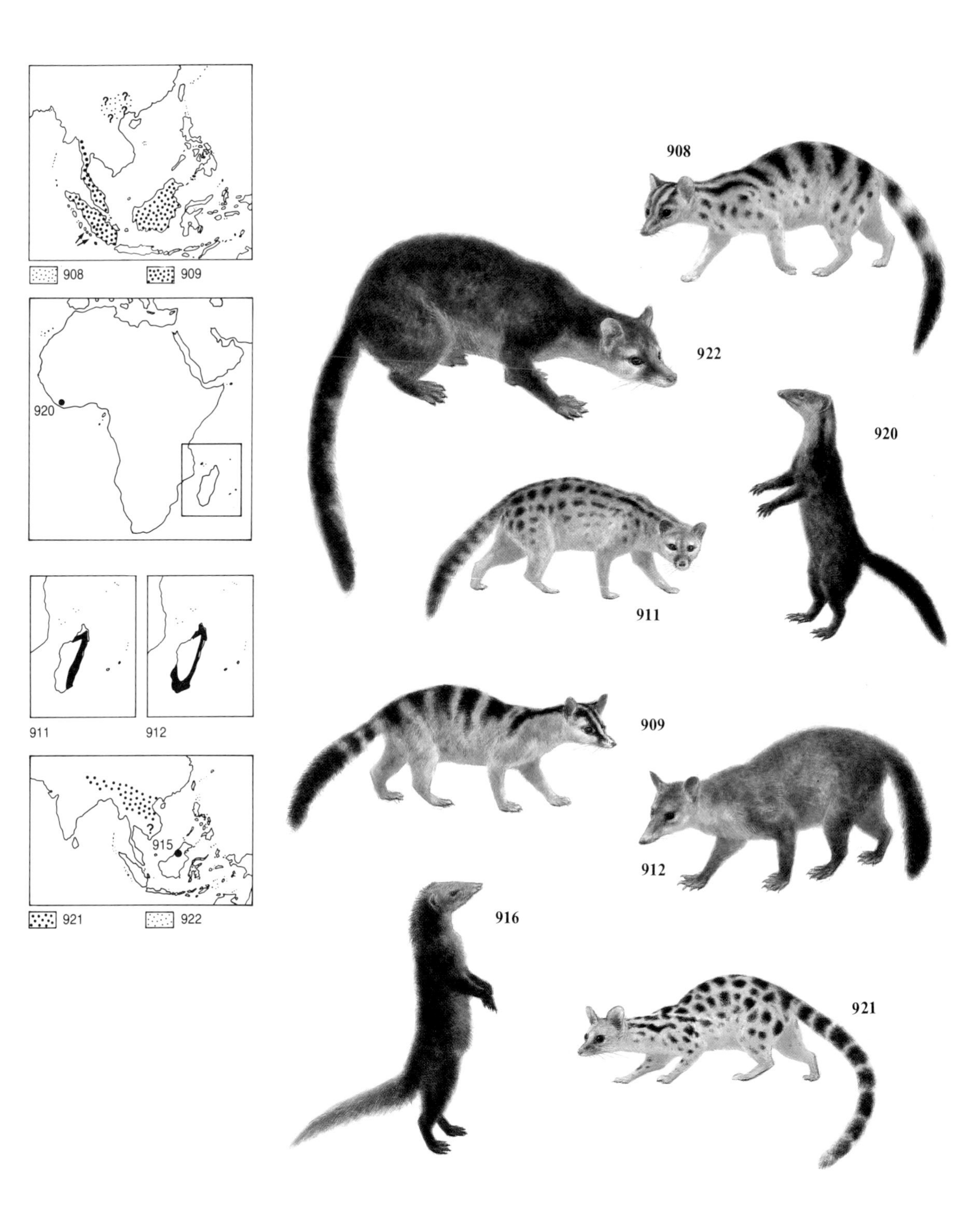
908
909
920
911
912
915
921
922
908
922
920
911
909
912
916
921

* THREE-STRIPED PALM CIVET

Arctogalidia trivirgata (**923**)

HB 16.9–20.9ins; T 20.1–26ins; WT 4.4–5.5lb.

Have three dark stripes along the back and a white stripe on the nose. Nocturnal and arboreal, they live in primary forests far from villages and other human habitations. Their range extends from NE India south and east to Peninsular Malaysia, Sumatra, Java and Borneo. In Thailand, and possibly elsewhere, numbers are thought to be declining through loss of habitat. Little is known of their precise status or distribution, particularly on the smaller islands of SE Asia.

* PALM CIVETS

Paradoxurus spp.

There is considerable variation in the palm civets and their classification is unclear. They occur on a large number of small islands throughout SE Asia, and although usually classified as subspecies of *P.hermaphroditus* (**924**), some may prove to be distinct species; virtually nothing is known of their status.

* AARDWOLF

Proteles cristatus (**925**)

HB 21.7–31.5ins; T 7.9–11.8ins; SH 17.7–19.7ins; WT 19.8–30.9lb.

Fur is long and soft, with relatively few guard hairs. General colour is yellowish-grey with black stripes. The bushy tail is black-tipped; along the back is a stiff crest. Found mainly in open country, usually making dens in an old Aardvark burrow. Mostly nocturnal feeding, almost exclusively on termites and other insects. Generally solitary but occasionally seen in small family groups, with litters of 1–5 young (usually 2–4). Although still widespread, occurring over much of eastern and southern Africa, in many areas they are thought to have declined due to hunting pressure, and loss of habitat. Transfixed by headlamps, they are also often killed by vehicles. Occur in many national parks and game reserves within their range. Aardwolves are sometimes kept in zoos, but although often long-lived, rarely breed.

* SPOTTED HYAENA

Crocuta crocuta (**926**)

HB 37–65ins; T 9.8–14.2ins; SH 27.6–36.2ins; WT 88.2–189.6lb.

Generally yellowish, with round brownish spots. The hair is coarse, and there is no mane. Females are usually longer than males but it is almost impossible to distinguish the sexes, as the external genitalia of the females are so like those of the males. Originally found in nearly all the more open habitats of Africa, south of the Sahara, at altitudes of up to 13,100ft. They feed mostly on medium-sized ungulates and zebras, and may hunt in packs of up to 25 animals. They are able to swallow prey and carrion very quickly and digest almost all the parts of an animal, devouring nearly 33lb at one time. Clans of up to 80 animals have their own den site, where each mother has a litter of 1–3 cubs. With the spread of agriculture, and in particular cattle and sheep ranching in southern Africa, Spotted Hyaenas have been exterminated from a large part of their former range. They have been hunted, poisoned and trapped, and earlier this century were often killed as vermin in game reserves. Most of the larger reserves and parks in suitable habitat in sub-Saharan Africa contain Spotted Hyaenas but, like many other larger predators in the more densely populated areas, they are unlikely to survive outside. Kept in many larger zoos, and regularly bred.

*** BROWN HYAENA

Hyaena brunnea (**927**)

HB 39–53ins; T 7.1–10.6ins; SH 25.2–34.7ins; WT 81.6–104.7lb.

Male slightly larger than female, and both sexes are generally dark brownish or blackish-brown, with the longer dirty-yellowish hairs on the upperparts. The fur is long and shaggy. Brown Hyaenas are usually nocturnal, living mainly in the arid areas of southern Africa. Found in desert areas as well as scrub and open woodland. Unlike the Spotted and Striped Hyaenas, they are not dependent on water supplies, relying on the water content of their prey, and also tsamma melons *Citrellus lanatus* and gemsbok cucumbers *Acanthosicyos nandianus*. They are mostly solitary hunters or live in small family groups, rarely killing large prey, but scavenging kills of their predators, killing small animals, taking birds' eggs, insects, etc. Over much of their range they have been persecuted for alleged killing of domestic livestock; they are also killed in traps set for jackals and Spotted Hyaenas. As a result of persecution, and also direct competition with Spotted Hyaenas in some areas, Brown Hyaenas are now extinct in much of southern Africa. Still widespread in Botswana and Namibia and relatively abundant in the Kalahari and Gemsbok national parks. Exhibited in some of the larger zoos and bred fairly regularly.

*** STRIPED HYAENA

Hyaena hyaena (**928**)

HB 36–47ins; T 10.2–18.5ins; SH 23.6–37.4ins; WT 55–121lb.

Both sexes are similar in size and colour. The background colour is greyish or pale brown with dark brown stripes on the body and legs. Along the spine is a mane of stiff hairs, up to 7.9ins long, the rest of the fur being rather long and shaggy. They live in open, often rocky country, up to an altitude of 10,800ft. Rarely found more than ca. 6 miles from fresh water, avoiding desert areas. They feed on mammals and carrion, up to the size of donkeys and antelopes, and also smaller vertebrates, insects and sometimes fruit; in some areas they are destructive to crops such as melons, grapes and peaches. There are 1–5 young in a litter. The Striped Hyaena was once widespread throughout N Africa, around the Sahara and as far south as Tanzania, and eastwards through the Levant, Arabia and the Middle East, as far east as India and Soviet Central Asia. In many parts of its range it is very much reduced in numbers and has disappeared from most areas close to human habitation. In N Africa it is confined to the montane areas of Morocco, Algeria and Tunisia. Occurs in relatively few national parks and reserves. Some of the larger zoos exhibit them, and small numbers are bred.

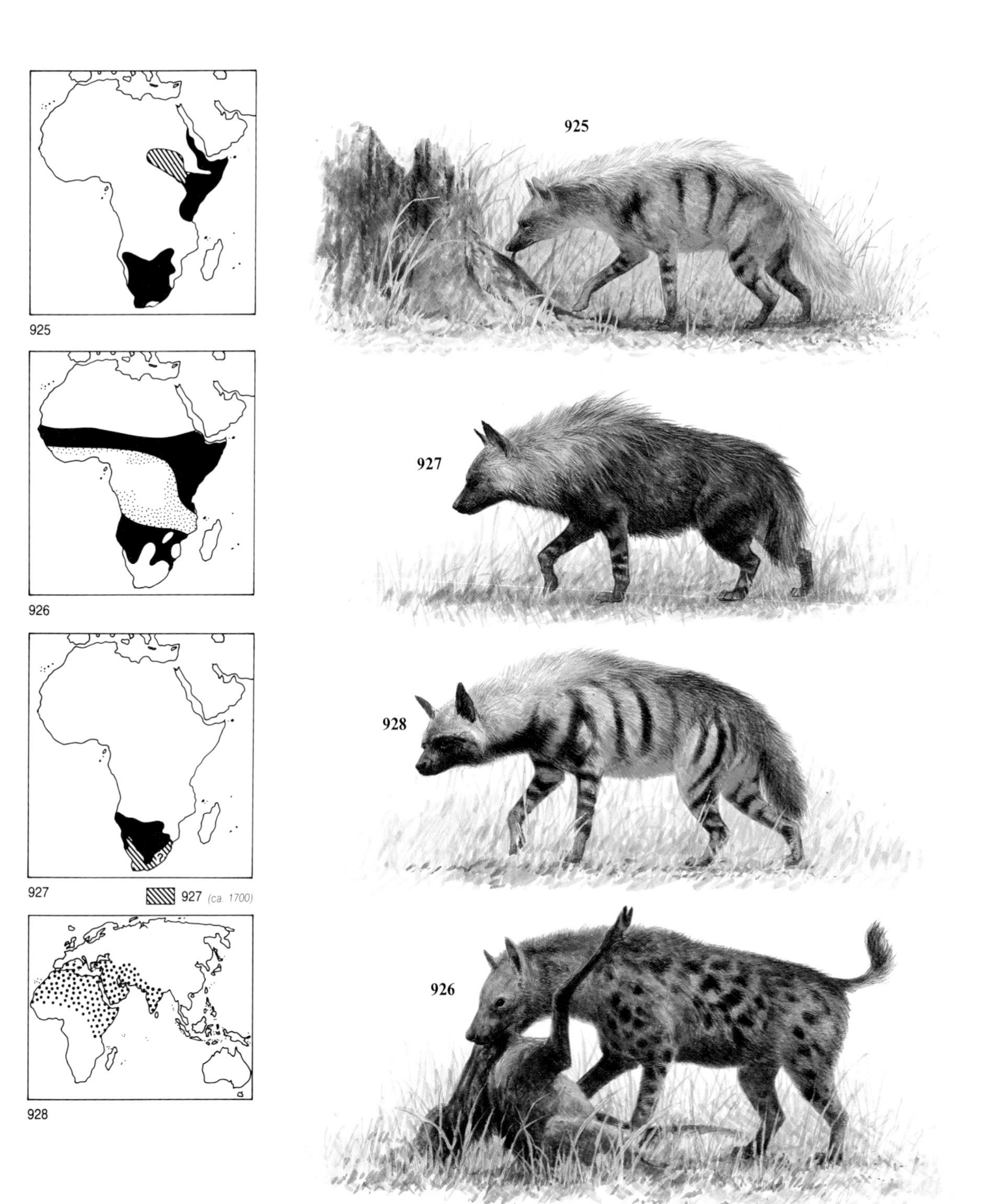
925
927
928
926
925
926
927
927 (ca. 1700)
928

*** LION

Panthera leo (**929**)

HB up to 98.5ins; T up to 41.5ins; SH ca.79ins; WT up to 551lb.

Female smaller than male. Some geographical variation, particularly in the extent and colouring of the mane of male. They live in groups (prides) based around related females and their young; males not defending a pride form batchelor prides or live alone; the average size of a pride is 15, max ca.40. There are 1–6 young in a litter. Lions often hunt as a group, feeding mostly on large mammals, and are also scavengers of hyaena and Hunting Dog kills; occasionally individuals or small groups take to preying on humans. Was one of the most widespread of all mammals, occurring at one time over most of the northern hemisphere and also parts of South America. Disappeared from the New World at the close of the Pleistocene, and from most of Europe with the spread of forests, but survived in SE Europe until around 2000 years ago. The decline of the Lion in Asia and Africa has been continuous within historic times and in the 20th century it has become extinct in the Middle and Near East, North and South Africa, most of Asia and West Africa. (In Asia *P.l.persica* only survives in the Forest of Gir in India, where it is at or near carrying capacity, and is artificially managed.) In East and Central Africa *P.l.leo* is becoming increasingly rare outside parks and reserves; it is unlikely that it will survive elsewhere by the end of the present century. In 1975 it was estimated that the Lion population of Africa had dropped from ca.400,000 in 1950 to 200,000, i.e. to ca.13,000 prides. Lions are common in zoos; they breed freely – and may actually produce a surplus. Listed on CITES App.II (*P.l.persica* on App.I), and protected in most countries where they still occur; they are still hunted in many areas.

*** LEOPARD

Panthera pardus (**930**)

HB 35–75ins; T 23–43ins; SH 17.7–30.7ins; WT 61.7–198lb.

Male generally larger than female. The colouring varies from greyish or pale yellow to rich buff or chestnut, with black spots which form rosettes over much of the body. Melanistic (black) leopards are common. In the colder, more northern parts of their range, the fur is generally longer and silkier. They are among the most adaptable and widespread of living mammals, occurring in forests, deserts, open grasslands, and even close to towns. They are normally but not always nocturnal, feeding mostly on mammals, such as antelope, deer, monkeys, pigs, and domestic livestock; occasionally they have become man-eaters. Leopards were formerly found over almost all of Africa, throughout the Near and Middle East, and S Asia, south to Sri Lanka, Malay Peninsula, Java and Kangean Islands. However they have been persecuted for centuries and are now extinct in most of N Africa, the Middle and Near East, and reduced to scattered populations over most of Asia and West Africa; in parts of Central and East Africa they remain fairly widespread and locally abundant. Several of the subspecies that have been described are particularly rare, including *P.p.orientalis* from Siberia, Manchuria and Korea; *P.p.nimr* from Arabia, north to Jordan and Israel; *P.p.tulliana* from Asia Minor; *P.p.jarvisi* from the Sinai Peninsula; and *P.p.panthera* from NW Africa. Protected in most of its range, and occurs in many national parks and reserves. Listed on CITES App.I. Large numbers are kept in zoos, and many of the rarer subspecies are being bred regularly.

*** TIGER

Panthera tigris (**931**)

HB 55–110ins; T 23.5–37.5ins; SH up to 43ins; WT 143.3–661.5lb.

The largest of the cats, with considerable variation in size; the smallest come from Bali, the largest from Siberia. Over their wide range, Tigers occur in many different habitats, including tropical and montane forests, mangrove swamps, arid grasslands and savannah, and rocky semi-desert. They are mainly nocturnal, hunting mostly medium to large mammals, including man. The tiger is one of the few animals that has regularly preyed on humans, and which is still a problem in some parts of their range. Densities of 1 per 14sq.mls have been recorded in India, with a home range of 20–385sq.mls; in Siberia ranges of up to 4050sq.mls have been recorded. They are normally solitary except when breeding; the litter is normally 1–6 young. Their present range is fragmented and reduced to scattered isolated populations; they are extinct in most of their former range. The main reason for this decline in the past has been hunting by man, as they have always been a threat to both humans and domestic livestock. Also a valued trophy animal, and skins have been extensively traded. Currently their main threat is probably destruction of the remainder of habitat. The Tiger is protected throughout most of its present range, and is listed on CITES App.I (except the Siberian, *P.t.altaica*, App.II). Most of its surviving populations are in national parks, and it is unlikely that many will survive outside parks and reserves. They breed freely in zoos, but the ancestry of the majority is mixed.

*** SNOW LEOPARD or OUNCE

Panthera uncia (**932**)

HB 39.5–51ins; T 31.5–39.5ins; SH 23.5ins; WT 55.1–165.4lb.

The fur is exceptionally thick and long; the colour varies from grey to cream above, with dark grey markings, and white below. Confined to montane areas of Afghanistan, east to Siberia and Tibet, at altitudes of 8850–19,700ft. There are 1–5 (usually 2 or 3) in a litter. The main cause of decline is hunting for its pelt which is highly prized in the fur trade. Although protected in most of its range, illegal trapping and hunting continued, and in 1985 furs were still being offered for sale in Kashmir. The displacement of its prey species (mountain goats, sheep, deer, boar etc) from alpine pastures by domestic livestock is also contributing to its decline in some areas. Exhibited in some zoos and breeding regularly. Listed on CITES App. I.

929
930
931
932
P. l. persica
929
♀
P. l. leo
P. l. leo
♂
P. pardus
P. pardus (melanistic)
930
P. t. altaica
931
P. t. sondaica
P. uncia
932

*** JAGUAR

Panthera onca (**933**)

HB 44–61ins; T 17.7–29.5ins; WT 79.4–348.4lb.

Colour varies from pale yellow to reddish-brown with black spots, which form rosettes; spots may merge to form lines and melanistic (black) animals are common, but the spots can usually be seen. Preferred habitats include forest and savannah, but also found in more arid areas and even deserts, though seldom far from fresh water. Good swimmers, and climb well. They feed on animals up to the size of peccaries, capybara, caiman and tapir. Jaguars are normally solitary, marking their territories with urine. They have 1–4 young. At the turn of the century they occurred as far north as southern USA, but now extinct or very rare in most of C America, and also in extreme south of range. Still widely distributed over most of South America, but locally extinct or endangered, and almost everywhere have declined. Persecuted as predators with the spread of cattle ranching, they have also been hunted for sport, and skins once commanded high prices in fur trade. Protected in most countries of range, listed on CITES App.I and found in most of the larger national parks and reserves in South America. Exhibited in larger zoos where they breed freely.

*** MARGAY or TREE OCELOT

Felis wiedii (**934**)

HB 18.1–31.1ins; T 13–20.1ins; WT up to 22lb.

Similar to Ocelot, but smaller and slimmer with proportionally longer tail. Found almost exclusively in forests in Mexico and C America, south throughout most of South America to Paraguay and N Argentina, but are now very rare or extinct in many parts of that range. They are arboreal and mainly nocturnal, feeding on birds and small mammals. Litters usually consist of 1 or 2 young. They have also been extensively hunted for pet trade and fur industry. Listed on CITES App.II with two C American subspecies (*F.w. salvinia* and *F.w. nicaraguae*) on App.I; protected in most countries of origin and occur in many national parks and reserves. Margays are common in captivity and regularly bred, but few of the captive populations are identified to subspecies.

*** JAGUARUNDI

Felis yagouaroundi (**935**)

HB 21.7–30.3ins; T 13–26ins; WT 9.9–19.8lb.

Jaguarundis have two colour phases: dark brownish-grey, and reddish-chestnut. They are slender, weasel-like cats, with rather short legs and a long tail. Found mainly in dense lowland forests, where they are less strictly nocturnal than other cats. Mainly terrestrial, they feed on small mammals and birds. They live in pairs or are solitary, with 1–4 in a litter. They are widely distributed from southern USA through C America, and over most of S America, but have declined markedly in many areas.

*** GEOFFROY'S CAT

Felis geoffroyi (**936**)

HB 17.7–27.6ins; T 10.2–13.8ins; WT up to 17.6lb.

Vary from greyish to bright ochre, with fine black spotting, and rings on tail; melanism is not uncommon. Live in open woodland and scrub up to an altitude of 10,850ft. They are nocturnal, feeding on small mammals, reptiles, fish etc. There are 2–3 young in a litter. Occur in Bolivia and SE Brazil, south to Patagonia. Like most other South American cats they have been heavily exploited by the fur trade, and are listed on CITES App.II. Found in most of the larger national parks within their range, and, although much depleted, are still the most abundant cat in most parts of range.

*** MOUNTAIN CAT

Felis jacobita (**937**)

HB up to 23.6ins; T up to 13.8ins; WT up to 8.8lb.

Mountain cats are variable in colour, but usually silvery grey with brown or yellowish spots or stripes. They are little-known, found mainly in dry habitats up to 1650ft, feeding on rodents, including viscachas and chinchillas (see p.128). Little is known about numbers or population densities, but rare throughout most of range. Listed on CITES App.I, protected in Peru. May occur in Lanca NP, Chile and in Pampa Galeras NP, Peru. Rare in captivity, and not recorded as having ever been bred.

*** OCELOT

Felis pardalis (**938**)

HB 21.7–39.4ins; T 11.8–17.7ins; WT up to 34.8lb.

Variable coloration, ranging from yellowish or whitish to tawny or greyish, with dark spots and stripes. Found in a wide range of habitats, including arid semi-deserts and tropical rain forests. Generally nocturnal, and arboreal, feeding on small birds, rodents, reptiles and other animals up to the size of small deer. Although they still occur over much of their original range, in the north they have disappeared from most of Texas, are extremely rare in Mexico, and have undergone massive declines in many other parts of their range. In addition to loss of habitat they have been extensively exploited for the fur trade; in the late 1960s and early 1970s several hundred thousand a year were involved in international trade. Now protected in most parts of their range and listed on CITES App.II (with some populations on App.I). They occur in many national parks and reserves throughout their range, and breed freely in captivity.

*** COUGAR, PUMA or MOUNTAIN LION

Felis concolor (**939**)

HB 39–75ins; T 20.9–36.2ins; WT up to 275lb.

Have two main colour phases: buff to brownish and greyish to slate-coloured. They vary considerably in size, the smallest occurring in the tropics. The most widely distributed mammal in the New World (except man), once found in almost all habitats except the completely barren. They feed on animals up to the size of deer. Generally solitary, their range may extend to 250sq.mls, depending on habitat. They have been hunted extensively since the arrival of European settlers, and regarded as a threat to livestock. Extinct or rare over most of their North American range, and considerably reduced in almost all other parts, where they are close to human settlements. In USA isolated populations occur in Florida and Texas. Survive in a large number of national parks and reserves, throughout their range, and are protected (at least partially) over most of their range; listed on CITES App.II (with some populations on App.I). They are common in zoos and breed freely. Claims have been made for the existence of the **Onza** – a cheetah-like cat – in Mexico, but these probably refer to Pumas.

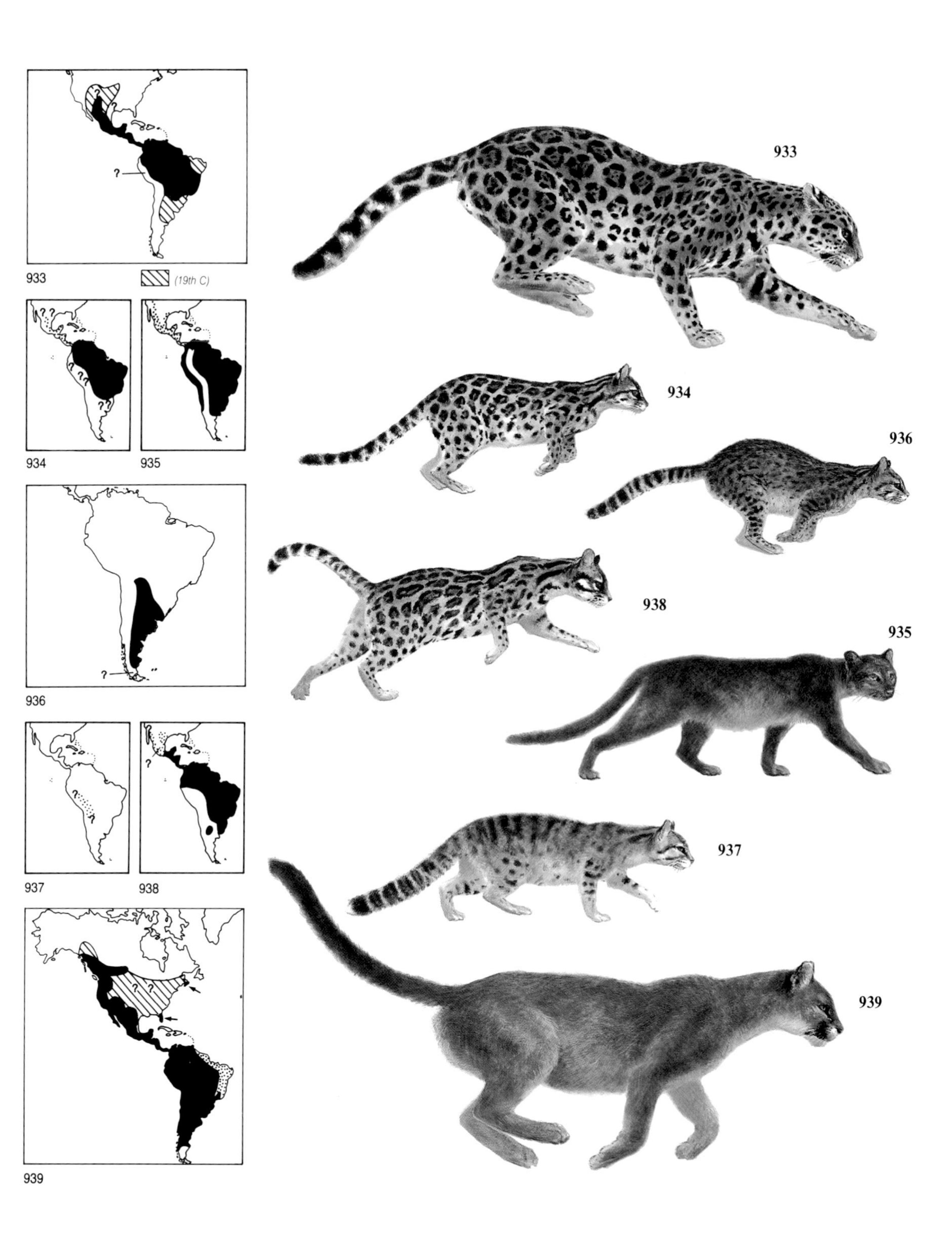
933
(19th C)
934
935
936
937
938
939
933
934
936
938
935
937
939

*** CHEETAH

Acinonyx jubatus (**940**)

HB 43.3–59.1ins; T 23.6–35.5ins; SH 27.6–37ins; WT 77.2–158.8lb.

Generally yellowish or buff, marked with black spots; very occasionally the spots join together to form longitudinal stripes – once regarded as a separate species, the King Cheetah *A. rex*, but now known to be a variety which occurs in a small area of Zimbabwe. The Cheetah is the fastest land mammal, hunting gazelles and other medium sized mammals by running them down at speeds of over 39 m.p.h. Once found in open habitats throughout Africa, the Near and Middle East, as far as southern USSR and NW India. Its range has been contracting for several hundred years; it is hunted as a predator on domestic livestock, particularly sheep and goats, for its skins, and is captured as a live animal for training as a 'Hunting Leopard', or for exhibition in zoos. Now extinct over most of the northern parts of its range. In Asia confined to reserves in the southern USSR and adjacent Iran, and possibly Pakistan. It became extinct in Arabia around 1950; in Africa north of the Sahara it is reduced to a few populations which may survive in Libya; in West Africa extinct over most of its range, and in East Africa mostly confined to national parks and reserves. It main stronghold is in southern Africa, particularly Namibia, where it is still fairly abundant. Kept on many private ranches in South Africa. Listed on CITES App.I, and the African Convention, and protected in most countries where it still occurs. They have been tamed for hunting for over 4000 years; commonly exhibited in zoos and bred in significant numbers.

*** SERVAL

Felis serval (**941**)

HB 26.4–39.4ins; T 9.5–17.7ins; SH 21.3–24.4ins; WT 19–39.7lb.

A long-legged, slender cat with dense spotting on a tawny or yellowish background colouring. There is considerable variation in the spotting, and in W Africa a darker form with small, dense spots was at one time regarded as a separate species, the **Servaline Cat** *F.servalina*. They live in open habitats, feeding on mammals up to the size of small antelope and birds up to the size of guinea fowl, as well as smaller prey. They are generally solitary, with a litter of 2–3 young. Formerly widespread and abundant in open grassy habitats throughout most of Africa, except in the very arid regions; they also occur in reed-beds and open woodlands and forests. Although still common in parts of their range they have disappeared from many areas particularly in the north and west, and extreme south. They are persecuted for their depradations on domestic fowl, and have also been exploited for fur. Listed on CITES App.II, and the African Convention; protected in some countries within their range and occurring in most national parks and reserves, particularly in E Africa.

*** LYNX

Felis lynx (**942**)

HB 31.5–51ins; T 2–9.8ins; SH 23.6–29.5ins; WT 11.2–83.8lb.

Extremely variable in size and coloration, but usually yellowish or buff with dark spots, and paler below. The fur is generally long and silky. Lynx are usually found in forested areas and feed on a wide variety of small to medium-sized mammals and birds, including hares and rabbits. Once widespread throughout most of mainland Europe, N Asia, the Near East and North America as far south as southern USA, their range is now considerably fragmented and it is extinct in most of the areas where human populations are widespread. The N American population *F.l.canadensis* is sometimes treated as a full species. The isolated Spanish population *F.l.pardina*, also sometimes treated as a full species, is one of the most threatened, with a population of 1000–1500; most of the isolated populations are vulnerable to change in habitat, declines in prey, such as rabbits and snowshoe hares, and trapping by man. Lynx have been successfully reintroduced into Sweden and other parts of Europe, and with protection are slowly recolonising some areas. Found in many national parks and reserves throughout their range; the Spanish Lynx occurs in the Coto Donana reserve. Regularly bred in captivity.

* SAND CAT

Felis margarita (**943**)

HB 17.7–22.5ins; T 11–13.8ins; WT ca.2.9lb.

Sandy to greyish above, whitish below. A reddish streak across the cheeks and the tail is ringed towards the black tip. The feet are densely furred on the soles, an adaptation to their habitat of deserts and sand dunes. They are nocturnal, hiding in burrows by day, and preying on rodents, such as jerboas, as well as birds, reptiles and other small vertebrates. They can live without drinking water. They have one or two litters a year of 2–4 young. Throughout their range they are generally rare, and have been exploited by the fur trade. Occasionally bred in captivity, and kept in some zoos. Listed on CITES App.II. They are scarce in Arabia; the subspecies *F.m.scheffeli* described in 1966, from Pakistan, is particularly threatened.

*** CARACAL

Felis caracal (**944**)

HB 23.6–36ins; T up to 12.2ins; SH 15–19.7ins; WT 28.7–41.9lb.

A rather tall, slender cat, generally sandy-coloured with only faint spotting on the sides. The ears are long and distinctly marked. Found in open habitats, including semi-desert, feeding on a wide variety of small animals,including small antelope, hyraxes, rodents, lizards and birds. Agile, springing up to 6ft high. Although their range in N Africa has probably contracted, elsewhere in Africa there is no evidence of major decline; however, in many parts of India and the Middle and Near East they have disappeared. Formerly tamed and trained for hunting, like the Cheetah, in India. Found in many national parks and reserves, mainly in Africa, and thrive in captivity. Listed on CITES App. I and II.

* BOBCAT

Felis rufus (**945**)

HB 25.6–43.3ins; T 3.9–7.5ins; WT 14.1–68.3lb.

Buff or brownish above, paler below with darker spotting. Ear tufts shorter than those of Lynx; they occur in a wider variety of habitats than Lynx. Although they have been heavily exploited for their fur and have undergone local declines, still occur over much of their original range, and are locally abundant. Being re-established in areas where they have been extirpated, such as New Jersey.

940
(19th C)
941
(2-3000 years ago)
942
943
945
944
F. l. canadensis
945
942
F. l. lynx
941
944
940

***** PALLAS'S CAT**
Felis manul **(946)**
HB 19.7–25.6ins; T 8.3–12.2ins; WT 5.5–7.7lb.
Generally yellowish, greyish or buff with white-tipped fur, giving a frosted appearance; the fur is particularly long and dense. Found in open steppe and desert at altitudes of up to 13,100ft and over, where they are usually nocturnal, feeding on pikas, and other small mammals. Considered threatened in the USSR, and Ladakh (India), but little known of status. Listed on CITES App. II and protected in USSR and India.

***** CLOUDED LEOPARD**
Neofelis nebulosa **(947)**
HB 23.6–39.4ins; T 21.7–35.5ins; WT 35.3–50.7lb.
Relatively large head with proportionally the largest canine teeth of any living cat. The markings are very variable, and melanistic animals occur. They are found in forests up to an altitude of 8200ft, where they are highly arboreal. They feed on birds, monkeys, deer and other similar-sized animals. Although they have a comparatively wide distribution, they are declining due to extensive loss of habitat, particularly in Thailand and Malaysia. They are also hunted for their skins, which until recently were often involved in international trade. Kept in most of the larger zoos; breeding fairly freely. Listed on CITES App.I, and protected over most of their range.

***** BAY (BORNEAN RED) CAT**
Felis badia **(948)**
HB 24.8–27.2ins; T 14.6–16.9ins.
Uniformly dark reddish or greyish, usually with three stripes on head, and white tip to the tail. One of the least-known cats, they are only known from a few localities in Borneo. Consequently presumed rare, also rare in captivity. Listed on CITES App.II.

*** FLAT-HEADED CAT**
Felis planiceps **(949)**
HB 17.3–21.9ins; T 5.1–6.7ins; WT 3.3–4.8lb.
Brownish with fine speckling of grey and buff; chin and chest white. Occurs from S Thailand south through Peninsular Malaysia to Borneo and Sumatra. A rare and elusive species, and very little is known of its precise distribution or status. Believed to frequent riverine habitats. Listed on CITES App.I.

******* IRIOMOTE CAT**
Felis iriomotensis **(950)**
HB ca.23.6ins; T ca.7.9ins.
Brownish with 5–7 rows of longitudinal spotting or stripes. Relatively short legs and tail. First discovered in 1967. Confined to the island of Iriomote Shima, between Taiwan and Japan, where it lives in lowland subtropical rain-forest, feeding on small animals. Although totally protected, it is still declining due to continued loss of habitat and the population has recently been estimated at 40–80.

*** ASIATIC GOLDEN CAT**
Felis temmincki **(951)**
HB 29.9–32.1ins; T 16.9–19.3ins; WT 26.5–33.1lb.
Rather variable in colour, normally tawny-brown, but may be golden, greyish-brown, or more rarely melanistic. Found mainly in dry forests, up to altitudes of more than 9850ft, where they are mostly terrestrial, feeding on animals up to the size of small deer, pheasants, lizards. They are rather sparsely distributed throughout most of their range, and have been hunted for skins and alleged magical properties of their fur. Protected over much of their range, which extends from S China and Nepal, south through Peninsular Malaysia to Sumatra. They are relatively rare in captivity, but occasionally bred. The closely related *** African Golden Cat** *F.aurata* **(952)** is confined to forests of W and C Africa, where, because of forest destruction, it may be threatened. It is more variable than *F.temmincki*, some populations being extensively spotted.

***** MARBLED CAT**
Felis marmorata **(953)**
HB 17.7–20.9ins; T 18.7–21.7ins; WT 4.4–11lb.
Rather like a small Clouded Leopard. Although still fairly widespread it is a little-known species, rare throughout its range and almost certainly declining due to loss of its forest habitat. Protected in India and Thailand, and probably occurring in several national parks throughout its range.

***** LEOPARD CAT**
Felis bengalensis **(954)**
HB 17.5–21.7ins; T 9.1–11.4ins; WT 6.6–11lb.
A small, well-spotted cat found in a wide range of habitats, often close to human settlements. They feed on a wide variety of small animals up to the size of small deer. They have a wide distribution from Amur, USSR, to the Himalayas, south to Indonesia. Occur on many small islands, and several subspecies have been described; these are not very clearly defined, and probably intergrade. Although still common and widespread, they have been extensively exploited for the fur trade, and are declining at least locally. Protected in most parts of their range, they occur in many national parks and reserves; they are listed on CITES App.II, with some populations on App.I. Kept in many zoos and regularly bred.

***** FISHING CAT**
Felis viverrinus **(955)**
HB 28.6–30.7ins; T 9.8–11.4ins; WT 15.4–24.2lb.
A relatively large, spotted cat with a rather fragmented distribution. Although still abundant in some areas, it is declining in others. Can survive close to human populations, but is also extensively trapped for fur – although the fur is of no particular value. Considered threatened in parts of India, where it is protected.

*** LITTLE SPOTTED CAT, TIGER CAT or ONCILLA**
Felis tigrinus **(956)**
HB 15.8–21.7ins; T 9.8–15.8ins; WT 3.9–5.9lb.
Rather variable in colour, but usually pale buff or rich ochre with rows of dark spots; about one in five are melanistic (black). Little-known in the wild, but usually found in forests where it is less arboreal than the Margay. Litters of 1–2 young recorded in captivity. Occurs in C America as far north as Costa Rica, and over most of S America; rare throughout its entire range. Has been hunted both for both pet trade and fur industry. Listed on CITES App.II with Costa Rican subspecies (*F.t.oncilla*) on App.I. Protected in many countries and probably occurs in several national parks and reserves. Rare in zoos, but a few have been bred.

All cats not listed on CITES App.I. are included on App. II.

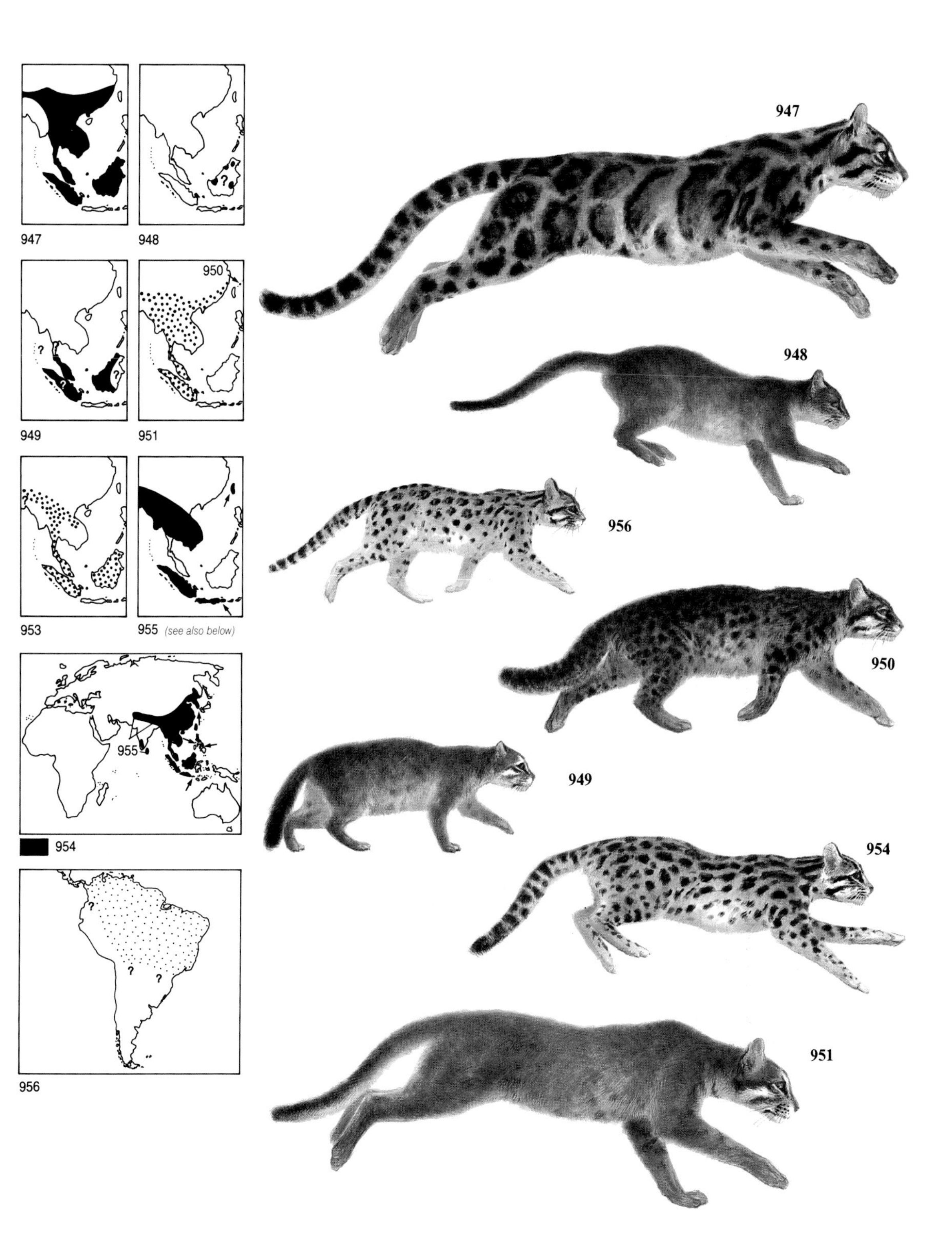
947
948
947
948
949
951
950
953
955 (see also below)
956
950
949
955
954
954
956
951

**** AUSTRALIAN SEA LION**
Neophoca cinerea **(957)**
Males: HB up to 11ft 6ins; WT up to 662lb.
Females: HB up to 6ft; WT up to 230lb.
They breed on beaches and rocky promontories. Often climb cliffs and have been known to occur 6 mls from the sea. Non-migratory, staying in the vicinity of their breeding grounds, where they feed on squid, crayfish, fish and penguins. One of the rarest pinnipeds in the world, with an estimated total population of 2000–3000. Probably always been fairly rare, and no evidence of any decline. Found along Great Australian Bight, from Kangaroo Island, to Houtman Rocks.

**** NEW ZEALAND SEA LION**
Neophoca hookeri **(958)**
Males: HB ca.10ft; WT ca.882lb.
Bulls are dark brown with a well developed mane of darker hair; females are much smaller, and lighter in colour. Breed on sandy beaches, and often wander inland for a considerable distance. They feed on crabs, mussels, small fish and penguins. Largely sedentary, they stay in the vicinity of the breeding ground throughout the year. Occurred on North Island, New Zealand, until ca.1000 years ago, but restricted to subantarctic islands. They were exterminated on Macquarie Island during the 19th century. The world population was estimated at ca.4000 in 1979. The main breeding colony, Auckland Island, is a nature reserve.

***** CALIFORNIAN SEA LION**
Zalophus californianus **(959)**
Males: HB 79–99ins; WT 440–660lb.
Females are much smaller (under 220lb).
The most noticeable feature of the male is the high forehead. The best-known of the sea lions, they are exhibited in many zoos and circuses since they are naturally playful, chasing each other and leaping clear of the water, and 'porpoising'. They feed on squid, octopus and a wide variety of fish. There are three quite separate populations: *Z.c.californianus* which breeds from San Miguel Island, California, throughout the Sea of Cortez, and south to Punta Estrada, Baja California, totalling about 35,000; *Z.c.wollenbaeki* breeds in the Galapagos Islands, and numbers about 20,000; the Japanese population, *Z.c.japonicus* is probably extinct. The Californian Sea Lion is protected throughout its range, and apart from the Japanese population, is stable or increasing.

***** NORTHERN or STELLER'S SEA LION**
Eumetopias jubatus **(960)**
HB 130–138ins; WT 1985lb. Females are much smaller than males.
The largest sea lion. It feeds on squid, octopus, crabs, clams and a variety of fish, and dives to 600ft. Breeds in isolated parts of the N Pacific, and some populations are considerably reduced. The world population is probably 200,000–300,000. From 1957 there has been a marked decline in the Aleutian population, which has fallen from 50,000 to 25,000 in 20 years. The Californian population declined generally from the 1930s onwards. In the 19th century they were extensively exploited; causes of the more recent declines are not fully known. Some are killed in fishing gear, and it may be that increased competition for food resources, both with other species of seal and with commercial fisheries, is affecting them.

***** SOUTH AMERICAN SEA LION**
Otaria flavescens **(961)**
Males HB ca.99ins; WT up to 750lb.
Females much smaller than males (under 330lb).
Although the world population was estimated in 1979 at nearly 275,000, there is cause for concern for a number of populations of the South American Sea Lion. They breed from the Isla Lobos de Tierra, Peru, south around most of the coast of S America to the Straits of Magellan and the Falkland Islands. Until recently exploited in many parts of their range, and still killed by fishermen on account of net damage; there is a general downward trend in their numbers, with many local extinctions. Even in areas where they are not hunted, such as the Falkland Islands, declines have been noted.

***** COMMON or HARBOUR SEAL**
Phoca vitulina **(962)**
HB 47–67ins; T 3.5–4.7ins; WT up to 550lb.
Greyish or brownish above with darker spots and blotches, spotted creamy white below. Mainly coastal and spend a lot of time hauled out on sand bars, rocks and beaches, basking. Although still widespread in the northern hemisphere, they have been subjected to centuries of exploitation and persecution, and are locally extinct or rare over much of their range, particularly close to human settlements. They are also increasingly threatened by oil spills and other forms of pollution. In the 1970s the world population was estimated at nearly 400,000. The E Pacific population is sometimes treated as a full species, *P.richardi* **(963)**, and the **Largha Seal**, *P. largha* **(964)** from the N.Pacific is closely related, and hunted in the Bering Sea.

*** RINGED SEAL**
Phoca hispida **(965)**
HB 33.5–63ins; T vestigial; WT 88.2–198.5lb.
Dark grey or blackish above, with oval-shaped white rings, pale grey on the underside. The Ringed Seal is closely related to the seals from the Caspian Sea and Lake Baikal, and several populations of Ringed Seals occur in land-locked lakes. One of the most abundant seals in the northern hemisphere with a total population of 6–7million. However in some areas it has declined and some populations may be threatened. The Baltic populations, in particular, are threatened by extensive pollution. *P.h.saimensis* occurs in Lake Saimaa, Finland; and *P.h.ladogensis* in Lake Ladoga, USSR, may be vulnerable.

*** GREY SEAL**
Halichoerus grypus **(966)**
HB 65–90ins; T vestigial; WT up to 684lb.
Male is substantially larger than female. Colouring variable, generally darker above. Feed on a wide variety of fish, together with cephalods and crustaceans. Have been extensively hunted since prehistoric times, both for their skins and also for meat and blubber. In more recent times they have been killed for alleged damage to commercial fisheries; some also drown in fishing tackle. They are also killed in oil spills. In the early 1970s, the world population was estimated at under 95,000, with over half in British waters, and ca.22,000 in Canada. The British population was expanding rapidly, but elsewhere it was declining; and in most of its European range, including parts of S Britain, it is now declining. Protected in most parts of its range, and where hunted usually subject to quotas.

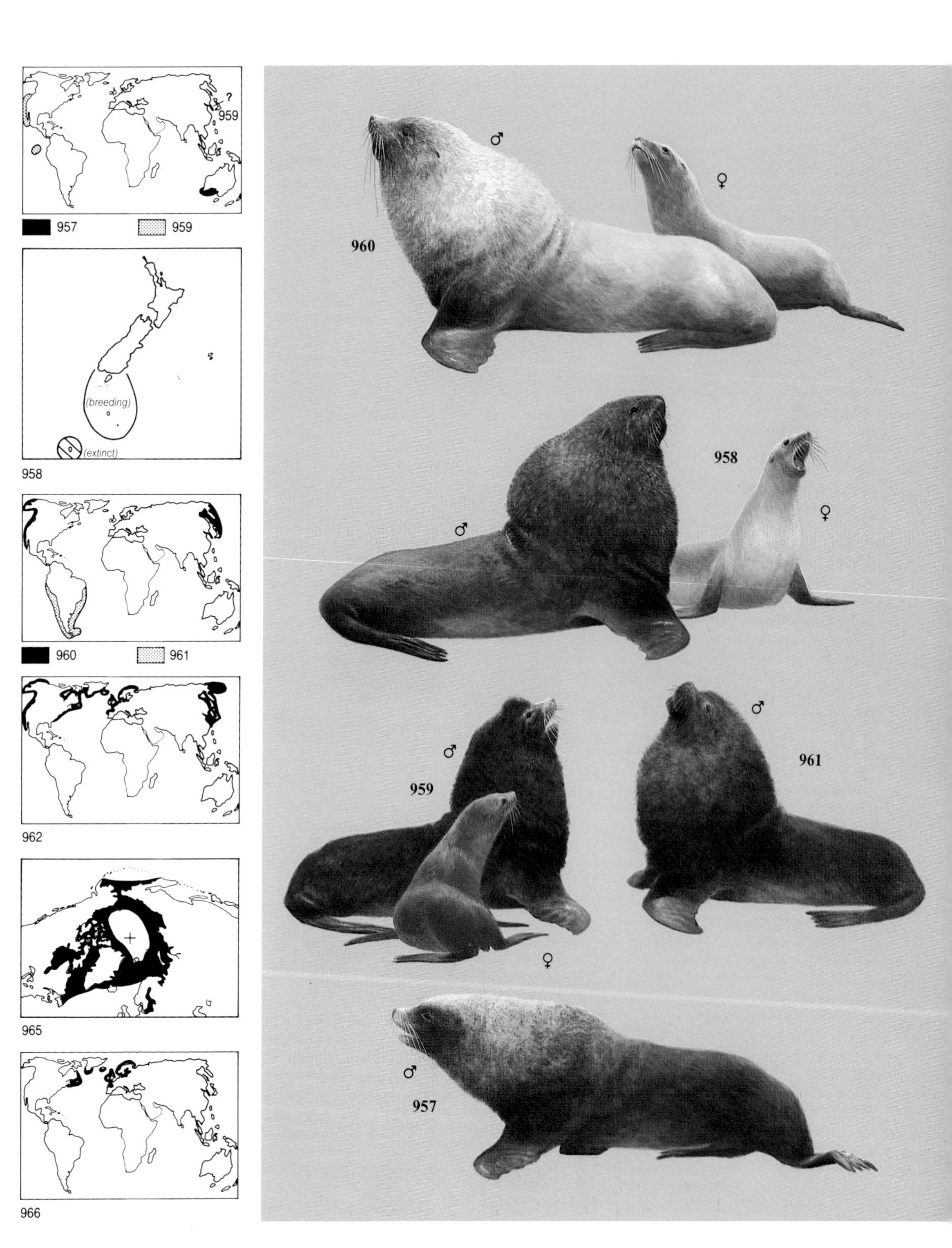

?
959
957
959
(breeding)
(extinct)
958
960
961
962
965
966
960
♂
♀
958
♂
♀
959
♂
♀
961
♂
957
♂

*** WALRUS**

Odobenus rosmarus (**967**)

HB 8ft 3ins–11ft 10ins; T vestigial; WT up to 1875lb; occasionally larger, but females much smaller.

Upper canines form large tusks, up to 60ins long. Sociable, gathering in mixed herds of up to 2000 or more during feeding and migration. When breeding older males defend harems. Found in the colder arctic waters, feeding on molluscs and crustaceans, diving to 350ft. Extensively exploited by subsistence hunters in the Arctic, and more recently commercially exploited for oil, and ivory. Under protection many populations are rebuilding, and most considered out of danger, but some may still be threatened by overhunting for ivory. Protected throughout their range and listed on CITES App.II.

******* MEDITERRANEAN MONK SEAL**

Monachus monachus (**968**)

HB up to 9ft 10ins; T vestigial; WT up to 660lb.

Dark grey or brownish above, with a pale belly. Pups born on land in late autumn; at birth have dark brown fur. Once widespread in the Mediterranean Sea, Black Sea and Atlantic waters to Madeira and the Canary Isles, and south along the W African coast to Cap Blanc. Formerly hunted for skins; more recently persecuted for its alleged damage to fisheries. By the 1950s total population estimated at ca.5000, but little known of its true status. Decline continued with some populations disappearing, until by the 1980s world population estimated at 500–1000, with few viable populations. Kept in captivity on many occasions but it is doubtful if a viable captive colony could be established. The **Caribbean Monk Seal** *M. tropicalis* is believed to have become extinct in the 1950s.

******* HAWAIIAN MONK SEAL**

Monachus schauinslandi (**969**)

Similar to *M.monachus* (968), but smaller, and greyer. Formerly widespread in the Hawaiian Islands, but by the time it was described in 1905, already depleted by sealers. From 1909 it was given protection and population recovered to ca.400 by 1924, and 1350 in 1958. Found mainly on the western Hawaiian Islands; they rarely stray to the main Hawaiian Islands. Have been exhibited in Hawaiian aquaria.

FUR SEALS

Arctocephalus spp.

During the 17th, 18th and 19th centuries, fur seals were intensely exploited, by commercial sealers, primarily for their skins, several species being reduced to near extinction. All rather similar in appearance, with a rich chestnut-brown underfur; the males have a heavy mane of black guard hairs, often tipped with white giving them a hoary appearance. The males defend harems on the sea shore.

***** JUAN FERNANDEZ FUR SEAL**

Arctocephalus philippii (**970**)

HB up to 79ins; T vestigial; WT up to 310lb; females markedly smaller.

Closely related to, and sometimes considered conspecific with, the Guadalupe Fur Seal *A.townsendi* (971). Confined to 3 islands of Juan Fernandez Archipelago, and San Ambrosio and San Felix Islands, off Chile. Once extremely abundant, perhaps ca.3.5 million, but hunting during the 17th, 18th and 19th centuries reduced them to apparent extinction by 1960s; in 6 years at the end of the 18th century 6 million seals were taken. However they were 'rediscovered' in 1965. The Juan Fernandez Archipelago was declared a national park in 1970 and by the late 1970s ca.2,500 were present. In 1978 the seals were given total protection. They are listed on CITES App.II.

*** GUADALUPE FUR SEAL**

Arctocephalus townsendi (**971**)

HB up to 79ins; T vestigial.

May have occurred on the Channel Islands of S California; now restricted to east coast of Guadalupe, ca.155 mls from the mainland of Baja California. Before the start of commercial sealing, they numbered possibly 200,000; by 1928 pronounced extinct. In 1949 a single bull was discovered, and in 1954 a colony of 14 found; other small isolated colonies are presumed to have survived. Population has slowly increased; by late 1970s had risen to 1300–1500 on Guadalupe and is growing.

*** NEW ZEALAND FUR SEAL**

Arctocephalus forsteri (**972**)

Now recovered from near extinction in 19th century in New Zealand, and by 1970s numbered ca.38,500 in New Zealand and the Antarctic Islands. However, the recovery of the Australian populations has been very much slower.

*** GALAPAGOS FUR SEAL**

Arctocephalus galapagoensis (**973**)

HB ca.71ins; T vestigial.

The smallest of the fur seals, it is closely related to the *** South American Fur Seal** *A.australis* (**974**), and often considered conspecific. Nearly exterminated in the 19th century, they are protected under Ecuadorian law, and have slowly increased in the last 50 years. In the late 1950s the population was estimated at 100–500, but it was probably higher; in the mid-1970s it was estimated at ca. 1000.

***** BAIKAL SEAL**

Phoca sibirica (**975**)

HB 47–55ins; T vestigial; WT 176–198lb.

A very small seal confined to Lake Baikal and Lake Oron in C Siberia, USSR. From 1930–1941, 5000–6000 were killed each year, but by the 1970s the annual kill had been reduced to ca.2000–3000, and the total population estimated at 40,000–50,000. There has been considerable concern over the effects of pollution from paper mills and other industrial effluents entering the Lake, but the population is believed to be stable or slightly increasing.

***** HOODED SEAL**

Cystophora cristata (**976**)

HB up to 99ins; T vestigial; WT up to 880lb.

Male larger than female, and both possess a bladder on the head, which can be inflated. Live in small groups, and are more gregarious during the breeding season when they gather on the edge of the ice. The main breeding areas are off Newfoundland, including the Gulf of St Lawrence, and around Jan Mayan Land. They gather off Greenland and elsewhere in summer to moult. Subject to extensive commercial exploitation, for both skins and oil. In the late 1970s ca.45,000 were killed each year, and it is believed that substantial numbers were being killed but sinking and lost. They are subject to quotas. The ***Harp Seal** *Pagophilus groenlandicus* (**977**), although depleted, is still relatively abundant.

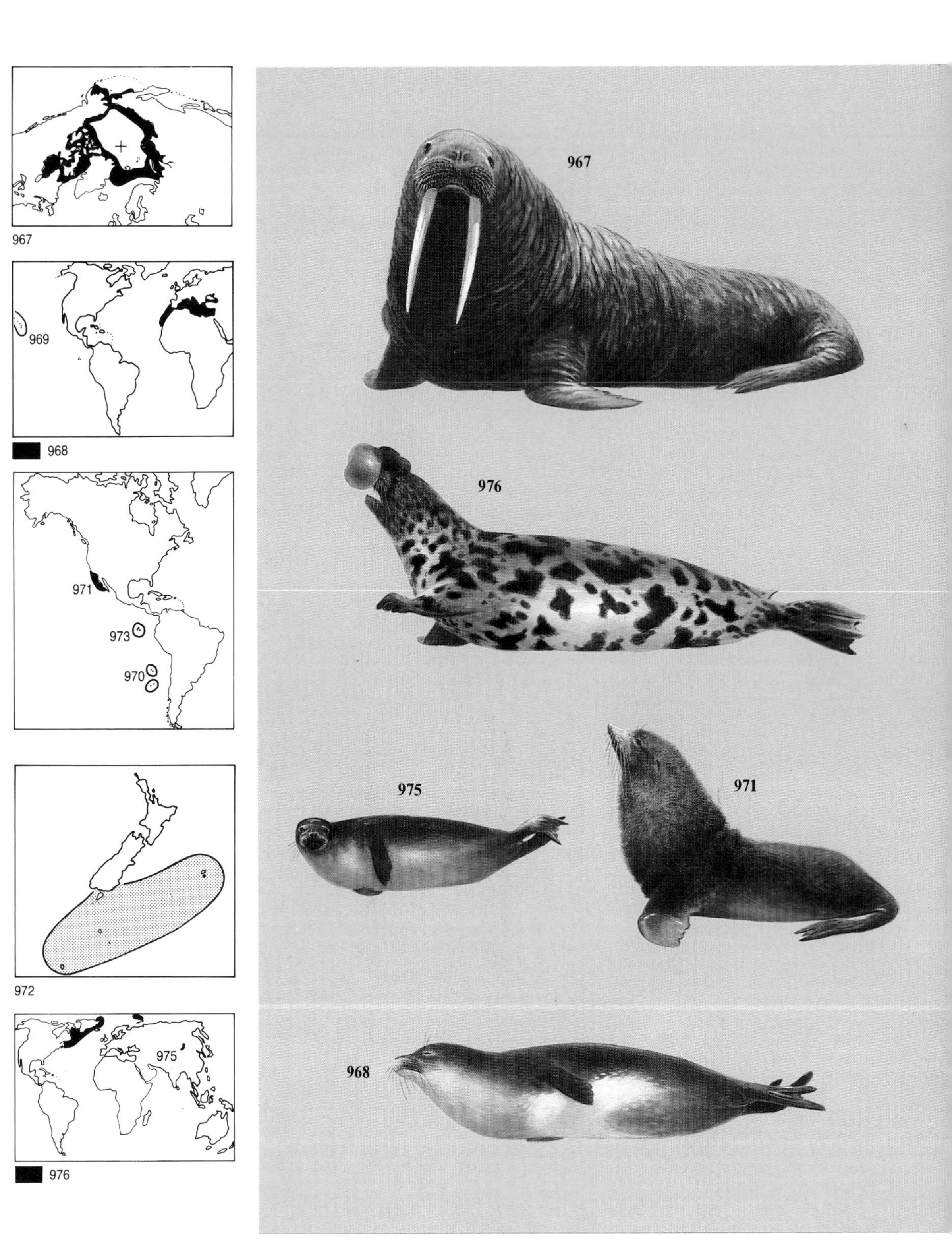

967
968
969
971
973
970
972
975
976
967
976
975
971
968

*** **ASIATIC ELEPHANT**
Elephas maximas (**978**)
HB 18–21ft (inc. trunk); T 4–5ft; SH 8ft–8ft 6ins; WT up to 11,025lb.
The Asiatic Elephant has considerably smaller ears than the African Elephant; it also has 4 hooves (3 in the African) on the hind foot, and a single 'finger' on the tip of the trunk (2 in the African). Females are generally smaller than males and usually lack prominent tusks; male tusks rarely weigh more than 100lb (a pair) although there are records of over 155lb per tusk and over 9ft 10ins length. They are found mostly in forests, but are adaptable and can occur in a wide range of habitats, at altitudes up to 11,800ft. They feed on grass and other vegetation, and also raid crops. In the past, Asiatic Elephants were found from Syria and Iraq eastwards across southern and Southeast Asia. In the 19th century they were still common over much of the Indian subcontinent, Sri Lanka and the eastern parts of their range. In the present century, expanding human populations and the destruction of forests in southern Asia fragmented and isolated their range until by the late 1970s a maximum of 42,000 were left; the population in Sumatra is probably the most endangered of all. They are still hunted for ivory. Listed on CITES App.I. Although frequently exhibited in zoos (and circuses) they are only rarely bred in captivity. Asiatic Elephants occur in many of the larger national parks and some forest reserves within their range. They also occur in Borneo and the Andaman Islands where they were probably introduced by man.

*** **AFRICAN ELEPHANT**
Loxodonta africana (**979**)
HB 20–25ft (inc trunk); T 3ft 3ins–4ft 3ins; SH 10–13ft; WT 1985–13,230lb.
The largest living land animal; skin greyish-brown, but may take on colour of its surroundings as they frequently wallow in mud. Both sexes carry tusks (enlarged incisors) which grow to a maximum of 11ft 6ins and weigh over 220lb; female tusks are much smaller than males. They once occurred in practically all habitats in Africa, from semi-desert to high montane forest up to 16,400ft, marshes and open savannah; they feed on a wide variety of vegetation, and also frequently raid crops. In the past they were found throughout most of Africa, except exteme deserts. By Roman times they were exterminated from most of N Africa. Throughout W Africa populations are fragmented, isolated and substantially reduced. In E Africa most of the once enormous population has been drastically reduced by ivory poaching, and only in C Africa are substantial populations believed to survive. Although extinct or critically endangered in many countries, in some areas they are still abundant enough to be a pest to agriculture. Elephant ivory is extensively traded, both legally and illegally, but is subject to special licence under CITES App.II. The African Elephant is protected in most countries where it occurs, and is in many national parks. It is only found in relatively few zoos and safari parks, and rarely bred in captivity.

*** **BLACK RHINOCEROS**
Diceros bicornis (**980**)
HB 12ft 6ins–14ft 3ins; T 28ins; SH 4ft 7ins–5ft; WT 220–3970lb.
Dark, with a long, pointed, protruding prehensile upper lip. Two prominent horns, the longest up to 48ins (average 20ins); occasionally a smaller third horn. They occur in bush country with thick cover, grasslands, and in open forest up to 11,500ft altitude, where they browse on an extremely wide variety of plants. The Black Rhino was formerly found in suitable habitat over most of Africa south of the Sahara. By the mid-1960s it still occurred throughout most of its range, but was already much rarer, being eliminated as part of tsetse fly control measures. By the mid 1970s its decline had accelerated, and poaching for its horn (sold to Asia for carving and medicinal purposes) led to its extermination in many places. By the late 1970s less than 30,000 remained in fragmented and isolated populations; since then they have continued to decline, to about 15,000 in 1980 and 9000 by 1984; seven subspecies are recognised, of which five have populations of less than 500. They are found in many national parks within their range, but even these are often poached. Listed on CITES App.I; exhibited in many zoos and breed frequently.

**** **WHITE RHINOCEROS**
Ceratotherium simum (**981**)
HB 11ft 10ins–16ft 5ins; T 35–40ins; SH 40–80ins; WT 5070–7940lb.
The two horns can grow to over 60ins, the front one being about three times the length of the rear. Readily distinguished from the Black Rhino by its square lip. It is found mainly in open grasslands and lightly wooded habitats where it grazes on grasses and herbage, in territories of about 0.8 sq.mls. White Rhinos occur in two separate populations, the Southern (*C.s.simum*) and the Northern (*C.s.cottoni*). Both subspecies have much reduced ranges. In 1882 the Southern White Rhino was believed extinct, but a small population was found in Umfolozi, Zululand, and at the turn of the century about 10 or 11 are thought to have survived; by the mid 1960s, they had increased to about 500 in the Umfolozi Game Reserve, and were being translocated to other parks and reserves. By the early 1980s there were over 3000 in South Africa and 600 elsewhere in Africa, and breeding groups in zoos and safari parks. In 1900 the Northern White Rhino was found from NW Uganda to Chad, and in the Belgian Congo (Zaire), French Equatorial Africa (Congo) and Sudan. At the time of their discovery they were far more numerous than the Southern, but by 1980 numbered about 1000, and by 1985 were believed to be restricted to a single population in the Garamba NP, Zaire, of less than 20 (400 in the 1970s). A small number exists in captivity. There are probably no White Rhinos now outside protected areas or zoos.

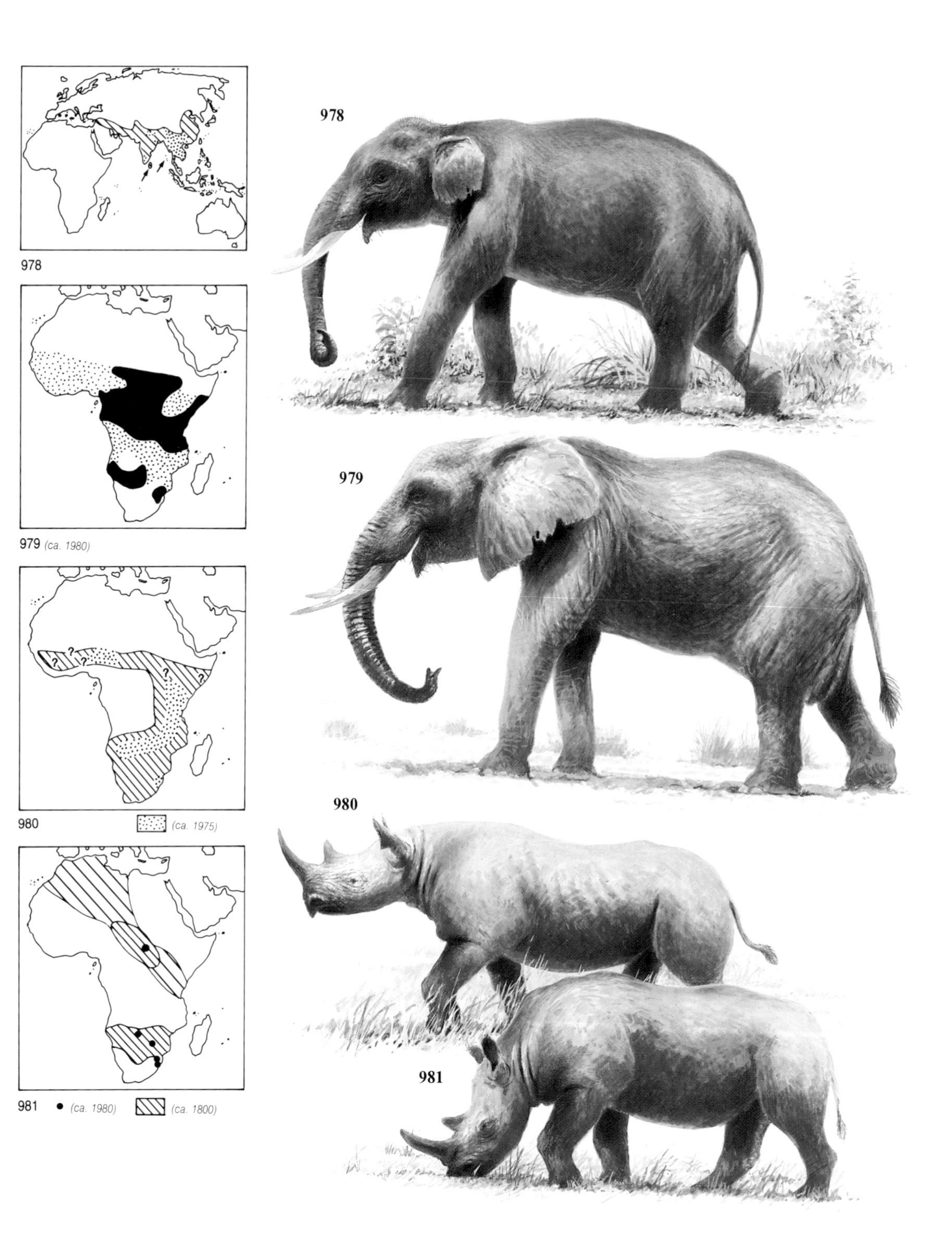
978
979 (ca. 1980)
980
(ca. 1975)
981
(ca. 1980)
(ca. 1800)
978
979
980
981

***** SUMATRAN or ASIATIC TWO-HORNED or HAIRY RHINOCEROS

Dicerorhinus sumatrensis (**982**)

HB 7ft 9ins–10ft 6ins; SH 3ft 7ins–4ft 11ins; WT 2200–4410lb.

A two-horned rhino with an 'armour-plated' appearance, covered in coarse hair. The horns are short, the rear one under 15ins and the front one smaller. Found mainly near water in secondary forest and often in hill country, as they climb well. They feed mainly at night on a wide range of vegetation including leaves, twigs, fruit and bamboos; during the day they often wallow in pools to avoid horseflies. At the end of the 19th century the Sumatran Rhino was still found over much of SE Asia from Assam and Bangladesh, south through Burma, Thailand and Vietnam, through the Malay Peninsula to Sumatra and Borneo. Hunting for its horn and almost all other organs (for alleged medicinal purposes) is largely responsible for its disappearance. There are now only scattered populations numbering a few hundred in total, isolated from each other. Has been kept in captivity but with none surviving. They are protected in most of their remaining range, and listed on CITES App. I.

*** GREAT INDIAN RHINOCEROS

Rhinoceros unicornis (**983**)

HB 13ft 9ins; T 30ins; SH 6ft 7ins; WT up to 4410lb.

Characteristic 'armour-plated' appearance, with large tubercles on the bare skin. The single horn may grow to 24ins, but is usually less than 8ins. Found in forests, swampy areas and in reed-beds; they are usually solitary, spending much time in wallows. They feed primarily on grasses, reeds and twigs; in some localities they feed in cultivated areas. Their former range was considerably greater than at present, though its exact extent is not known. They were certainly found throughout the foothills of the Himalayas from N Pakistan, east through India and Nepal, to Assam and Bengal. They may also have occurred in Burma, Thailand and other parts of SE Asia until the Middle Ages. By the 1900s the population was considerably reduced in India; the British authorities banned hunting in 1910 and established a series of sanctuaries. By late 1950s a total of about 400 survived, the majority in Kaziranga Sanctuary, Assam, with others in Bengal and Assam, and about 300 in Nepal. By the early 1980s the total was estimated at over 1000. The main threats to rhinos in the sanctuaries are competition from domestic stock grazing within reserves, and that surplus animals will leave the boundaries of the sanctuary. Small numbers are maintained in about 30 zoos, and an increasing proportion are captive-bred. The Great Indian Rhino is listed on CITES App.I and strictly protected throughout its range.

***** JAVAN or LESSER ONE-HORNED RHINOCEROS

Rhinoceros sondaicus (**984**)

HB 63–69ins; T ca.28ins; SH ca.71ins; WT 1.6–2.2 tons.

Like the Great Indian Rhino, the Javan is characterised by loose folds of skin, but in this species one of these continues across the midline of the back. The single horn is fairly small, normally less than 10ins, and is often only a low protuberance. Found mostly in forested, hilly areas up to an altitude of 6550ft, where they browse on shrubs and bushes, often in secondary habitats. Until the middle of the 19th century, the Javan Rhino was widespread and often abundant from Bengal, east through Burma, SW China, to Vietnam and south through Thailand, Laos, Cambodia and Malaya, to Sumatra and Java. Although it once roamed over most of lowland Java, the human population, which increased from 3–4 million in 1800 to 57 million by 1958, had pushed them into a few remote areas and by the 1930s they were confined to the Udjung Kulon Reserve in western Java. By 1940 it was presumed extinct in Sumatra, but there continue to be scattered unconfirmed reports elsewhere in mainland Asia, including the Thai/Burma border. In 1950 it was still present in the Sunderbans, the Brahmaputra Valley, in the Chittagong Hills and several other localities, but by 1960 it was extinct. The present world population is around 50. It is listed on CITES App. I.

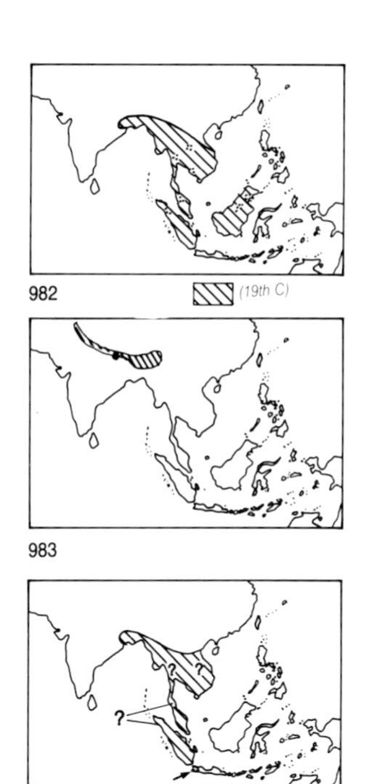
982
(19th C)
983
?
984
(19th C)

983

984

982

*** **MOUNTAIN TAPIR**
Tapirus pinchaque (**985**)
HB ca.71ins; T vestigial; WT ca.496lb.
The smallest tapir, and the only species with thick, bristly hair. The snout and upper lips form a short, fleshy proboscis. Found in montane forests often shrouded in mists, at altitudes of 6500–14,800ft in scrub habitats, dominated by stunted trees and shrubs. They feed on vegetation including ferns, bamboos and other plant shoots. Confined to the Andes, their range is much fragmented, and they are subject to hunting pressure as well as loss of habitat. Protected in Colombia, Peru and Ecuador but because of their remote range, this is difficult to enforce. There are very few in captivity, and the first was bred as recently as 1977; in the 1960s and 1970s many were captured, with high mortality, for export to zoos. Listed on CITES App.I.

*** **BAIRD'S TAPIR**
Tapirus bairdii (**986**)
T vestigial.
Uniform brownish, with a low, narrow mane. They live in a wide variety of habitats from sea level to 11,000ft, including mangrove swamps, woodlands, tropical moist forest, and even above the tree-line. Generally nocturnal, they feed on a wide variety of vegetable matter, including fallen fruits, leaves, flowers and twigs and have a single young. Found from S Mexico south to Ecuador, but declining throughout their range, which is becoming increasingly fragmented. Extinct in El Salvador, and extremely rare in Mexico, but there is little precise information on status. They occur in several protected areas, including Corcovado NP, Costa Rica. The main threats are loss of habitat and hunting for meat. Listed on CITES App.I, and protected throughout range. They are rare in zoos, but small numbers are now being bred.

* **BRAZILIAN TAPIR**
Tapirus terrestris (**987**)
HB ca.71ins; T vestigial; SH ca.31.5ins.
The adults are a fairly uniform dark brown, but the young have very contrasting longitudinal white stripes, blotches and spotting, almost identical to those of young Malayan Tapir. Probably the most widespread and locally abundant tapir, hunted for meat in many parts of its range, despite legal protection. Destruction of habitat is fragmenting its range and also leading to local extinctions. It occurs in many protected areas and is listed on CITES App.II.

*** **ASIAN (MALAYAN) TAPIR**
Tapirus indicus (**988**)
HB 87–94ins; T vestigial; WT 550–827lb (max.1190lb).
The adults' markings are very striking, but apparently offer excellent camouflage at night, when shafts of moonlight on dense vegetation produce a similar black-and-white pattern; the young are similar to *T.terrestris* (987) young. They are usually solitary, and mainly nocturnal, feeding on a wide variety of forest vegetation. Their range extends from S Burma and Thailand, south through the Malay Peninsula and also Sumatra. Although fossils have been found in Borneo, there is no evidence that they survived into historic times. Their range is now considerably fragmented, and they are increasingly confined to the more remote mountain tops; they occur in most of the larger forest reserves within their range. Protected throughout their range, and not hunted in Muslim areas. Listed on CITES App.I. Exhibited in many zoos, where they breed freely.

*** **DUGONG**
Dugong dugong (**989**)
HB (inc.T) 8ft 3ins–13ft 2ins; WT 794–2240lb.
A large marine mammal with a horizontal tail fluke similar to those of whales; the forearms are modified into paddles. They are entirely vegetarian and their range follows very closely the distribution of their principal food plants, *Potamogetonaceae* (pondweeds) and *Hydrocharitaceae* (frogbits). Once found throughout the warmer waters of the W Pacific, Indian Ocean and Mediterranean Sea, they are now extinct in the Mediterranean, and declining rapidly in most other parts. Although protected in many parts of their range, significant increases in their population have only been reported in Australia. They have been hunted extensively for meat and oil (over 110 pints from a single animal) and also for their incisor teeth which are used as amulets in SE Asia. Currently the greatest threat comes from fisheries and the increasing use of nylon fishing nets; Dugongs are frequently caught and drowned in these. Dugongs are thought to be the origin of sailors' tale of sirens and mermaids. They occur in only a few protected areas and do not flourish in captivity.

*** **MANATEES**
Trichechus spp.
Total length: up to 14ft 9ins; WT up to 794lb.
Grey or blackish with a few bristles scattered over the hide (often over 2ins thick). The 3 species are found in warm, coastal waters where they feed, eating up to 33lb of aquatic plants per day. They swim with a vertical movement of the tail. The **Amazon Manatee** *T. inunguis* (**990**) is found throughout the Amazon Basin, and possibly the Orinoco, and coastal waters of Brazil. From the late 18th century until about 1960, up to 10,000 were killed annually for oil, meat and hides, but in the 1960s the catch declined and in 1973 they were given protection in Brazil. Hunting continues, despite protection throughout their range, and they are now extremely rare in Peru and other countries on the edge of their range. They occur in several protected areas, and are listed on CITES App.I. Relatively few in zoos, but they can be maintained under semi-captive conditions in closed waters such as lakes and canals. The **American Manatee** *T.manatus* (**991**) which was formerly widespread and abundant in the waters of southern USA, the Caribbean and South America, has undergone a massive decline. Although protected over most of its present-day range, it continues to decline in many areas from more recent threats, such as pollution, drowning in fishing gear and damage from the propellors of powerboats. Listed on CITES App.I; small numbers are maintained in captivity. The **African Manatee** *T.senegalensis* (**992**) is found in coastal waters and rivers from Senegal to Angola. The threats, past and present, are similar to the previous species, but it has probably declined less and still occupies a large part of its original range. Listed on CITES App.II and class A of the African Convention.

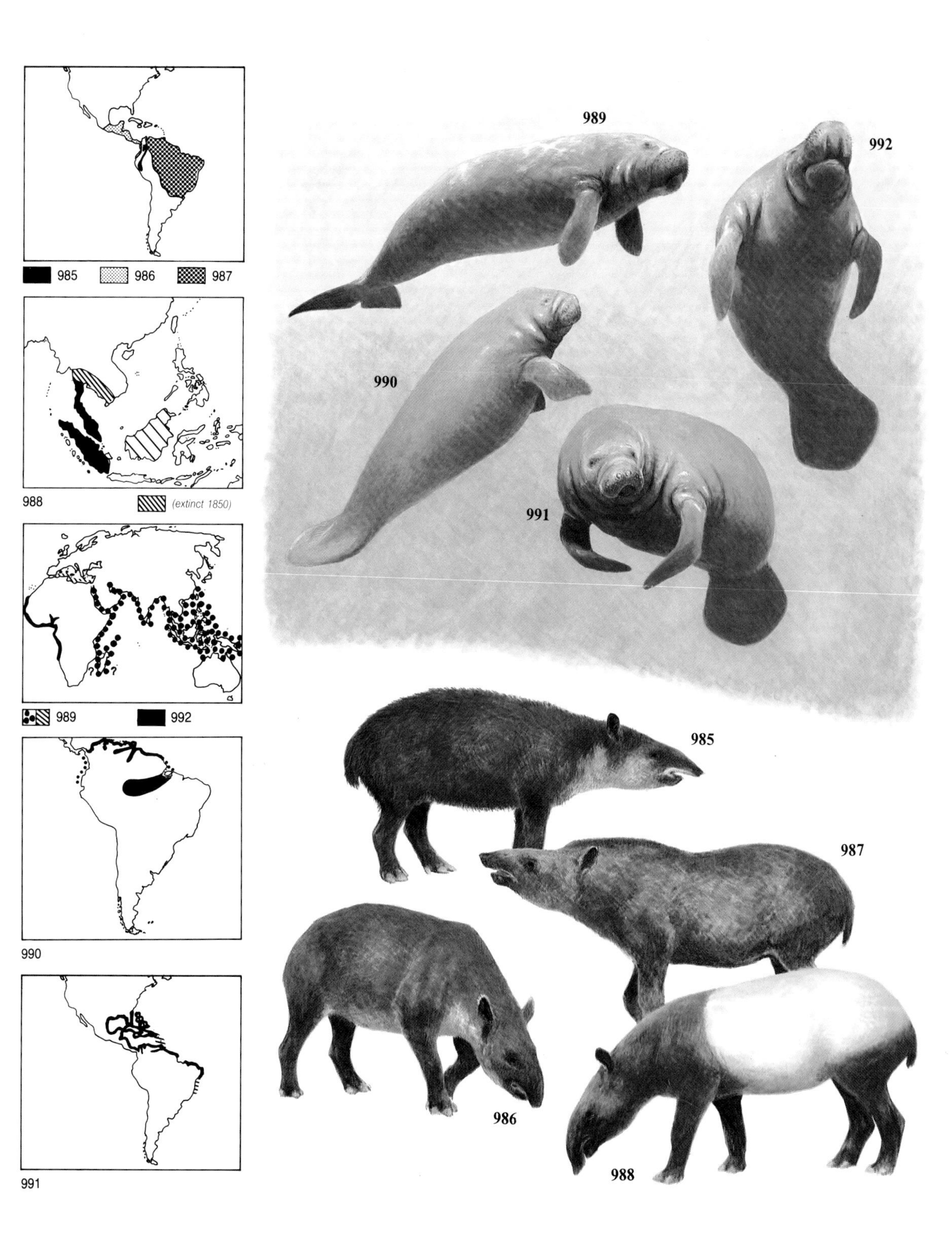
985
986
987
988
(extinct 1850)
989
992
990
991
989
992
990
991
985
987
986
988

*** KIANG

Equus kiang (**993**)

SH ca.55ins; WT up to 882lb.

The largest of the asses, characterised by its large head, and 'Roman' nose. The tail is tufted at the tip. In summer the coat is bright rufous, in winter it is more brown, and long and thick. They live at altitudes of 15,750ft in the arid steppes of Tibet, in herds of 5–10, or up to 40. Because of the inhospitable and inaccessible nature of their habitat, they are less threatened than other equids. Their present distribution extends from China (Xizang, Qinghai, Sichuan) into Ladakh in the Indian State of Jammu and Kashmir. Small numbers are kept in captivity, mostly in USSR, all of which are presumed to be captive-bred.

*** ASIATIC WILD ASS

Equus hemionus (**994**)

SH 43.3–55.2ins.

Smaller than *E. kiang* (993), with narrower head, shorter mane, and longer ears. Lighter-coloured and less rufous in winter. Once found in lowland deserts and steppes of much of Asia. Found in Anatolia 3000 years ago, and on European side of the Urals in 18th Century. At the beginning of this century their range, although reduced at the edges, was continuous from the Gobi and Lake Baikal, west to Kazakhstan, Palestine, and south to the Thar Desert of India, and N Arabia; now reduced, largely by hunting for sport and food, to 4 separate populations. The only populations which are reasonably numerous are the **Dziggetai** *E.h. hemionus* from Mongolian Sinkiang (Xinjiang), China; in 1982 an estimated 358 were recorded in the Kalamaili Mountains Wildlife Reserve, Sinkiang. The **Khur** *E.h. khur* is confined to the Rann of Kutch in India, where ca.870 survived in 1962, dropping to 368 in 1969, but increasing to ca.2000 by 1984. The **Onager** *E.h.onager*, which occurs in Iran, was increasing in the 1960s and 1970s but its current status is unknown. The **Kulan** *E.h. kulan*, although now confined to Turkmenistan, has a reasonably healthy and well-protected population, both in the Badkhyz Reserve, and as an introduction on the island of Barsa Kelmes in the Aral Sea. Asiatic asses are kept in many zoos; the most commonly kept is the Kulan, followed by the Onager, with very few Dziggetais or Khurs.

**** AFRICAN WILD ASS or DONKEY

Equus asinus (=africanus) (**995**)

SH up to 49.5ins.

The ancestor of the domestic donkey, but a more uniform grey, and generally larger. Now confined to remote, arid parts of Africa, where they feed on almost any vegetation.North of Sinai it is only known from Neolithic remains and Biblical references. The western population, *E.a. atlanticus* became extinct ca.300AD, though isolated populations may have survived more recently. By early 1970s the world population was estimated at ca.3000; with ca.1500 *E.a. africanus* in the Sudan, and ca.1500 *E.a. somalicus* in Somalia and Ethiopia, including ca.700 in the Yangudi-Rasa NP, Ethiopia. Military activity and drought in much of their habitat may have had an adverse effect, and little is known of their current status. Small numbers are kept in captivity, notably at Hai Bar Reserve, Israel. Listed on CITES App.I and Class A of the African Convention; they are theoretically protected throughout their present-day range.

**** WILD or PRZEWALSKI'S HORSE

Equus caballus (=*przewalski* or *ferus*) (**996**)

HB ca.49.5ins; SH 49.2–59.1ins; WT 550–662lb.

The ancestor of domestic horses and ponies. Once widespread across Europe and N Asia and extensively hunted by Paleolithic (Stone Age) men. Their already reduced range continued to contract, and by the 13th century they were extinct over all of W Europe, though surviving in E Europe and steppes of the Ukraine until the 18th century, and even until 1918; by the beginning of the 20th century were more or less confined to Mongolia and adjacent China. By the 1970s they were probably extinct in the wild, and it is likely that most, if not all, wild horses had interbred with domestic horses for the past century or more. Western populations were known as **Tarpans**, and lived in forests and open steppes. A Tarpan, bred from primitive Polish Konik ponies, has been reintroduced into the Bialowieza Forest, Poland. The eastern or Przewalski's Horse is kept in considerable numbers in captivity, and it is planned to reintroduce them into the wild.

*** GREVY'S ZEBRA

Equus grevyi (**997**)

HB 8ft 3ins–8ft 6ins; T 27.5–29.5ins; SH 59ins; WT 770–948lb.

Strikingly marked with narrow black stripes on a white background over most of the upperparts, and distinctive striping on the hindquarters; the mane is tall and erect. There are three rather isolated populations, confined to rather open dry habitats in Kenya, north of the Tana River; Ethiopia on the east side of Omo River to Lake Zwai, and Somalia. Believed to have declined drastically in the 1970s, and the remaining herds are estimated to total 10,000–15,000. There are self-sustaining populations in zoos. Listed on CITES App.I.

*** MOUNTAIN ZEBRA

Equus zebra (**998**)

HB 87–102ins; T 29.5–31.5ins; SH up to ca.59ins; WT up to ca.741lb.

Distinguished from other zebras by 'gridiron' pattern on the rump. Confined to montane areas of southern Africa, from S W Angola, through Namibia to Cape Province. There are 2 distinct subspecies: *E.z. zebra* from the Cape, and *E.z. hartmannae,* Hartmann's Zebra, from Namibia and Angola. The Cape population was protected as early as 1656, but this was relaxed when the British took over in 1806. In 1937, when the Mountain Zebra NP was established, the population had been reduced to ca.47 but by the 1980s there were over 200. Hartmann's Zebra is more widespread and numerous, with ca.15,000 recorded in 1960. They are also kept in several ranches and some zoos. The Cape Mountain Zebra is listed on CITES App.I; Hartmann's on App.II.

*** COMMON or BURCHELL'S ZEBRA

Equus burchelli (**999**)

HB 75–96ins; T 16.9–22.5ins; WT 385–783lb.

Striping variable, but always broader than other living zebras; often considerably paler on hindquarters. The most widespread zebra, formerly occurring from Ethiopa, Sudan and Somalia south to South Africa, Angola and Namibia. Range now much reduced, particularly in the south where *E.b. burchelli* became extinct ca.1910. Still numerous in parts of E Africa, but increasingly confined to national parks and other protected areas. Exhibited in most larger zoos.

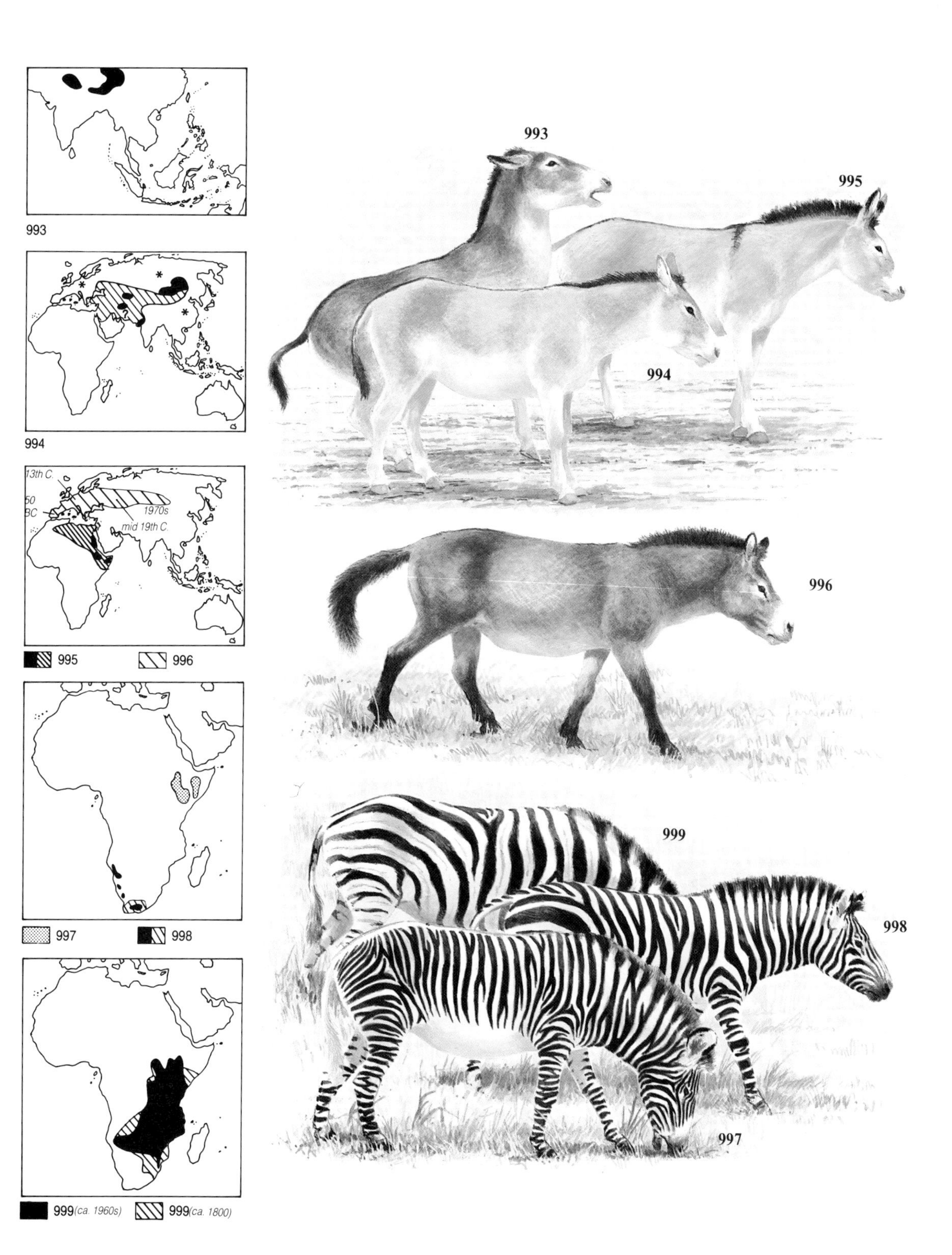
993
994
13th C.
50
BC
1970s
mid 19th C.
995
996
997
998
999(ca. 1960s)
999(ca. 1800)
993
995
994
996
999
998
997

*** WART HOG

Phacochoerus aethiopicus **(1000)**

HB 39.4–59ins; T 13.8–33.5ins; SH 25.6–33.5ins; WT 106–330lb.

The greyish-brown hide is sparsely covered with bristles which also form a yellowish or brown mane. They live in family groups, a pair with 2–4 young and feed on grass, roots, tubers, and also carrion. The 'warts' are protection in the fights between males. Widespread over most of Africa south of the Sahara, at altitudes of up to 8200ft, but declining in many areas, as their rooting and wallowing behaviour is incompatible with agriculture. In the more densely settled parts of Africa they are now more or less confined to parks and reserves.

***** PYGMY HOG

Sus salvanius **(1001)**

HB 19.7–23.6ins; T ca.1.2ins; SH 9.8–11.8ins; WT up to 22lb.

The smallest pig, living in small family groups, and feeding on roots, tubers and other vegetable matter, also insects and other invertebrates. They used to live in grassy swamplands in the foothills of the Himalayas of Nepal, Sikkim, Bhutan, NE India and Bangladesh; thought to have become extinct in the 1960s, and today survive only in Assam. Their existing habitat is threatened by burning in the dry season, and increasing pressure from human settlements. In the late 1970s the total population was estimated at less than 150. Known to occur in several forest reserves and also the Manas Wildlife Sanctuary, but their populations are now extrememly small. Have been bred in captivity, but as yet are not self-sustaining. Fully protected in India and listed on CITES App I.

*** BABIRUSA

Babyrousa babyrussa **(1002)**

HB up to 43.5ins; SH up to 31.5ins; WT up to 220lb.

The male's upper canines grow upwards and backwards, through the flesh of the snout. Their preferred habitat is undisturbed forest, but they also occur in secondary forests, where they feed mostly on vegetation and fallen fruit, and do not root about like other pigs. Found in Indonesia on the islands of Sulawesi, Buru, the Togian and Sula Islands. Habitat destruction is an important cause of decline; they are protected but still hunted; they occur in some reserves including Lore Kalimantan. Listed on CITES App.I; small numbers are being bred in captivity particularly at Surabaya Zoo, Indonesia.

*** CHACOAN PECCARY

Catagonus wagneri **(1003)**

HB up to ca.43.5ins; T 2.8–3.9ins; WT ca.81.5lb.

Generally brownish-grey with a pale collar. Found in arid thorn forests, where they are active by day feeding on cacti, roots, seeds and other vegetable matter. Confined to the Gran Chaco region of Argentina, Bolivia, Paraguay, and possibly Brazil. The species was described from sub-fossil remains in 1930 and although well-known to the local Indians, not known to scientists as a living species until 1974. They live in small groups of 5 or 6, which stay together even when they being attacked and killed. Their main threats come from habitat clearance for cattle ranching, together with hunting, and also other forms of disturbance from oil exploration, road-building, etc. Protected in Paraguay, and occur within the 3600 sq.ml Defensores del Chaco NP, Paraguay. Small numbers have been captured for captive breeding.

*** WHITE-LIPPED PECCARY

Tayassu pecari **(1004)**

SH ca.21.7ins.

Similar to preceding species, but lacking any pale collar. Formerly abundant from Oaxaca and Veracruz in Mexico, south through C America to W Ecuador, Brazil and NE Argentina. Although still locally abundant in S and C America, extinct over much of its former range, and there are only isolated populations in Costa Rica and some other parts of C America. Although protected in much of its range, still extensively hunted. Found in many national parks and reserves over their range; rare in captivity, with few bred.

*** GIANT FOREST HOG

Hylochoerus meinertzhageni **(1005)**

HB 51–83ins; T 11.8–17.7ins; SH 33.5–39.4ins; WT 286.6–606.4lb.

Male larger than female. The skin is greyish, with long, black hair. Usually nocturnal, hiding by day in dense thickets. They occur in a wide variety of mainly forested habitats, and also on plains with stands of trees and shrubs. They live in family groups, with their 1–4 young, feeding mainly on vegetable matter. Widespread in forests from W Africa, east to Ethiopia, Kenya and Tanzania, but their range is increasingly fragmented and isolated, particularly in the west.

*** BEARDED PIG

Sus barbatus **(1006)**

HB 47–63ins; T 6.7–10.2ins; SH ca.35.5ins; WT 125.7–330.7lb.

Colour variable, but generally blackish when young, adults paler. Normally live in groups of 4–5, but also in larger groups, and have litters of up to 11 young. The piglets are striped horizontally with orange/brown. They tend to hide by day emerging at night to feed on crops. Five subspecies are found from the Malay Peninsula south to Sumatra, Borneo, Palawan and the Philippines. They are extensively hunted in many parts of their range, particularly in areas where there are no religious taboos against pigs, such as Visayan Isles, Philippines. In many areas, despite having undergone massive declines, they are still widely regarded as pests. Although there is no captive breeding programme as such, there is no reason why they should not breed easily in captivity. The * **Javan Warty Pig** *S. verrucosus* **(1007)** is confined to Java, Madura and Bawean, where it has been shot and poisoned as an agricultural pest. They do not occur in any reserves and little is known of their status. Although declining, they are still abundant in some areas. The * **Sulawesi Warty Pig** *S. celebensis* **(1008)**, which may be conspecific with *S. verrucosus*, has been domesticated; its wild populations are declining.

* WILD BOAR

Sus scrofa **(1009)**

HB up to 71ins; T ca.11.8ins; SH up to 43.5ins; WT up to 772lb.

The wild ancestor of the domestic pig. Formerly distributed throughout Europe and Asia, including the British Isles, Japan, Sri Lanka, some Mediterranean Islands and N Africa. Its distribution in SE Asia has been extensively modified by man. Extinct in British Isles, much of Europe and N Africa; range fragmented; in past 25 years dramatic increase in some parts of Europe, due to changes in agriculture, and shifts in the human population. Hunted extensively. Found in many reserves and national parks. Introduced into many parts of the world, including USA, New Zealand and Hawaii.

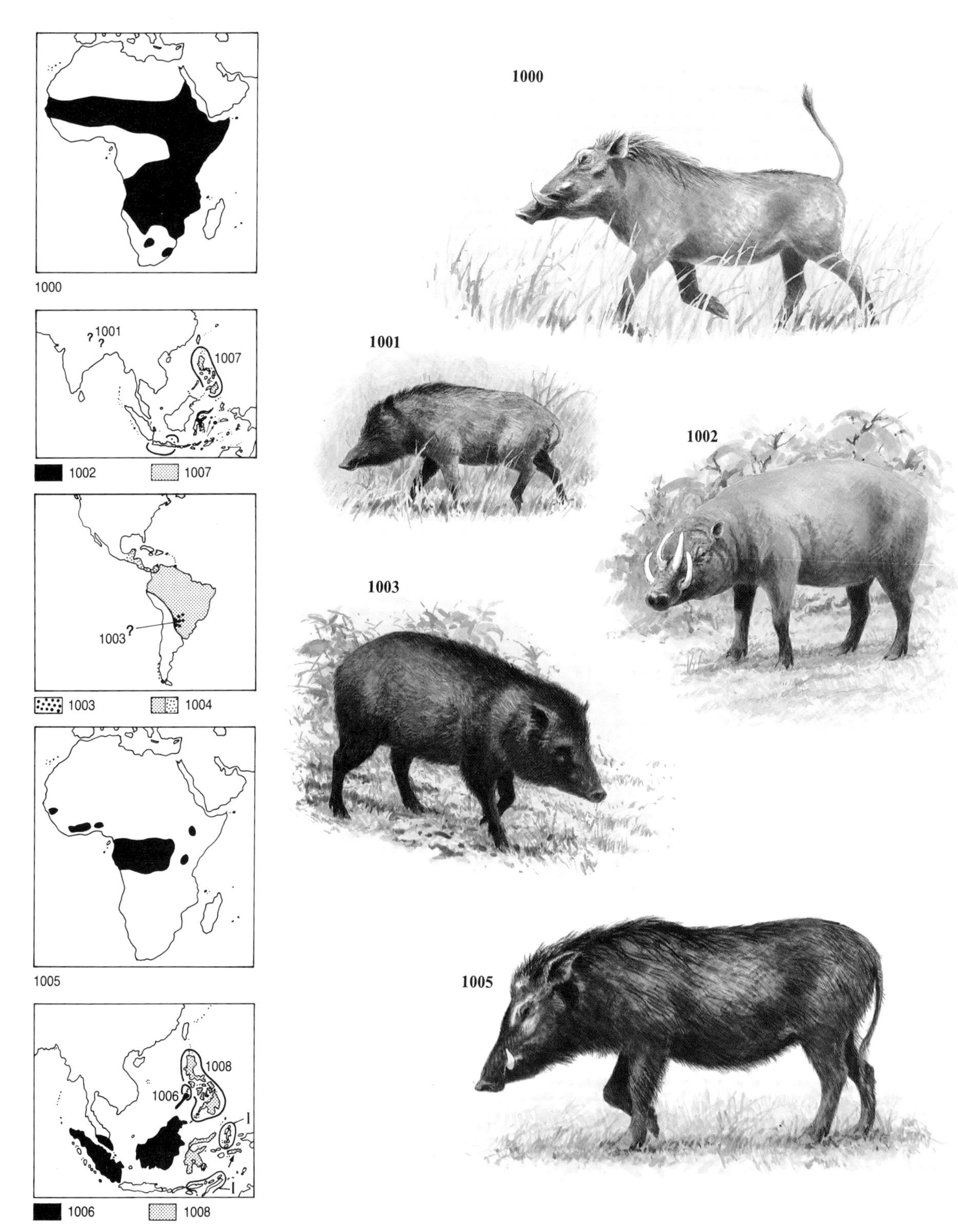
1000
1001
1007
1002
1007
1003
1003
1004
1005
1008
1006
1006
1008
1000
1001
1002
1003
1005

*** HIPPOPOTAMUS

Hippopotamus amphibius **(1010)**

HB up to 14ft 9ins; T ca.19.7ins; SH 4ft 11ins–5ft 5ins; WT up to 9920lb.

The skin is almost naked with a sparse covering of hair. Both the upper and lower canines are enlarged into tusks up to 2ft long, weighing up to 6.6lb. Most of the day is spent resting in or close to water, emerging at night to graze, mostly on grasses. Sociable animals, with a mature male dominant over a group of females and young. Once found over almost all of Africa where there were water temperatures of 18–36°C. Within historic times, have been exterminated over the entire northern part of their range, the last one disappearing from Egypt ca.1816. Also extinct or reduced to isolated remnant populations in the west and south of their range. In the Upper Nile Valley and other parts of East Africa large populations still survive, particularly in national parks. In some areas numbers are sufficiently large as to be a pest to agriculture, but overall it appears that they are continuing to decline both in numbers and range. Hippopotamuses have been hunted for the ivory from their teeth, for meat and fat, and also for trophies; in areas close to human settlement they damage crops by trampling and eating them. They can be aggressive and cause many human deaths. Protected in most countries; listed in the African Convention; exhibited in many of the larger zoos where they breed regularly.

*** PYGMY HIPPOPOTAMUS

Choeropsis liberiensis **(1011)**

HB 59–69ins; T 5.9–8.3ins; SH 29–39ins; WT 352.8–595.3lb.

Like a small Hippopotamus in general shape and appearance, but with less protruding eyes and well separated toes. They are less aquatic than the Hippopotamus, found in wet forests and swamps, always close to water, feeding on leaves, shoots, fruits, tubers etc. They are normally solitary, and nocturnal. Have probably always been sparsely populated, but have become increasingly rare, due to illegal hunting and widespread habitat destruction. Listed on CITES App.II and the African Convention. Kept in many zoos where they breed well.

**** BACTRIAN CAMEL

Camelus bactrianus **(1012)**

HB 7ft 5ins–11ft 3ins; T ca.21.7ins; SH 5ft 11ins–6ft 11ins; WT 992–1521lb.

Colour of fur varies from dark brown to sandy or greyish; when moulting it is shed in large lumps. They possess a number of adaptations to living in sandy desert and other arid environments, including thick eyelashes, nostrils which can be closed to keep out sand, and broad feet. They feed on almost any vegetation and can survive without water for extended periods, drinking up to 125 pints in order to replenish. Formerly found in herds of up to 30; a single calf is born every other year. The Bactrian Camel is known to have been domesticated by about the 4th century BC, remaining widespread in the wild, and locally fairly common until the latter part of the 19th century. Now restricted to the Gobi Desert on the borders of Mongolia and China, where 300–500 are thought to survive. They are protected and occur within reserves. Domesticated Bactrian Camels are common in zoos.

**** ONE-HUMPED CAMEL

Camelus dromedarius **(1013)**

Extinct in the wild, although large feral populations exist in Australia, and small populations in Asia. Believed to have become extinct in the wild ca.2000 years ago, originally occurring in the Near and Middle East.

*** GUANACO

Lama guanicoe **(1014)**

HB ca.71ins; T ca.10.6ins; WT up to 165lb.

Dark brownish above, pale buff or white below; the fur is dense and woolly. Found mainly in fairly arid country, in mountains and plains from S Peru and E Argentina, throughout Patagonia south to Tierra del Fuego. Formerly widespread and abundant within this range. In 1978 population estimated at 50,000–100,000, and declining due to hunting pressures, in Argentina; reduced to a few scattered groups and rarer than the Vicuna (1015) in Peru; in Chile once-large populations in the north are now reduced to scattered groups on the coastal range and parts of the Andes, while the southern population was estimated at 3000 on the mainland, and 7000 on Tierra del Fuego. The Guanaco is either closely related to or the ancestor of the Llama and Alpaca, which were domesticated ca.4500 years ago. They occur in several reserves within their range, notably Tierra del Fuego and Los Glaciares NP in Argentina and the O'Higgins NP, and adjacent reserves in Chile. Frequently kept in zoos where they are self-sustaining. Protected in Peru, Chile and Argentina, and listed on CITES App.II.

*** VICUNA

Vicugna vicugna **(1015)**

HB 50–75ins; T 5.9–9.8ins; SH 27.6–43.3ins; WT 77– 143.3lb.

Rich brownish above, paler below. Similar to the Guanaco, but smaller and lighter in colour. The incisors are unique among hoofed animals as they grow continuously. They usually live in small groups consisting of a male, several females and their offspring, or in batchelor herds, in territories of up to 0.12sq.mls. They live in arid montane grasslands and plains, at altitudes of 11,500–18,850ft and once occurred widely in the Andes from S Ecuador, south to Argentina and N Chile. During the Inca period they were rounded up and sheared, but subsequent to European colonisation, they were slaughtered for skins and meat. It is estimated that there were up to 1.5million at the time of the collapse of the Incas. By 1940s there were still ca.400,000, but by 1967 this had dropped to ca.10,000, caused by demand for skins, and the spread of livestock into their grazing grounds thereby competing for food. Strict conservation measures have allowed numbers to rebuild, and there are now estimated to be more than 80,000 in the wild. Small numbers are bred in zoos. They are subject to a conservation agreement between Peru, Chile, Argentina and Bolivia, and are listed on CITES. App. I.

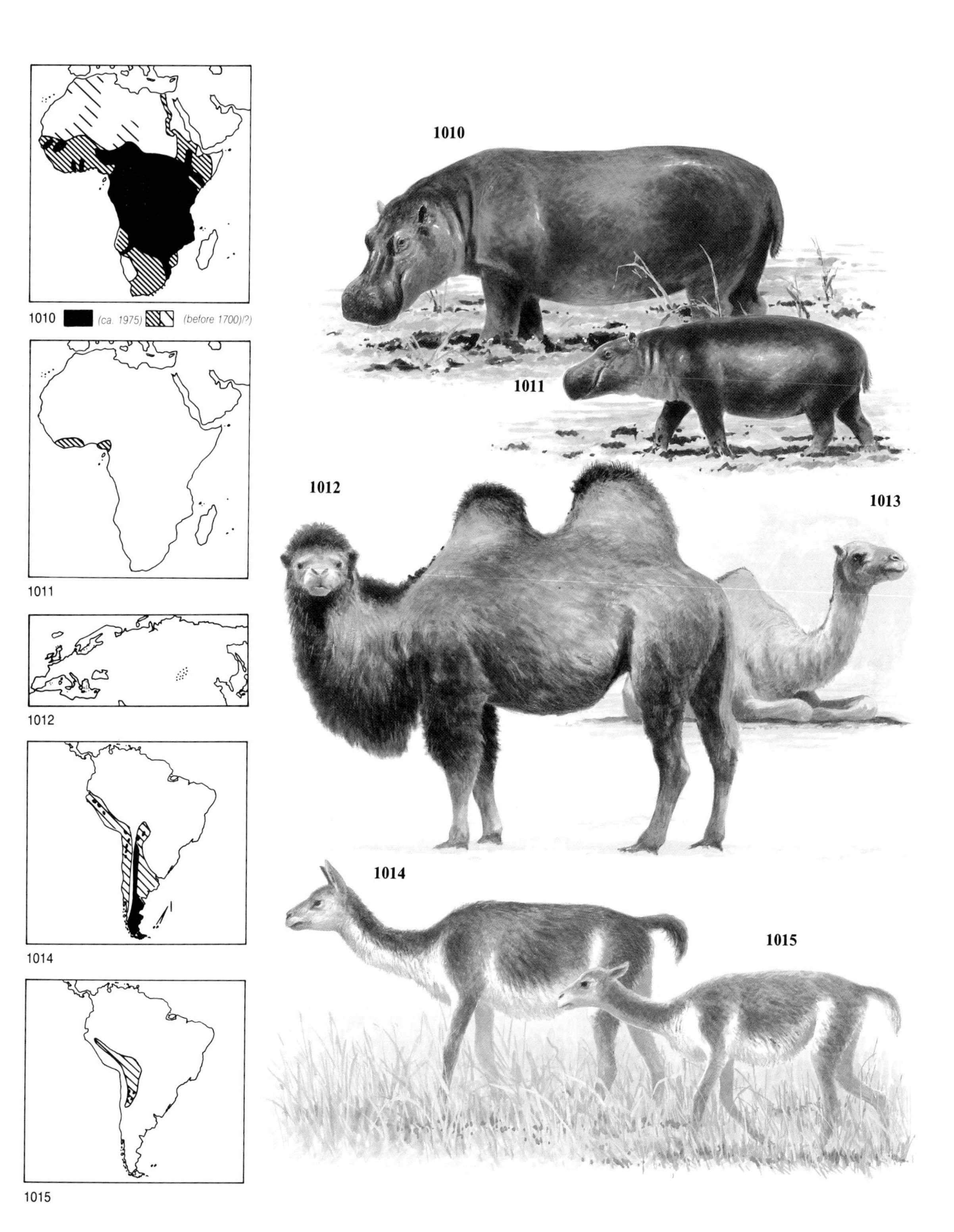
1010
1011
1012
1013
1014
1015
1010 (ca. 1975) (before 1700)/?)
1011
1012
1014
1015

*** **MUSK DEER**
Moschus spp.
HB 31.5–39.4ins; T 1.5–2.4ins; SH 20.1–24ins; WT 15.4–37.5lb.
The three species are closely related and very similar in general appearance. There is no agreement among biologists as to the definition of the species, and there is an enormous amount of confusion in their taxonomy.They lack antlers, but the upper canines of the males are developed into tusks about 3ins long. Their name is derived from the gland in the abdomen of the males, which secretes a waxy substance used in the manufacture of perfumes and other toiletries. Musk deer live at altitudes of 8550–11,800ft in the south of their range, lower in the north; *M. moschiferus* (**1016**) is found in Siberia, Mongolia, NE China, Korea, and Sakhalin Island; *M. sifanicus* (**1017**) is found from E Afghanistan and N Pakistan to China and N Burma; *M. chrysogaster* (**1018**) is found in the Himalayas of India, Nepal, Bhutan and Tibet. Musk deer have been extensively hunted for their musk glands, and are now locally extinct in many parts of their range. During the 1930s an average of 10,000–15,000 were killed each year for their musk. However, it is now possible to extract the musk without killing the deer, and because of the high value of musk, ranches and farms have recently been established in China. Only small numbers kept in zoos. They occur in national parks and reserves, and are listed on CITES App.II, with some populations on App.I.

*** **FEA'S MUNTJAC**
Muntiacus feae (**1019**)
HB ca.34.5ins; T ca.3.9ins; WT ca.48.5lb.
Fea's Muntjac is restricted to a small area in Thailand and adjacent Burma, where it is hunted extensively. Habits are said to resemble those of the more widespread Common Muntjac; they are confined to evergreen forests in upland areas. (Closely related to *M. rooseveltorum*, which is only known from Muong Yo in Laos; this species has not been recorded for over 50 years and is presumed extinct.) Only occasionally exhibited in zoos and there are no self-sustaining populations.

*** **BLACK or HAIRY FRONTED MUNTJAC**
Muntiacus crinifrons (**1020**)
For nearly a century after its discovery, the Black Muntjac was only known from 3 specimens, but in the 1980s it was rediscovered with a restricted range in EC China.On Borneo two species of Muntjac occur: the * **Bornean Yellow Muntjac** *M. atheroides* (**1021**) has only recently been effectively distinguished from the Bornean population of the Common or Red Muntjac *M.muntjak pleiharicus*. Little is known of their status, ecology or distribution.

*** **FALLOW DEER**
Cervus (=Dama) dama (**1022**)
HB 51–69ins; T 5.9–9.1ins; SH 31.5–43.3ins; WT 88.2–220.5lb.
The most variable of all deer, with 4 main varieties: brown above with white spots in summer, greyish-brown in winter; pale fawn, heavily spotted with white, all the year; dark, almost black, with faint spotting; white – but not albino. The males carry broad, palmate antlers, up to 37ins long. They live in herds, and during the breeding season males establish territories. The range of the Fallow Deer has been considerably altered by man, but it was probably originally found in open woodland habitats around the Mediterranean region. It is extinct in N Africa, became extinct in Greece in the 19th century, and in Sardinia in 1950s. The eastern population, often regarded as a separate species, *C.d. mesopotamicus*, the Mesopotamian Fallow Deer, became increasingly rare in the 1940s, and was believed extinct; however a small population of less than 50 was discovered in the 1960s in marshes on the Iranian/Iraq border. The fate of this population in what is currently a war zone is unknown. Fallow Deer are numerous in parks, zoos, and as introductions in most parts of the world; however most of these populations are partially domesticated. The Mesopotamian Fallow Deer is being bred in small numbers in zoos; the largest group is in Hai-Bar, Israel. This subspecies is listed on CITES. App. I.

** **PRINCE ALFRED'S SPOTTED DEER**
Cervus alfredi (**1023**)
The size of a small Sika Deer (1032) and often considered only a subspecies of the Sambar *C. unicolor*, it has a row of whitish spots on each side of the back, and a scattering of spots elsewhere. It is confined to the Philippines, and extinct on the islands of Cebu and Guimaras; its status on Leyte, Masbate, Panay, Negros and Samar islands is poorly known.

*** **MARIANA SAMBAR**
C. mariannus ((**1024**)
HB 39.4–59ins; T 3.1–4.7ins; SH 21.7–27.6ins; WT 88.2–132.3lb.
Smallest of the *Cervus* deer; it is little known, and confined to the Mariana Islands, Philippines, including Guam Rota and Saipan. It is probably only a subspecies of the Sambar, and introduced to the islands by man, but is sometimes regarded as a full species.

***RUSA**
Cervus timorensis (**1025**)
Rather like a small Red Deer, this species has a wide distribution from Java east to Timor and the Moluccas, which has been undoubtedly modified by man. Populations in New Guinea, Australia and New Britain are certainly introduced by man. Several populations are probably threatened or seriously endangered, such as the Javan Rusa *C.t. russa*, but this subspecies has been introduced into Madagascar, Borneo and Sulawesi, and is thriving. Similarly, little is known of the status of many Sambar subspecies (*C. unicolor*).

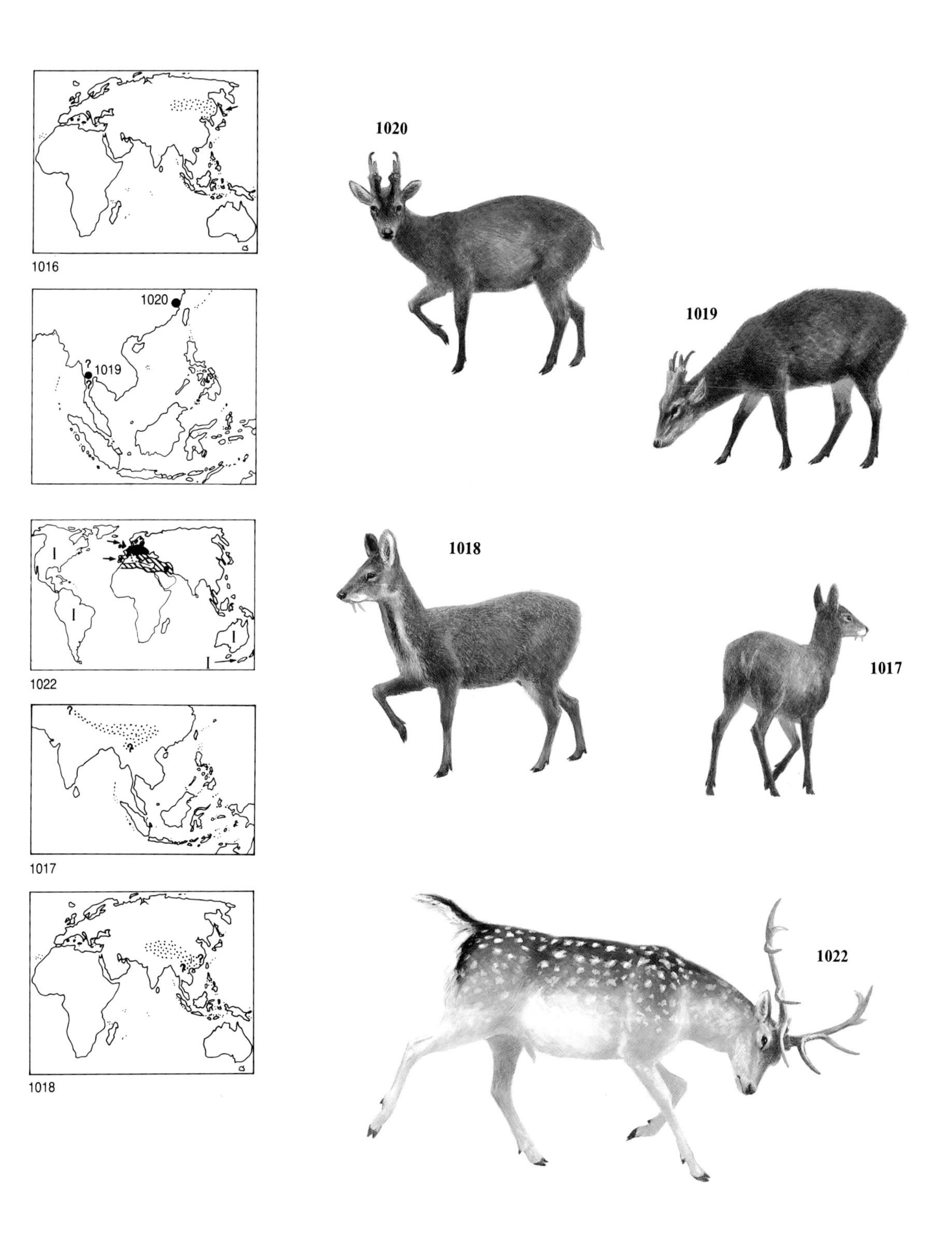
1016
1020
?
1019
1022
I
I
I
I
1017
?
1018
1020
1019
1018
1017
1022

***** SWAMP DEER or BARASINGHA**
Cervus duvauceli **(1026)**
SH up to ca.53ins; WT up to ca.397lb.
The antlers of the stags grow to a maximum of ca.41ins. Colouring is rather variable, from brownish to yellowish, often with small white spots on back and sides. The shape of the hooves also varies, with the swamp-dwelling *C.d. duvauceli* having large splayed hooves. Now confined to India and Nepal, occurring in a wide range of habitats, including marshes, where they are often extremely aquatic, and grassy areas close to water; also forest. *C.d. duvauceli* is restricted to the Terai of Uttar Pradesh, Nepal, Assam, and the Sunderbans, and *C.d. branderi* is found in drier habitats in Madhya Pradesh. By the mid 1960s the total population was estimated at less than 4000, with several populations occurring in protected areas. There are substantial populations in zoos, particularly in the USA, but most are not identified to subspecies level. Listed on CITES App.I. They were unsuccessfully introduced into Australia in the 1860s.

***** THOROLD'S DEER**
Cervus albirostris **(1027)**
SH up to 47.5ins.
Brownish above, creamy below with nose, lips chin and throat white; the hairs on the withers point forwards giving a humped appearance. It is confined to E and N Tibet, the upper reaches of the Yangtze River and Nan Shan, China. It has been depleted by hunting for the antler trade. Little is known of its current status, but there are small numbers in captivity.

*** HOG DEER**
Cervus porcinus **(1028)**
HB 55ins; T 6.7–8.3ins; SH 25.6–28.4ins; WT 154.3–242.5lb.
Their name originates from their way of running through the undergrowth, head held down like a wild pig. Found in riverine habitats, marshes and swamps with tall grasses. Where they are undisturbed they are often diurnal and form small herds; elswhere they are mainly nocturnal and solitary. Although locally abundant, particularly in parts of India, they are much reduced in range and numbers in the Indo-Chinese part of their range, and believed to have become extinct in Bangladesh during the 1971 war. The Bawean Island population ** *C. kuhli* **(1029)** is sometimes treated as a full species; in 1969 the population was estimated at 500. To the west of the main Philippine Islands, the Calamian Islands are inhabited by **** *C. calamianensis* **(1030)**, which numbered less than 900 in total by the mid-1970s; They only occur with a protected area, but poaching is rife. A captive herd has been established in Britain.

***** BROW-ANTLERED DEER or THAMIN**
Cervus eldi **(1031)**
HB 59–67ins; T 8.7–9.8ins; SH 47.3–51.2ins; WT 209.5–330.7lb.
The antlers of the stags grow to over 3ft. Stags are dark brown or blackish, hinds much lighter in colour. Three populations are recognised: the **Thamin** *C.e. thamin* from Tenasserim (Burma), Thailand, and formerly the Malay Peninsula; the **Sangai** *C.e. eldi*, restricted to the Keibul Lamjao NP, at the south of Logtak Lake in Manipur, where the population fell from ca.100 in 1969 to 14 in 1975; and the **Thailand Brow-antlered Deer** *C.e. siamensis* from Thailand east to Hainan, which is now extinct or close to extinction over most of its range. They have declined mainly through overhunting, and also through loss of habitat and competition with domestic livestock, but respond well to protection. The Sangai increased due to protection from 1977 on, when Keibul Lamjao was declared a NP and by 1977 there were 49 in captivity in Indian zoos. The status of the Thamin in the wild is poorly known but there are many in zoos where they breed regularly. Nominally protected through most of their range, and listed on CITES App.I.

***** SIKA DEER**
Cervus nippon **(1032)**
HB up to 59ins; SH up to 31.5ins; WT 88.2–143.3lb.
A very variable species both in colouring and size. Formerly restricted to the Far East, from Ussuri in Siberia through Korea to N Vietnam and Japan, and Ryukyus and Taiwan. They have been extensively introduced in many parts of the world, and have sometimes become pests. Some of the original populations, particularly those on islands are threatened, mostly by habitat loss and overhunting; these include *C.n. taiouanus* from Taiwan; *C.n. keramae* from the Ryukyu Islands; *C.n. mandarinus* from N China, *C.n. grassianus* from C China and *C.n. kopschi* from EC China. Of these, only the Taiwan Sika is commonly seen in zoos, where in the early 1980s the population was reaching 500, all captive-bred.

***** RED DEER or WAPITI**
Cervus elaphus **(1033)**
HB up to 98ins; T up to 7.5ins; WT up to nearly 1100lb.
Considerable variation in size, and numerous subspecies described, often on the basis of antler shape; the antlers grow to more than 5ft and weigh over 28lb. Although mainly a woodland species browsing on leaves and other vegetation, they are extremely adaptable and are found in a wide variety of habitats including semidesert, parkland and open grasslands. Their original distribution is confused by numerous introductions; also what was once a more or less continuous distribution throughout much of N America, Europe and N Asia is now extremely fragmented. Among the more threatened populations are the **Hangul** *C.e. hanglu* confined to Kashmir, which once numbered as few as 500 or 600, but is rebuilding its numbers in Dachigam NP; *C.e. bactrianus* from Central Asia and *C.e. yarkandensis* from Sinkiang, are also considered to be in some danger; *C.e. corsicanus* became extinct on Corsica, but survived on Sardinia and is being reintroduced into Corsica; *C.e. barbarus* is the only deer to occur naturally in Africa, and is now confined to a small area in the Atlas Mountains. In N America a large number of subspecies of Wapiti have been described, but are difficult to separate. Several have become extinct since European colonisation, and, particularly in the east, their range is often fragmented; in the late 1960s a total of 440,000 were estimated in N America. Red Deer thrive in captivity and have been extensively introduced. *C.e. hanglu* and *C.e. bactrianus* are listed on CITES App.I and II respectively.

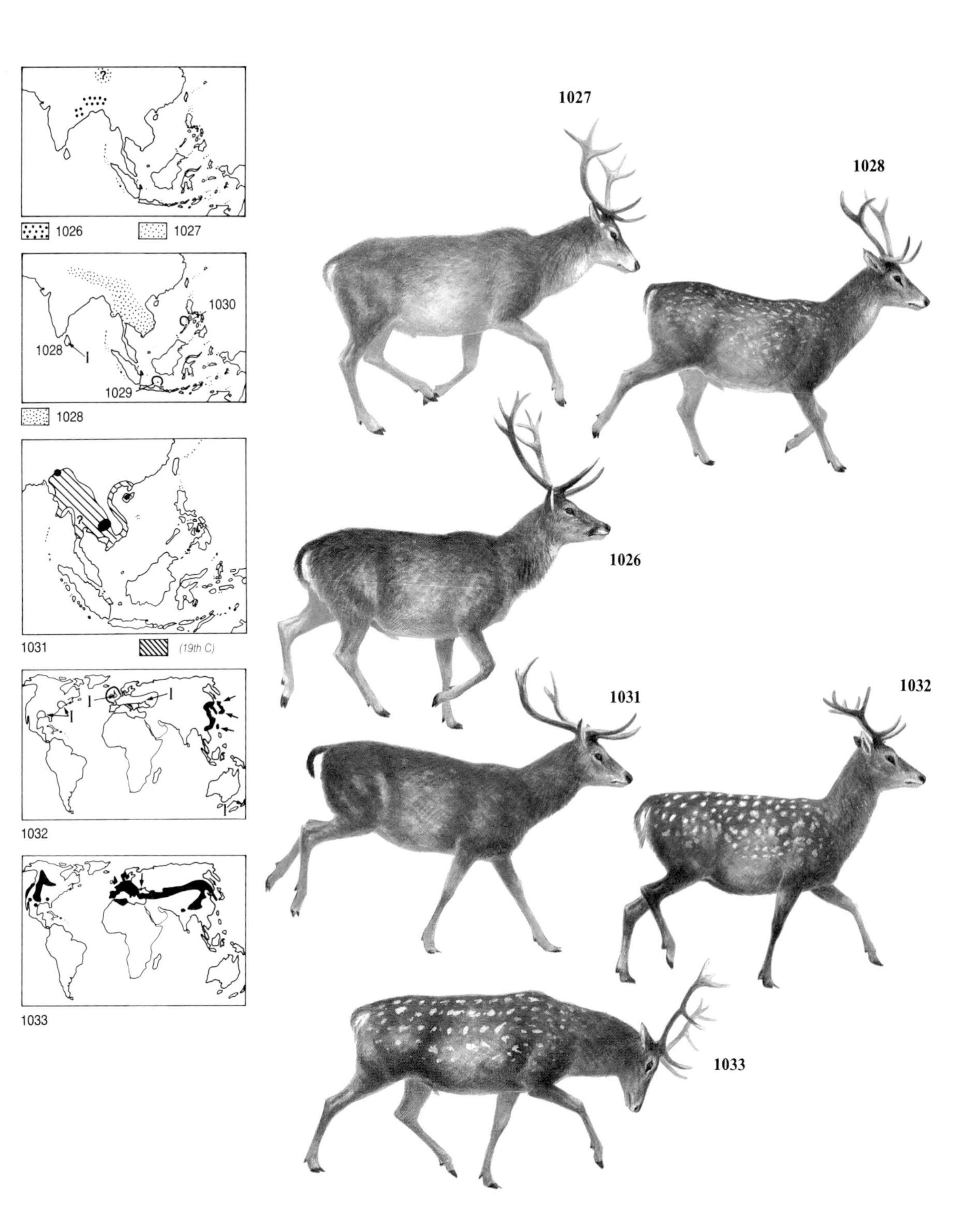
?
1026
1027
1030
1028
1029
1028
1031
(19th C)
1032
1033
1027
1028
1026
1031
1032
1033

****** PERE DAVID'S DEER**
Elaphurus davidianus **(1034)**
HB ca.59ins; T ca.19.7ins; SH ca.45.5ins; WT 330–440lb.
The tail is long, the hooves large and spreading. Pere David's Deer is thought to have originally occurred in marshy habitats in NE China, but became extinct in the wild ca.1800 years ago. Captives survived in the Imperial Hunting Park until the end of the 19th century, and animals from here were sent to Europe by Father Armand David shortly before they were exterminated in China. European stocks were gathered together at Woburn, UK, by the Duke of Bedford and as the stock increased herds were established in most major zoos. By the 1980s there were over 1000, worldwide, and a semi-wild group was being established in China.

***** MARSH DEER or GUASU PUCU**
Blastocerus dichotomus **(1035)**
HB 71–77ins; T 3.9–5.9ins; SH 43.4–47.3ins; WT 220–331lb.
The largest South American deer, bright chestnut in summer, duller in winter; the legs are blackish and tail black below. The male's antlers usually fork twice. The hooves can be splayed very widely and are bound together by membrane, enabling them to walk on very soft ground. They live in wet, wooded areas, often going into water, and lying up during the day, emerging at night to feed on grasses, reeds and other plants. Unlike most other deer, the breeding of Marsh Deer does not appear to be markedly seasonal, with antlers being grown and shed throughout the year and the single calf being born at any time of the year. Numbers have declined drastically in many parts of its range, through loss of habitat to agriculture and drainage, and also by excessive hunting pressures. Believed to have become extinct in Uruguay by about 1929. Found in NE Argentina, Bolivia, Brazil south of the Amazon, Paraguay, SE Peru where they occur in some national parks and reserves, and protected in all parts of its range, though the law is rarely enforced. Listed on CITES App I, though not extensively traded. Relatively few are held in captivity, and so far captive breeding has not been successful.

*** MULE DEER**
Odocoileus hemionus **(1036)**
HB 57ins; T 4.5–9.1ins; SH ca.37.5ins; WT 70.6–397lb.
Antlers grow to maximum spread of 47ins. The summer coat is reddish, greyish in winter, and fawns are spotted. They are very adaptable and occur in a wide range of habitats in the western half of North America, but mainly in dry, open forest, shrublands, and prairies. Seven subspecies are generally recognised. Those from the coastal area in N California are known as **Black-tailed Deer** and an estimated 5 million existed in pre-Columbian times.Following settlement by Europeans numbers declined, but subsequently increased, in the early part of the 20th century, due to beneficial changes in vegetation; in the 1960s and 1970s numbers decreased again – the reasons for this are not fully understood. The **Cedros Island Mule Deer** *O.h. cerrosensis* is a poorly defined subspecies confined to Cedros, off Baja California, Mexico; it is possible that a few hundred exist.

***** PAMPAS DEER**
Ozotoceros bezoarticus **(1037)**
HB 43.3–51ins; T 3.9–5.9ins; SH 2.8–3ins; WT 66.1–88.2lb.
Reddish-brown or fawn above, with a darker face and paler underparts; the tail is white below and dark brown above. Originally widespread in the pampas and other open habitats of Brazil, south and east to Bolivia, Paraguay, Uruguay and C Argentina. They live singly or in small herds. In the south of the range, breeding is seasonal, with calves born in April, although they are not seasonal nearer the equator. When disturbed, the mother with young, may feign injury to distract the intruder. The Pampas Deer has declined throughout most of its range due to loss of habitat and uncontrolled hunting, and the Argentine population *O.b. celer* is particularly threatened. In 1975 this population was believed to have fallen to about 100, but with protection had risen to 400 by 1980. Protected throughout its range, though in many parts this is not enforced. Listed on CITES App.I and occurs in many parks and reserves. Small numbers are kept in captivity, where they could be self-sustaining and provide animals for reintroduction.

***** WHITE-TAILED DEER**
Odocoileus virginianus **(1038)**
HB up to 94ins; T 5.9–13ins; SH up to 39.5ins; WT up to ca.423lb (usually less).
Distinguished from the Mule Deer by its long bushy tail, white on the underside. The males carry antlers, which grow to a spread of 35ins. They are one of the most adaptable and widespread mammals in the Americas, occurring from S Canada through most of the USA, Mexico and south to Peru and NE Brazil; some 125,000 were hunted under licence in Canada, and 1,875,000 in USA in 1978, with similar numbers in recent years. However some populations are threatened. The **Key Deer** *O.v. clavium*, a small race, is confined to the islands in the Lower Florida Keys. In 1949 they were on the verge of extinction, with less than 30 surviving, but by the late 1970s had built up to 300–400 or more. The **Columbian White-tailed Deer** *O.v. leucurus* formerly occurred throughout west Washington State, and Oregon, but now confined to a small area along the lower Columbia River, part of which is a reserve. This population has increased under protection and is now considered out of danger, but their main threat now is hybridisation with the **Columbian Black-tailed Deer** *O. hemionus columbianus*.

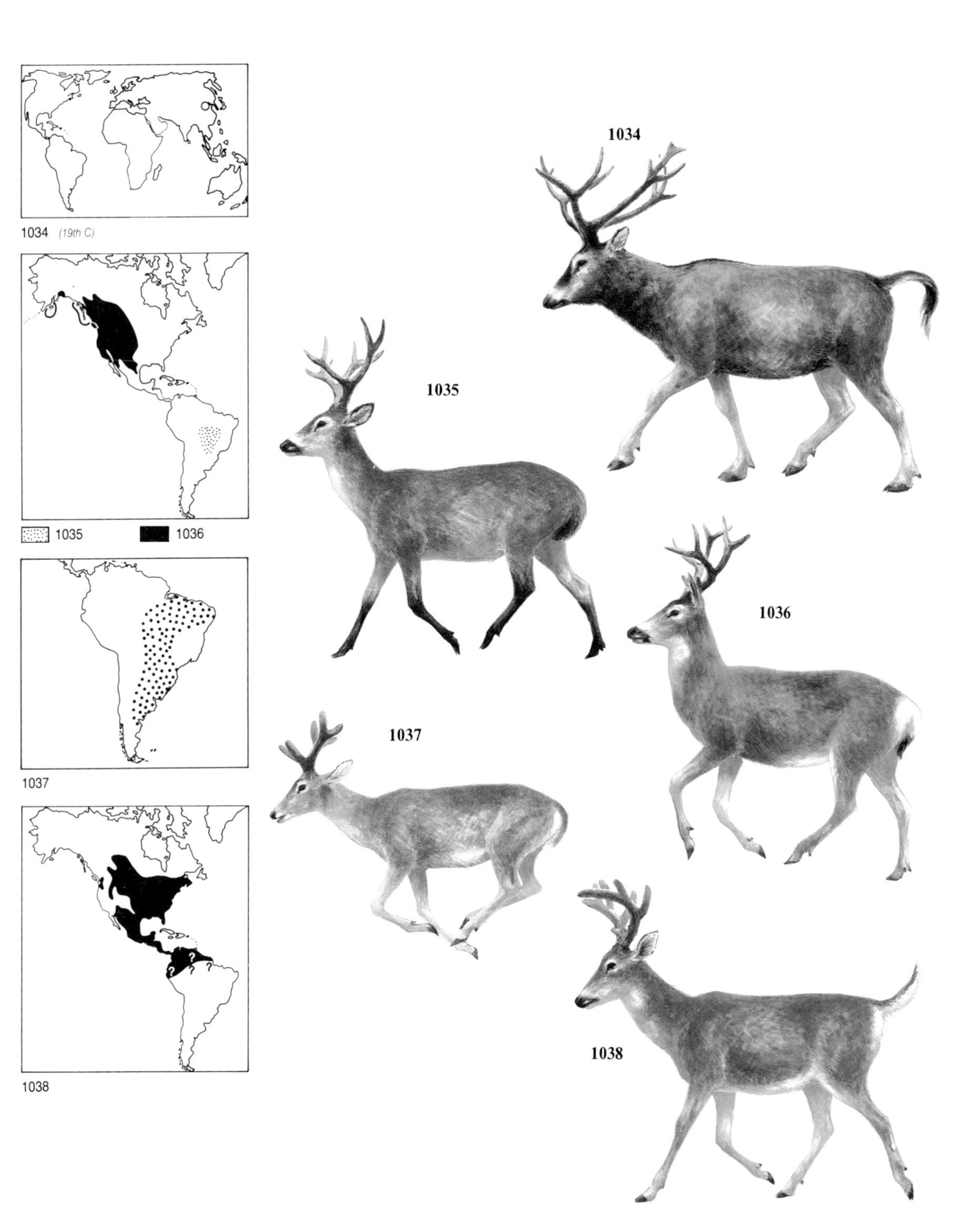
1034 (19th C)
1035
1036
1037
1038
?
1034
1035
1036
1037
1038

***** NORTHERN PUDU**
Pudu mephistophiles (**1039**)
HB ca.35.5; T ca.1.2ins; SH ca.13.8ins; WT up to 22lb.
Generally a rich brown, with blackish face and legs, and long white hairs inside the ears. They have short spiky antlers. Found in dense forest from 6550–13,100ft in the Andes. Very little is known of habits, distribution or status, as they are extremely difficult to observe. Their range is probably fragmented, but includes the Andes of Colombia, Ecuador and N Peru. They are extensively hunted, and their habitat suffers from fires. Only rarely seen in zoos. Listed on CITES App.II.

*** SOUTHERN PUDU**
Pudu pudu (**1040**)
Similar to the Northern Pudu, but generally browner; fawns have 3 rows of spots on the back. They live at lower altitudes than the Northern Pudu, ranging from sea level to 5550ft, and occur in S Chile, including Chiloe Island; there are also small numbers in SW Argentina. Although suffering from loss of their forest habitat, they are not thought to be in any immediate danger. Listed on CITES App.I. Small numbers are kept in zoos, where they breed regularly.

***** PERUVIAN HUEMUL**
Hippocamelus antisensis (**1041**)
HB ca.59ins; T ca.4.7ins; SH up to 31.5ins; WT up to 143lb.
Males have small spiky antlers. Found in forests and grasslands at altitudes of 10,850–16,400ft, feeding on grasses, sedges and other herbage. They live in small groups of females and their young, the males being solitairy outside the breeding season. They are still widespread, but declining, in the Andes of Peru, W Bolivia, NE Chile, and NW Argentina. In some areas they are reported to be increasing with protection. They occur in Lauca NP, Chile; Ulla Ulla Fauna Reserve, Bolivia and several other protected areas. Often found in areas inhabited by Vicuna (1015) and have benefited from the protection given to them. Rarely exhibited in zoos and there are no records of captive breeding. Listed on CITES App.I.

******* CHILEAN HUEMUL**
Hippocamelus bisulcus (**1042**)
Small spiky antlers, carried by the males. Very similar both in size and in coloration to the Peruvian Huemul (1041), but paler underneath, with a brown spot on the rump and brown under the tail. Found in forests and shrublands, mostly at higher altitudes, but very little is known of their current distribution and status. By the early 1980s its total population was estimated at under 2000, spread from about 37°S in Chile, south through the Andes of Chile and Argentina, and possibly some of the larger islands. They occur in several national parks, including Los Alerces and Los Glaciares in Argentina, and Rio Simpson NP, Chile. The main threats come from hunting, as well as loss of habitat, competition from domestic livestock, and disease. Protected by law, but not easily enforced in their rather remote habitat. Listed on CITES App.I.

***** REINDEER or CARIBOU**
Rangifer tarandus (**1043**)
HB 47.3–87ins; T 2.8–8.3ins; SH 31.5–55.2ins; WT 132.3–701.2lb.
The fur is thick, with woolly underfur and coarse guard hairs. Colour is variable, but mostly brownish above, paler below. The only deer where both sexes carry antlers, up to 51ins long in males. Formerly found in the tundra and adjacent boreal forests of the northern hemisphere, and also in some montane areas further south. Feed on a wide variety of plants including lichens, leaves, berries, twigs and shoots. Herds congregate for seasonal migrations and once numbered many thousands. In N America 30 discrete migrating populations have been identified. Before European colonisation, the Caribou population of N America has been estimated at 3.5million, and in the mid-1970s, at 1.1million; they are threatened locally by overhunting, as well as pressures associated with oil developments. Numerous subspecies have been described from N America, but most are dubiously distinguishable. The Queen Charlotte Island population *R.t. dawsoni* became extinct about 1910; a distinct white population *R.t. pearyi* has become extinct in Greenland, and declined to 10,000–15,000 in the rest of its range in the Canadian Arctic Archipelago. In Europe, Reindeer were formerly much more widespread, occurring as far south as Germany and Poland until Medieval times; they became extinct in the wild throughout Europe, except USSR. In the USSR the total population was estimated at ca.800,000 in the mid-1970s. In the 1950s they began to recolonise Finland, and by the early 1980s there were 400–600 in Kuhmo, Finland. Interbreeding with domesticated Reindeer, and hunting, have been responsible for its extinction in Europe. Although often stated to have occurred in Scotland in historic times, the remains are probably those from beasts brought from Scandinavia. Reindeer are frequently exhibited in zoos, but most are interbred with domestic forms. They have been introduced into S Georgia, Kerguelen Island, Iceland, Scotland and many other places, with a total domestic population of ca.3 million in the late 1970s.

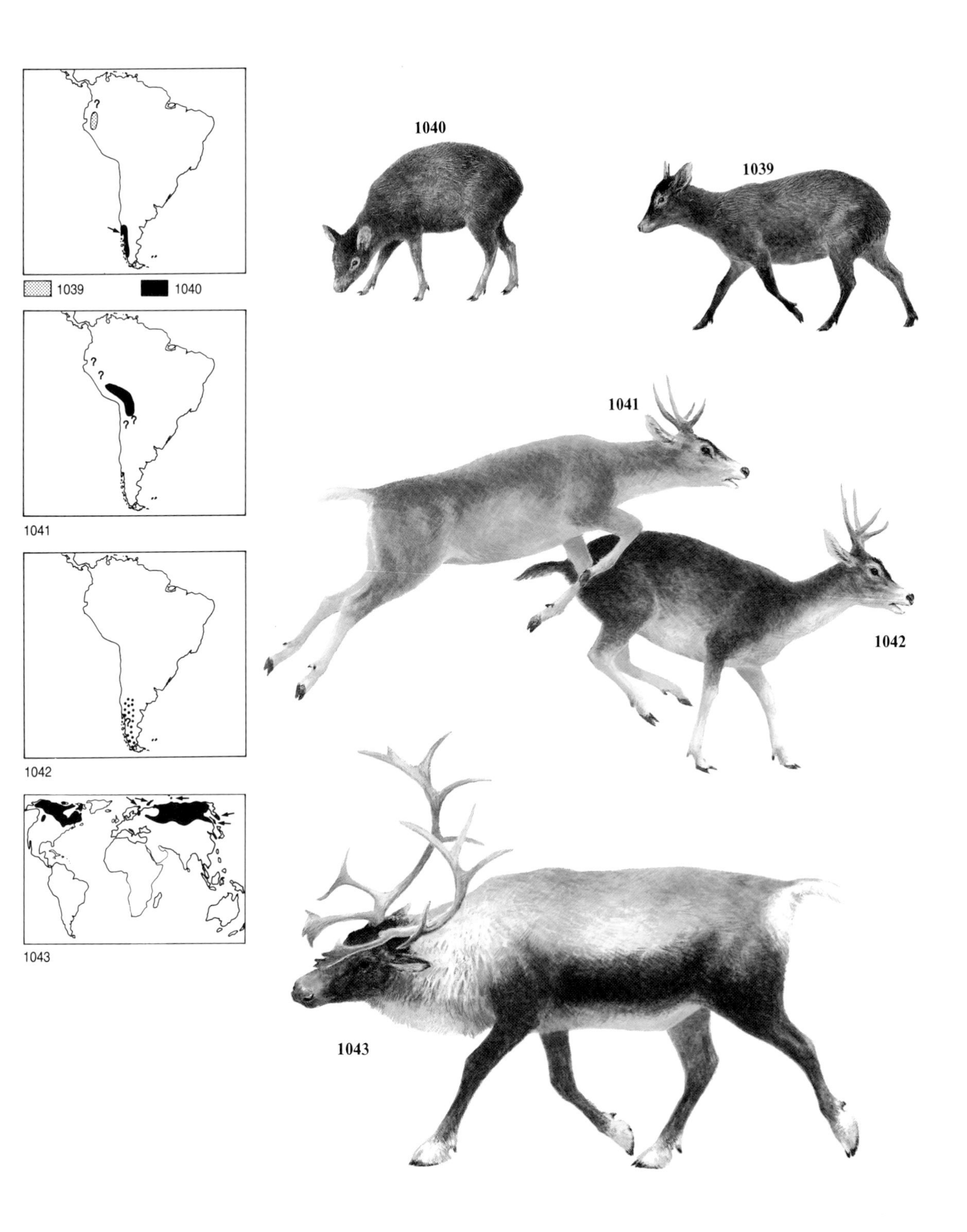
1039
1040
1041
1042
1043
1040
1039
1041
1042
1043

***** COMMON ELAND**
Taurotragus oryx (**1044**)
HB 6ft 7ins; T 19.5–29.5ins; SH 49.5–69ins; WT 660–2095lb.
Horns on the male grow to ca.39.5ins, on females to ca.27.5ins. They are rather ox-like animals, with a small hump and a dewlap on the lower part of the throat. Formerly widespread in open habitats from SE Sudan, throughout E Africa, and most of southern Africa. They live in herds, usually small, but occasionally up to 1000 or more. Their numbers have declined drastically, particularly in southern Africa, both as a result of the spread of agriculture and hunting, and also from disease, particularly rinderpest, which exterminated them in many parts of South Africa in the 1890s. They have since recolonised many areas, including the Kruger NP; they are found in most of the national parks of E Africa, and have been domesticated in Africa and Askania Nova, USSR. Frequently exhibited in zoos where they breed freely.

****** GIANT ELAND**
Taurotragus derbianus (**1045**)
HB up to 11ft 4ins; T up to 35.5ins; SH up to 71ins; WT up to 1.1 tons.
Both sexes carry horns which, in the male, grow to maximum 47ins. The largest living antelope, closely related to, and often considered conspecific with, *T.oryx* (1044). Confined to open, wooded country where they are mainly browsers. Like the Common Eland, their flesh is highly prized, and their hides are valuable. They have been exterminated over most of their former range, which included a wide belt across sub-saharan Africa from Senegal and the Gambia to the SW Sudan and NE Uganda. The western population, *T.d. derbianus* is now extremely rare, but protected and stable in Senegal; the eastern population *T.d. gigas* is more numerous and widespread, but has undergone a substantial decline. Found in the Garamba NP, Zaire, Niokolo-Koba NP, Senegal, and several other protected areas. Only a few Giant Eland are kept in captivity, mostly captive-bred. Listed in Annex A of the African Convention.

*** MOUNTAIN NYALA**
Tragelaphus buxtoni (**1046**)
HB 75–102ins; T 7.9–9.8ins; SH 35.5–53.2ins; WT 330–662lb.
Only the males carry horns, up to 47ins long. Dark greyish-brown, with white markings. A little-known species not described until 1910, found in forests and tree-heaths at an altitude of 8850–13,800ft, browsing on grasses, leaves and shoots. They live in small herds of up to about 15. Confined to Ethiopia, with populations estimated in the mid-1970s at less than 11,500 in Arussi Mountains, and less than 980 in Bali Mountains. They are threatened by habitat destruction and disturbance, as well as hunting. Other related species such as the * **Sitatunga** *T. spekei* (**1047**) are also suffering local declines, but cannot be regarded as immediately endangered. The *** **Bongo** *T. euryceros* (**1048**) is rather rare, even within the areas where it is known to occur, and several populations are isolated by unsuitable habitat. In West Africa and Kenya they are particularly vulnerable.

***** PRONGHORN**
Antilocapra americana (**1049**)
HB 39.4–59ins; T 3–7.1ins; SH 31.5–41ins; WT 79.4–154.3lb.
Permanent horns are carried by males and most females; they grow to about 10ins in males, about 5ins in females. Found in a wide variety of open habitats up to an altitude of 11,000ft, where they are both browsers and grazers feeding on grasses, herbs, bushes, cacti and other plants; if necessary they can survive without water. During autumn they gather into large, loose, mixed herds of up to 1000; in spring these herds break into smaller groups and the sexes separate. The young are born in spring: usually a single calf the first year, and from then on twins or occasionally triplets. Before the colonisation of North America by Europeans there were more than 35 million Pronghorns. Overhunting and the spread of ranching reduced them to less than 20,000 by the 1920s. Conservation measures have allowed the population to rebuild to a level where hunting is now allowed in many parts of their range. The Mexican population continues to decline. *A.a. sonoriensis* and *A.a. peninsularis* are listed on CITES App.I and *A.a. mexicana* on App.II. Rarely seen in zoos outside N America.

***** GIRAFFE**
Giraffa camelopardalis (**1050**)
HB 11ft 6ins–17ft 5ins; T 30–39.5ins; SH 6ft 7ins–12ft 2ins; WT 1213–4256lb.
The tallest mammal, reaching a maximum of ca.19ft. The colour and pattern consists of varying shades of brown blotches on a buff or creamy background. Both sexes have 2–4 knobbed horns. They live mainly in open woodlands and dry savannah, browsing on trees and bushes. Herds usually consist of females and young, but bachelor herds also occur; bulls are usually solitary, wandering from herd to herd. The range of the Giraffe has been drastically reduced within historic times; they once ranged in the area which is now the Sahara Desert and thoughout most of Africa where suitable habitat existed. Now extinct north of about 15°N and reduced to isolated fragments, mostly in national parks and reserves in the west and south of their range; even in E Africa, they are becoming increasingly confined. While not likely to become extinct, their future outside reserves is doubtful. Exhibited in most of the larger zoos and frequently bred, often through several generations. Protected in most countries in which they occur. Around 20 subspecies of Giraffe have been described, but their status is far from certain; however several are now isolated and occur in discrete populations with little chance of interbreeding with others. **Rothschild's Giraffe** *G.c. rothschildi* from W.Kenya, has a restricted range and small numbers, but the **Reticulated** *G.c. reticulata* is still fairly common, even outside parks and reserves.

*** OKAPI**
Okapia johnstoni (**1051**)
HB 77–85ins; T 11.8–16.5ins; SH 59.1–67ins; WT 463–551lb.
Dark, almost black with a reddish or purplish sheen; the hindquarters and legs are transversely striped. The ears and eyes are large and the tongue is so long that it is used to clean the eyes. Found in dark, dense equatorial forests where they are primarily browsers, feeding on a wide variety of leaves and fruits of trees. They live singly or in small groups, and the single young is born during the rainy season. The Okapi was unknown to science until 1900, and has remained a little-known species because of its habitat. Extinct in Uganda, and now confined to the forests of Zaire, where it has been protected since 1933. Main threats are hunting, and in some areas destruction of habitat, but they are probably not in any immediate danger. Exhibited in only a few zoos, but most are now captive-bred, except those in Zaire.

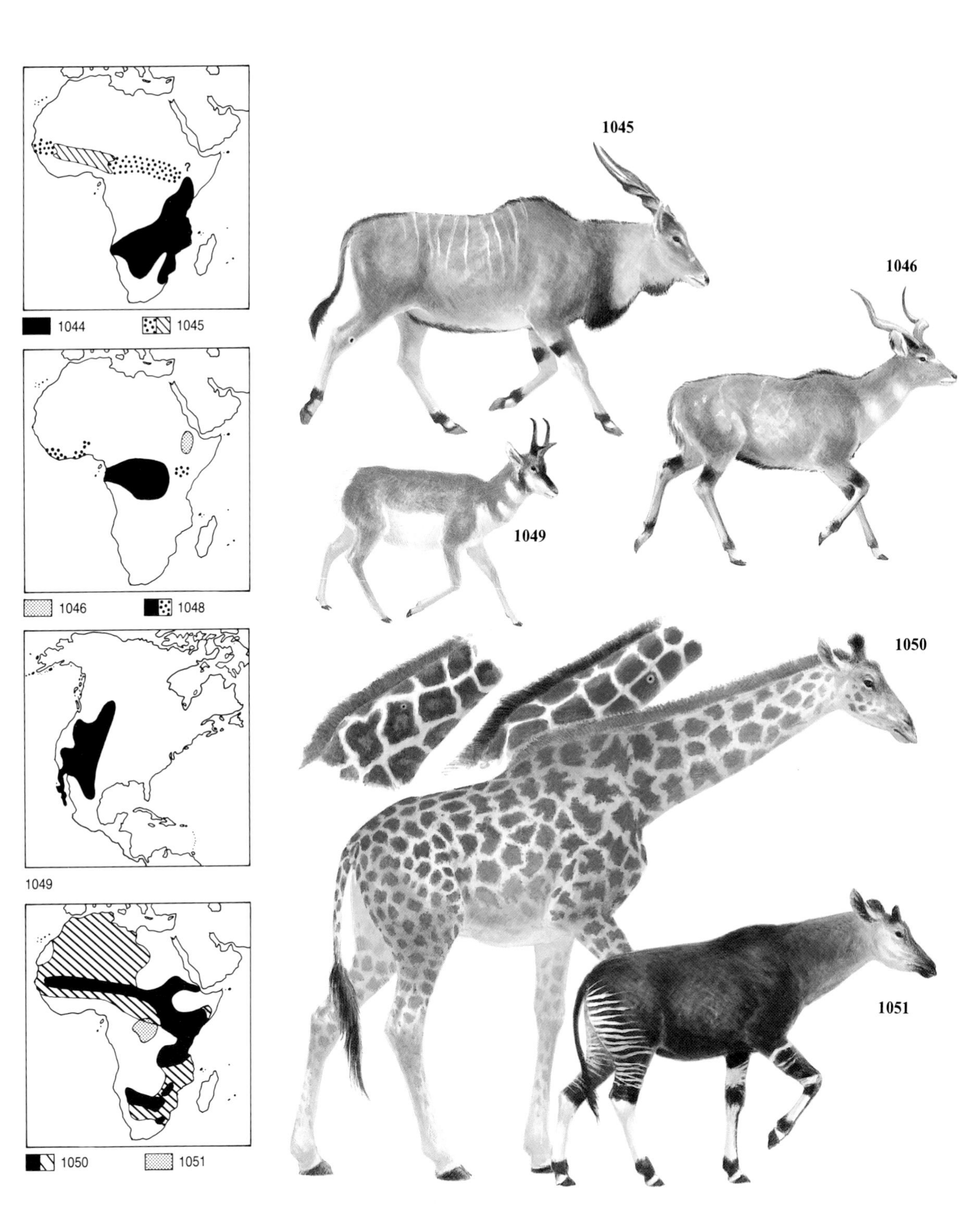
1044
1045
1046
1048
1049
1050
1051
1045
1046
1049
1050
1051
?

*** AFRICAN BUFFALO

Syncerus caffer (**1052**)

HB 6ft 11ins–9ft 10ins; T 29.5–39ins; SH 39.5–67ins; WT 1103–1985lb.
Both sexes carry horns, which in the males grow to 59ins. The African Buffalo varies considerably in size, with some of the forest populations half the size of those from the plains and savannah. Some of the largest are in South Africa, the **Cape Buffalo** *S.c. caffer*, which now have a very restricted range. At one time *S. caffer* occurred in almost all habitats south of the Sahara, preferring those near water, and with some tree or scrub cover. They are gregarious, occurring in herds of up to several thousand. In most parts of their range births occur throughout the year, and there is a single calf. In the wild they are known to have lived up to 18 years. Their range has been considerably reduced by habitat changes and they have also declined due to hunting; in the west of their range, populations are now considerably fragmented, and only occur as isolated groups. Even in parts of East Africa they are becoming increasingly confined to national parks, and in South Africa almost entirely confined to parks and reserves. The rinderpest epidemic of the 19th century was largely responsible for the extinction of the African Buffalo in southern Africa, and with the spread of European farmers, it was unable to recolonise once the epidemic had passed. Exhibited in most of the larger zoos and regularly bred.

*** WATER BUFFALO

Bubalus bubalis (**1053**)

HB 7ft 11ins–9ft 10ins; T 23.5–39.5ins; SH 59–75ins; WT 0.8–1.3 tons.
Generally greyish or blackish. Both sexes carry horns, which curve upwards and inwards, growing to 47ins along curve – the largest horns of any living cattle (their record is nearly 79ins). Found in wet grasslands and swamps, near to pools and wallows, where they cake themselves with mud as a defence against insect bites. They live in herds of females and their young, the males being in separate bachelor herds outside the breeding season. Water Buffalo were once widespread from E Nepal and India, east to Vietnam, and south to Malaysia, but are now extinct over a large part of their range. They have been displaced by the spread of agriculture, together with competition and diseases from domestic cattle; they are also hunted. They occur in many protected areas within their range and are bred in small numbers in captivity. *B. bubalis* is the ancestor of the domestic Water Buffalo, which is now found in most tropical and subtropical regions, both as a domesticated and, in many areas, a feral species.

***** TAMARAW

Bubalus mindorensis (**1054**)

SH ca.39.5ins; WT ca.662lb.
The horns are short and thick, growing to maximum of ca.20ins. They live in dense forest, grazing in open glades, close to water for wallowing; they feed on grasses, bamboo shoots and aquatic vegetation. Confined to the island of Mindoro in the Philippines, where by the mid-1970s the population was estimated at less than 200–280 in Mt Iglit Game Reserve, with some others in reserves on Mt Calavite and Mt Sablayan. Listed on CITES App.I.

*** LOWLAND ANOA

Bubalus depressicornis (**1055**)

HB ca.67ins; T ca.10.2ins; SH ca.39.5ins.
The horns grow to ca.14ins, and are triangular in section. Very small cattle, dark brownish, often with white markings on the legs; the young are a yellowish-brown. They live in small family groups or pairs, but little is known of their habits in the wild. Confined to undisturbed forests on the island of Sulawesi, Indonesia. There are small numbers in captivity, but an effective captive management programme is lacking.

*** MOUNTAIN ANOA

Bubalus quarlesi (**1056**)

Very closely related to the Lowland Anoa, and until recently often regarded as conspecific. Confined to the uplands of Sulawesi, but little is known of its status or habits. Listed on CITES App.I. Small numbers in captivity, with a few being bred, mostly in Germany.

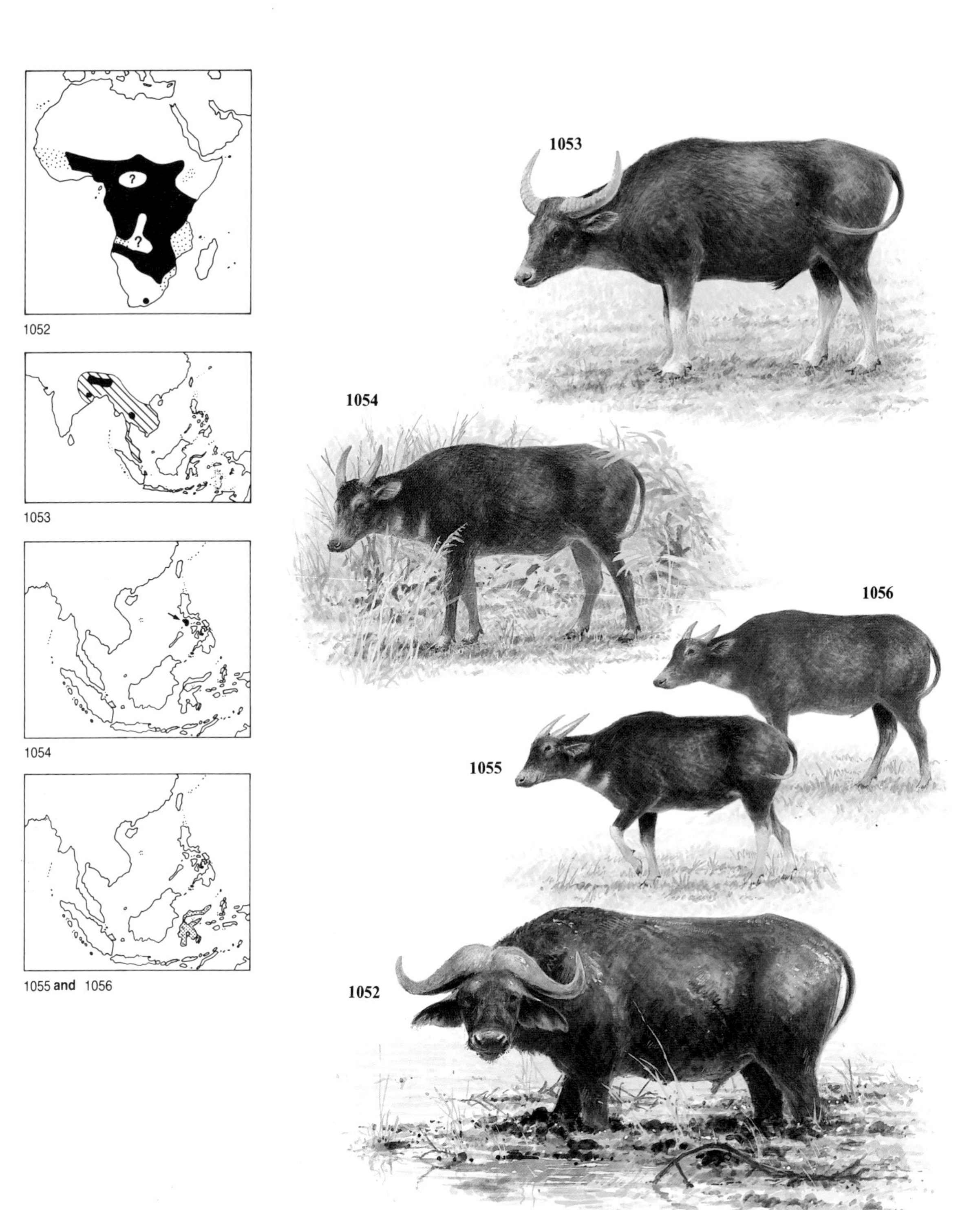
1053
1054
1056
1055
1052
1052
1053
1054
1055 and 1056

*** BANTENG

Bos javanicus (**1057**)

HB 77–88ins; T 23.6–27.6ins; SH 51–63ins; WT 1102–1985lb.

Colour variable, but often more bluish-black than Gaur; has white stockings and rump patch, and both sexes carry distinctive horns. Habitat dependent on dense forest, but often found around clearings. Largely nocturnal, in areas close to human habitation, hiding by day. During monsoons may move to higher ground. They feed on a variety of herbage as well as bamboo shoots and grasses. Groups of up to 40 animals live together led by a single adult male; other males are solitary or live in bachelor herds. They have 1 or 2 calves each year. They have disappeared from many areas through loss of habitat, as well as increased hunting pressure. They occur from Burma, Thailand and Indo-China, south to Java and Borneo, at altitudes of up to 6550ft. Most populations are endangered, although in Indonesia the Banteng is domesticated. Protected in many of its range states, and occurs in national parks and reserves. Exhibited in some zoos, most of these being captive-bred. Introduced Banteng (domestic) live ferally on Coburg Peninsula, Australia.

*** GAUR

Bos gaurus (=frontalis) (**1058**)

HB 8ft 2ins–10ft 10ins; T 27.5–39.5ins; SH 63–86ins; WT 1433–2205lb.

Horns up to 39ins long. Dark reddish- or blackish-brown with white stockings. The male has a shoulder hump. Found in hill forests (up to 5900ft) with grassy clearings, normally near water for drinking. Diurnal in undisturbed habitats, where they browse and graze during the cooler hours of the day; near human habitation they are normally nocturnal. They live in herds of up to 40 animals led by a single male. The single calf is usually born during the cooler months. Gaur populations have been drastically reduced, though they are still scattered over a wide area. Threatened by habitat loss, hunting, and by cattle diseases. In 1981 estimated that there were less than 400 in Malaysia; large populations still exist in some Indian reserves such as Mudumalai and Kanha NP. The domestic form of the Gaur, the **Gayal** or **Mithan**, which is smaller, sometimes occurs ferally with some 50,000 in Arunachal Pradesh, India. The Gaur is protected in most of its range, and occurs in many national parks and reserves. Not very common in zoos, where most animals are captive-bred.

***** KOUPREY

Bos sauveli (**1059**)

HB 86–88ins; T 39.4–43.3ins; SH 67–75ins; WT 1543–1985lb.

Greyish; underparts being pale in females and young, and dark brown in males. They have pale greyish stockings and a rather long tail, with a bushy tip. Males have a dewlap, which may almost reach the ground with age. The horns of the female are lyre-shaped, those of mature males widespread and frayed at the tips. Found in low hilly areas, with patches of forest grazing in open grassland. They are diurnal, hiding in the hotter parts of the day. They live in herds of up to 20, which may contain more than one male. Since their discovery in 1937, when a young bull arrived at Paris Zoo, they have remained one of the least-known mammals. Most of their range is in an area subject to war in Vietnam and Kampuchea; they have also been allegedly sighted in Thailand. There were estimated to be 1000 in 1940; 500 in 1951; and 100 in 1969. Since then they have hovered on the brink of extinction, with occasional sightings. None is currently in captivity. It has been suggested that they may be resistant to rinderpest, but it is equally possible that their rarity is due to susceptibility to the disease; there is no evidence for either hypothesis. One of the world's rarest mammals, and one of the largest to be described this century.

*** WILD YAK

Bos mutus (=grunniens) (**1060**)

HB up to 10ft 6ins; SH 79ins; WT up to 2205lb.

Blackish-brown, with long hair; females are only about one third of the size of males. Found in steppe and mountain areas of 20,000ft, around the snowline, particularly during the summer months. In winter they survive temperatures as low as -40°C. They graze on grasses, herbs and lichens. Yaks gather in herds of 20–200, occasionally up to 1000 or more, females and young; males are solitary or live in small bachelor herds. A single calf is born every two years. Although once very numerous and widespread, numbers have declined drastically, largely due to uncontrolled hunting. The Yak is domesticated, and is now widespread in the higher regions of the Himalayas and the mountains of Central Asia. Although found in some zoos, these are usually of the domesticated variety, which is smaller and has more varied coloration.

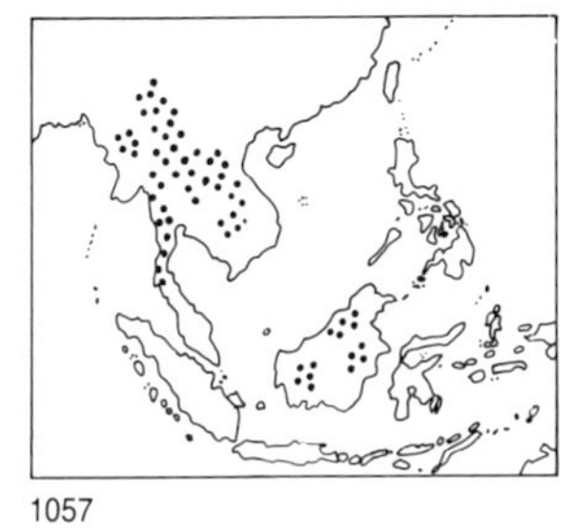

1057

1058

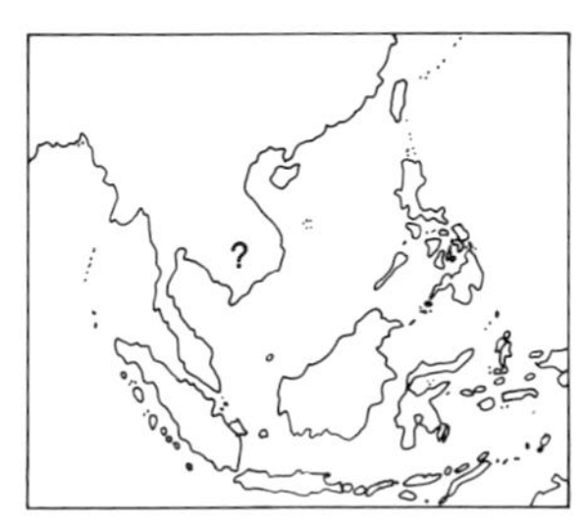

1059

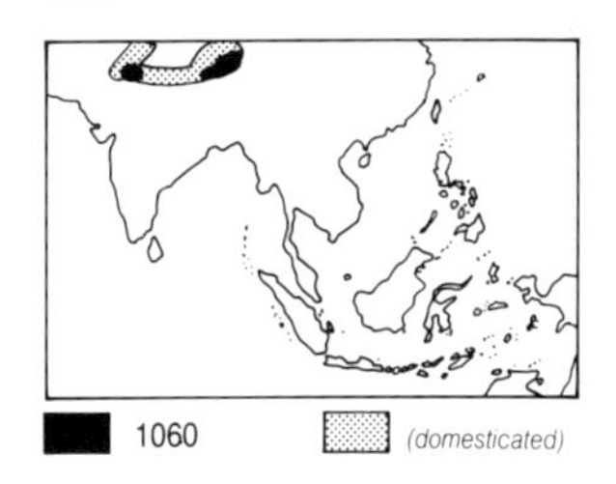

1060 (domesticated)

ANTELOPES

Most antelopes found in Africa and Asia have declined within historic times, and many continue to do so. They have been hunted extensively, and modern firearms, and the human population explosion over the past 150 years, has led to rapid declines and fragmentation of their range. Most species deserve some form of listing, at least locally.

*** JENTINK'S DUIKER

Cephalophus jentinki (**1061**)

HB ca.53ins; T ca.5.9ins; SH ca.31.5ins; WT up to 154.5lb.

Both sexes carry horns up to 6.9ins long. Like most other duikers, occur in dense forest, but little known about their habits; only a handful of specimens in museums, and few observations in the wild. Known from scattered localities in Liberia and possibly the Ivory Coast.

*** BLUE DUIKER

Cephalophus monticola (**1062**)

HB up to 35.5ins; T up to 5.1ins; SH up to 15.8ins; WT up to 22lb.

Both sexes have small horns; those of the males' up to 4ins, the females' up to ca.1.6ins, or sometimes absent. Closely related to Maxwell's Duiker *C. maxwelli*, which is one of the most widespread species in W Africa. The Blue Duiker occurs in forest regions of Nigeria, east of Cross River, east across sub-Saharan Africa to Kenya and S Africa. Although still locally abundant, over most of its range it is hunted extensively and has disappeared from many areas of dense human populations. They probably occur in many protected areas within their range. In S Africa, after declines in the 1940s, numbers remained low until 1964 Tsitsikamma NP was declared, and poaching became strictly controlled. In the mid-1970s numbers increased rapidly, continuing into the1980s. Small numbers are held in zoos, where they are breed regularly. Listed on CITES App.II. Several other species and subspecies of duiker may be threatened, but being relatively small and often nocturnal, living in thick vegetation, they are little studied. W African species, where loss of forest is particularly marked, are most at risk and many have increasingly fragmented and isolated ranges; these include the strikingly marked *** **Zebra** or **Banded Duiker** *C. zebra* (**1063**), confined to a few high forests in Sierra Leone, Liberia and Ivory Coast; and the *** **Yellow-backed Duiker** *C. sylvicultor* (**1064**) which, although it has a wide range in C Africa, is rare and fragmented in W Africa; most other species need careful monitoring, in particular *** **Ader's Duiker** *C. adersi* (**1065**) from Zanzibar and E Africa, and * **Abbott's Duiker** *C. spadix* (**1066**) from Tanzania.

*** COMMON REEDBUCK

Redunca arundinum (**1067**)

HB 47–63ins; T 7.1–11.8ins; SH 25.6–41.4ins; WT 110.2–209.5lb.

Males have horns up to 18ins long. Although they still have a wide range in E and S Africa, they are declining in many areas, particularly close to human settlements. Once common in S Africa, now rare except in Natal and Transvaal. They depend on access to fresh water. Rare in zoos, only occasionally bred. The * **Mountain Reedbuck** *R. fulvorufula* (**1068**) has isolated populations in N Cameroon, NE Africa, S Mozambique and S Africa, not immediately threatened. Little is known of the status of the * **Bohor Reedbuck** *R. redunca* (**1069**) in W Africa, but it is likely to be declining with human pressure.

*** LECHWE

Kobus leche (**1070**)

HB 51–71ins; T 11.8–17.7ins; SH 33.5–43.3ins; WT 132.3–286.6lb.

In areas where they are hunted by man they are usually nocturnal; they live in marshes and swamps and around lakes, grazing in water up to shoulder-deep. Their breeding behaviour is complex and they often congregate in large numbers. Lechwe occur in localised discrete populations, and do not appear to recolonise after local extinctions. Four subspecies are generally recognised: the **Black Lechwe** *K.l. smithemani*, found in the area around Bangweolo Lake, N Zambia; formerly occurred further north, but became extinct, and were reintroduced in mid-1970s, when the total population was estimated at 25,000. The **Red Lechwe** *K.l. leche* occurs from Victoria Falls, to the Lower Zambesi, Okavanga Delta and Chobe; several local extinctions have occurred, but in the mid-1970s total population was ca.30,000. The **Kafue Lechwe** *K.l. kafuensis* from the Kafue Plains of S Zambia, is greatly reduced – ca.40,000 in mid-1970s – and much of their habitat is threatened by development. *K.l. robertsi* from N Zambia is extinct. Listed on CITES App.II. Exhibited in many zoos and breed freely.

*** ROAN ANTELOPE

Hippotragus equinus (**1071**)

HB 75–94ins; T 14.6–18.9ins; SH 49.6–57ins; WT 491.7–661.5lb.

A very large antelope, surpassed only by Eland. Both sexes carry curved horns, which in the male grow to ca. 39ins; the females' are smaller. They are greyish-brown, sometimes tinged reddish, particularly in W Africa. The ears are long (up to 12ins) and white on the inside. They formerly had a wide range in wooded grasslands, where they appear to fill the niche often occupied by horses and zebras, and consequently competed with zebras. Although their range is much fragmented, particularly in W Africa, and populations throughout their range are generally considerably lower than in the past, they flourish in a large number of parks, including Kruger NP, S Africa; Waza, Cameroon; and many E African parks and reserves. Their preferred habitat is often that most suitable for agriculture, but they are still widespread in areas infested with tsetse fly. A few zoos maintain small herds and they are self-sustaining. Listed on CITES App.II.

*** SABLE ANTELOPE

Hippotragus niger (**1072**)

HB 77–83ins; T 15–18.1ins; SH 46–55ins; WT 450–580lb.

Both sexes carry horns which grow up to 64ins in the males. The young are brownish, gradually becoming russet, then shiny black in adult males, with black on the forequarters in older females. They live in small herds, which may join together to form groups of 100 or more. Found in woodland, or grassland close to woods, usually near drinking water. Have been extensively hunted for trophies; their main protecting factor at present is probably the presence of tsetse fly in much of their range. The **Giant Sable** *H.n. variani*, a subspecies confined to Angola, is reduced to a population estimated at 2000–3000 in 1970. It occurs in reserves but may have suffered during civil wars in the area. Sable occur in many parks including Kruger NP in S Africa and the Shimba Hills Reserve in Kenya. They are kept in many zoos and also on ranches in S Africa. The Giant Sable is listed on CITES App. I.

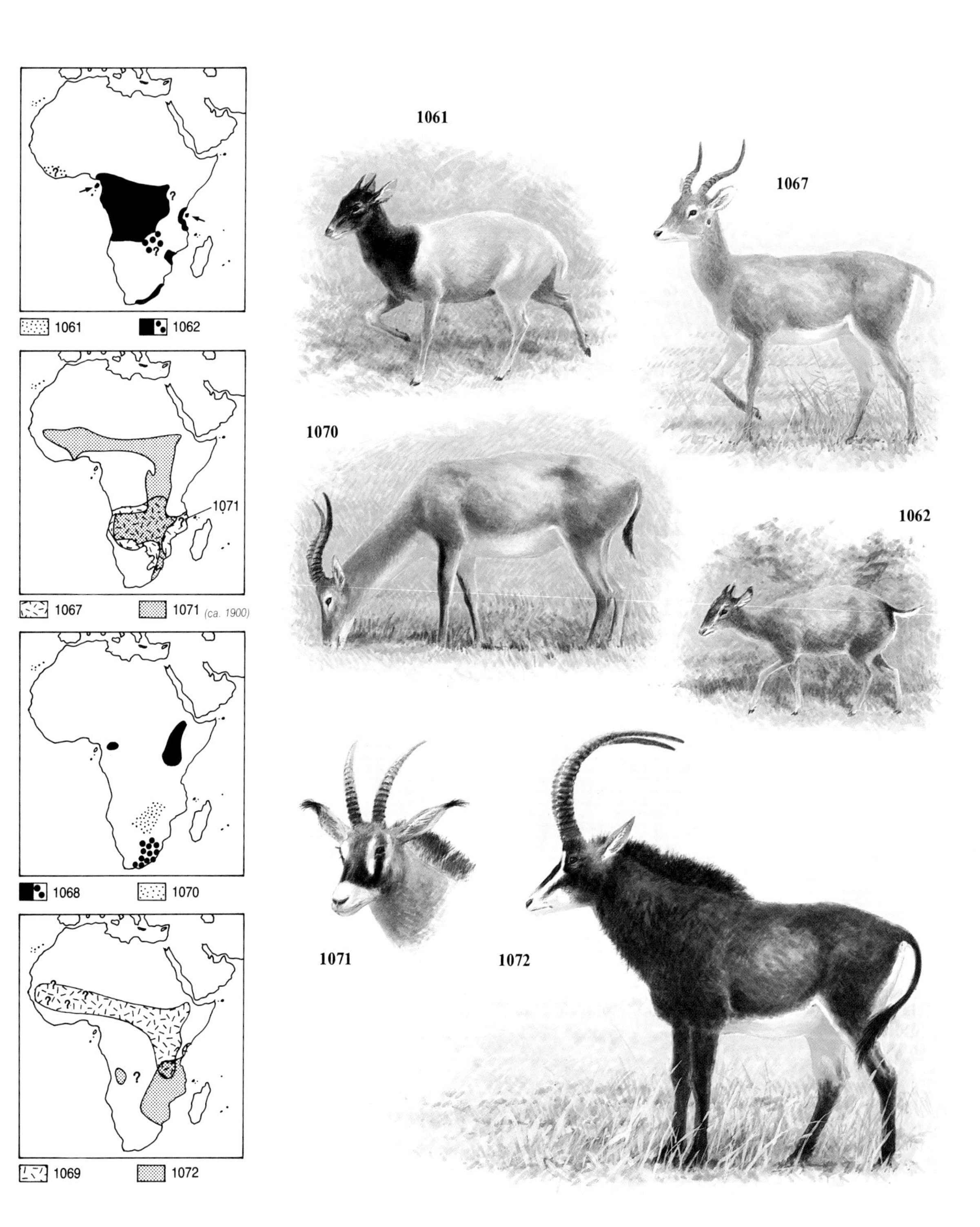
1061
1062
1067
1071 (ca. 1900)
1068
1070
1069
1072
1061
1067
1070
1062
1071
1072

****** ARABIAN or WHITE ORYX**

Oryx leucoryx (**1073**)

HB up to ca.55ins; T ca.11ins; WT ca.121lb.

The horns, carried by both sexes, are up to 28ins long. Formerly occurred in rocky deserts, dunes and other arid habitats, from the Levant, S through Arabia and E to Iraq. By 1960s motorised hunting had reduced them to scattered populations in S Arabia. In 1962 some of the last wild Oryx were captured, and captive breeding commenced in the USA. The last wild one was probably killed in 1972. In the early 1980s the captive population was over 150 and some were reintroduced into fenced reserves in Jordan, Oman and Israel; later those in Oman were released into the wild. In addition to the descendants of the US captives, there have always been significant numbers in captivity in Arabia. Although still rare in captivity they are being dispersed to zoos throughout the world. Listed on CITES App.I and totally protected wherever they occur.

****** ADDAX**

Addax nasomaculatus (**1074**)

HB 59–67ins; T 9.8–13.8ins; SH 37.4–45.3ins; WT 132.3–275.6lb.

Horns up to 35ins along the curves. The most desert-adapted of all antelope, can exist without drinking water, relying on moisture in vegetation and dew. They live in herds of up to about 20; in the past these joined to form larger groups. Are easily exhausted when pursued, and motorised hunting has reduced them to the edge of extinction. They were also hunted for their leather for making sandals and shoes. At the turn of the century their range extended over most of the Sahara and surrounding arid areas, surviving today in small numbers only in Niger and adjacent countries. Fortunately have flourished in captivity and there are sufficient for a reintroduction programme when when a suitably protected reserve can be identified. Listed on CITES App.I.

****** SCIMITAR ORYX**

Oryx dammah (**1075**)

SH ca.47.5ins; WT ca.440lb.

The horns, carried by both sexes, are long and curved, and grow to over 40ins. Like other oryx, found in arid habitats and capable of surviving without drinking water, using the moisture from vegetation. They live in herds of 20–40, and move over very large distances in search of food; herds may join together; groups of more than 1000 have been recorded in the past. Their range formerly extended over the arid semi-desert zones around the Sahara, from Morocco and Senegal, east to Egypt and Sudan, but by the 1950s their range was fragmented, though substantial numbers survived in some areas. By late 1970s they only survived in scattered populations, and by mid-1980s were on the brink of extinction. Substantial numbers survive in captivity and in 1986, a herd was established in Tunisia for eventual reintroduction. They are nominally protected in most of their range, and may occur in the Air NP, Niger. Listed on CITES App.I. The * **Fringe-eared Oryx** or **Gemsbok** *Oryx gazella* (**1076**) is still widely distributed in E and SW Africa, but declining locally.

****** BONTEBOK/BLESBOK**

Damaliscus dorcas (**1077**)

HB 55–63ins; T 11.8–17.7ins; SH 33.5–39.4ins; WT 132.3–176.4lb.

The two subspecies are well marked – the **Bontebok** *D.d. dorcas* is generally rich dark brown, with purplish sheen to the darkest areas, and a white patch on the buttocks. The **Blesbok** *D.d. phillipsi* has the white facial blaze divided with a narrow band between the eyes. At the time of the arrival of Europeans in southern Africa, they were widespread and abundant, although separated by about 186mls. Hunting brought both close to extinction, but in 1837 the Van der Byl family created a reserve for Bontebok near Bredasdorp. In 1931 the Bontebok NP was created, and by 1969 there were ca.800. Introductions have since been made to reserves and private ranches. Blesbok never dropped to such low levels, but are now only found in ranches and reserves. Both subspecies are kept in major zoos, where they breed regularly.

***** HUNTER'S HARTEBEEST**

Damaliscus hunteri (**1078**)

HB 47–79ins; T 11.8–17.7ins; SH 39.4–49.3ins; WT 165.5–353lb.

Horns up to 28ins. Uniform sandy colouring, which may become greyish in old males. They are confined to a 62mls-wide strip, north of the Tana River, Kenya. They are apparently declining through competition with cattle; from 1973–77 numbers dropped markedly. Relatively few in captivity. The *** **Topi** *Damaliscus lunatus* (**1079**), is sometimes considered conspecific with Hunter's Hartebeest and because it is variable in appearance with a wide geographical range, has also been thought to represent more than one species. Widespread and locally abundant, but threatened in many parts of its range, particularly in the north and west. Found in open country from Senegal to Ethiopia and south to northern S Africa.

***** HARTEBEEST or KONGONI**

Alcephalus busephalus (**1080**)

HB 59–96ins; T 11.8–27.6ins; SH 43–59ins; WT 220–496lb.

The horns grow to 27.6ins. Extremely variable in appearance with a large number of subspecies described, 7 of which are distinct and do not overlap: **Bubal Hartebeest** *A.b. buselaphus* from N Africa is extinct. Earlier this century the **Western Hartebeest** *A.b. major* was probably the commonest antelope from Senegal east to the Congo, but now rare or extinct over most of W Africa, although surviving in reasonable numbers in some areas. The overall range of *A.b. cokei* and *A.b. lelwel* remained more or less unchanged until recently; numbers are reduced in many areas, and declines noted around the edges of their range. The **Red Hartebeest** *A.b. caama* is abundant in Botswana, but much diminished since the 17th century, when it occurred in much of S Africa. In recent years it has been reintroduced on ranches there. **Swayne's Hartebeest** *A.b. swaynei* from Ethiopia and Somalia has been extensively hunted and is severely threatened, as is **Tora Hartebeest** *A.b. tora* from a small area in Sudan and NW Ethiopia. The species is found in many national parks, and kept in zoos, but there is no programme to preserve the threatened populations.

***** LICHTENSTEIN'S HARTEBEEST**

Alcelaphus lichtensteini (**1081**)

HB 63–78ins; T 15.8–20.1ins; SH 47–53ins; WT 275–450lb.

Distinguished from the Kongoni by a chestnut area on the back. Like Kongoni are found in grassy plains and open woodlands. Although still widespread are much reduced in numbers and many populations are isolated and fragmented. Occur in many national parks and reserves, but rare in zoos.

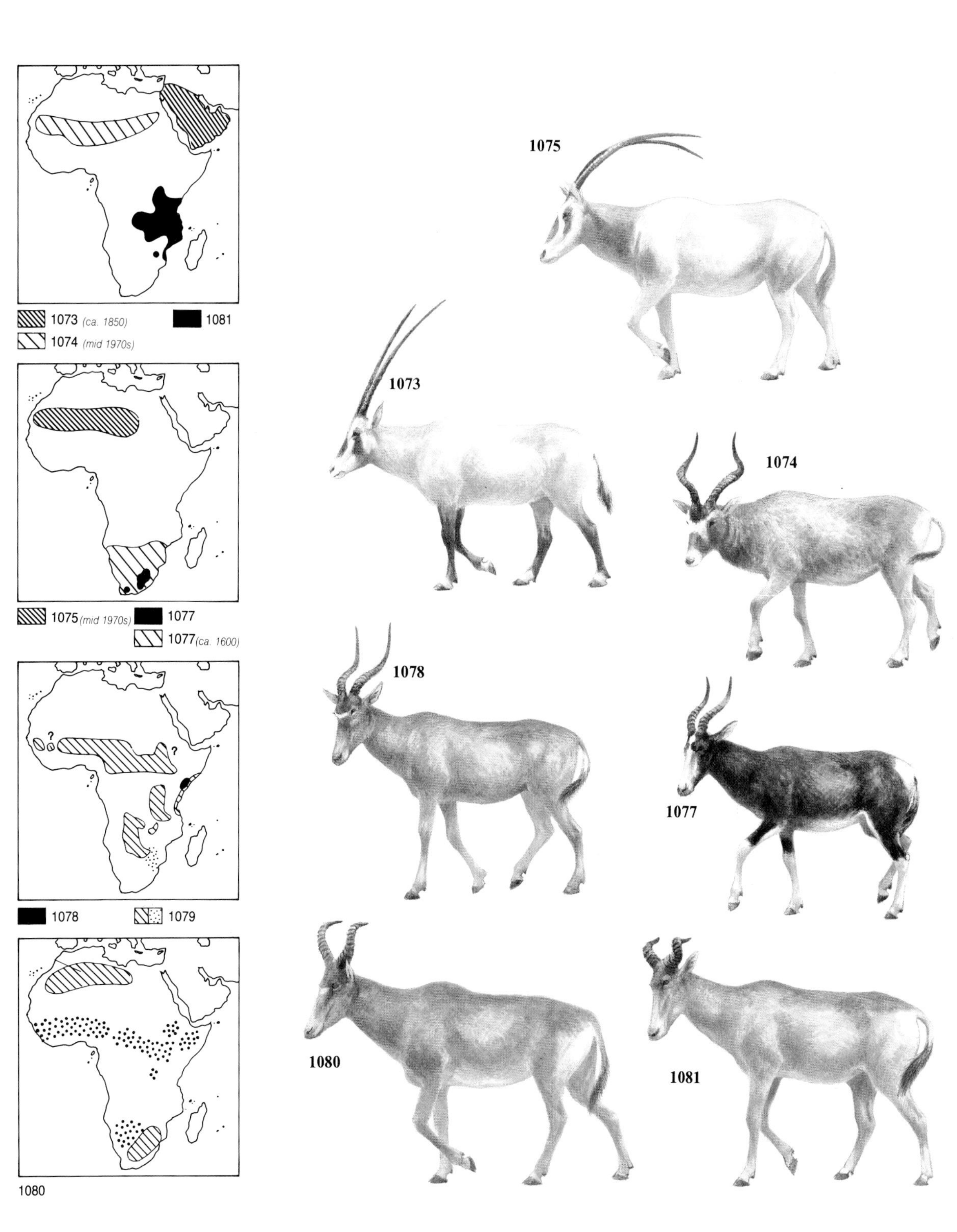
1073 (ca. 1850)
1081
1074 (mid 1970s)
1075 (mid 1970s)
1077
1077 (ca. 1600)
1078
1079
1080
1075
1073
1074
1078
1077
1080
1081

****** AMERICAN BISON**
Bison bison **(1082)**
HB up to 11ft 6ins; T up to 23.5ins; SH up to 79ins; WT up to 1.1 tons.
Generally larger than the European Bison, with longer hair on the neck and shoulders. However they are very closely related and sometimes regarded as merely populations (subspecies) of the same species. In America they are often known as Buffalo; although traditionally associated with the prairies and grasslands of North America, they also formerly occurred extensively in woodlands, open forests and montane habitats. Formerly the **Plains Buffalo** *B.b. bison* existed in herds of hundreds of thousands, with a total population of perhaps 50 million. They were hunted commercially, and also to help subdue the native Indian populations, who were dependent on them. By the early 19th century they were extinct east of the Mississippi, and by 1900 only 1000 survived on the entire North American continent, mostly in Canada. Subsequent conservation efforts by private individuals and government agencies have allowed them to recover to over 50,000; many of the populations were of mixed ancestry. In 1957 a herd of pure **Wood Buffalo** *B.b. athabascae* was discovered in N Canada, and is protected in the Wood Buffalo NP (nearly 18,000sq.mls) in Alberta. American Bison are frequently exhibited in zoos where they breed freely; they are also crossed with domestic cattle and ranched commercially. The Wood Buffalo is listed on CITES App.I.

****** EUROPEAN BISON**
Bison bonasus **(1083)**
HB ca.98ins; T ca.19.7ins; SH 75ins; WT up to 1985lb.
Male larger than female and both sexes carry short horns. In general appearance, they are similar to domestic cattle (to which they are very closely related) with a thick shaggy mane on neck and shoulders. Generally a fairly uniform dark brown colour. The bulls are usually solitary outside the breeding season, but females live in herds of up to 30. The cows leave the herd in summer to give birth to a single calf. They are mainly browsers feeding on a wide range of deciduous trees and shrubs, and on heathers and evergreen shrubs and trees in winter. By the beginning of the 20th century they were restricted to the Bialowiecza Forest and the Caucasus in Russia. At the outbreak of World War I the Bialowiecza population numbered over 700, but was completely destroyed by 1919. The Caucasus population (which was considered a separate subspecies) became extinct in 1925, but the Bialowiecza herd was re-established in Poland from animals which survived in British and other zoos; subsequently other herds have been re-established in the USSR and Romania under semi-natural conditions. They have been calculated to live at densities of ca.3 per square mile. Exhibited in some of the larger zoos, particularly in Europe.

****** BLACK WILDEBEEST or WHITE-TAILED GNU**
Connochaetes gnou **(1084)**
HB 67–87ins; T 31.5–39.4ins; SH 35.5–47ins; WT ca.397lb.
Both sexes carry horns which grow to nearly 31ins. Generally buff-brown in colour, with old males almost black; the tail is dark at the base, with long whitish hair covering the remainder and almost reaching the ground. They are animals of the open plains, and formerly ranged in hundreds of thousands across much of S Africa. With the spread of European colonists and agriculture, together with overhunting, they were reduced to the point of extinction; in 1965 the entire population was estimated at 1700–1800 in reserves and private ranches. Since then a coordinated conservation programme has allowed numbers to rebuild, and by 1970 numbered over 3000; reintroductions have subsequently taken place. They also breed in many zoos outside South Africa.

***** BLUE WILDEBEEST or BRINDLED GNU**
Connochaetes taurinus **(1085)**
HB 76–82ins; T 17.7–22.1ins; SH 50–55ins; WT 260–595lb.
Horns grow up to 29ins. The general colouring is greyish-silver with some brownish markings and a black mane, face and tail. They feed almost exclusively on short grasses, and their range is, to a large extent, determined by the existence of suitable pasture, and also the availability of water in areas where the grazing is dry. They undertake seasonal migrations to and from suitable pasture, and live in herds of 10–1000 individuals. One of the largest populations occurs in the Serengeti; in 1957 it was estimated at 101,000 rising to 220,000 in 1961, 500,000 in 1970, and 1.3 million by 1977. However elsewhere they are declining, and locally extinct in many other parts of their range. They are subject to rinderpest, and are important prey for Lions, Hunting Dogs, hyaenas, and many other large predators. The introduction of extensive control fencing in Botswana reduced them from 250,000 to around 30,000 in a few years, by cutting migrating herds off from water. Often kept on ranches and zoos where they breed freely.

1082

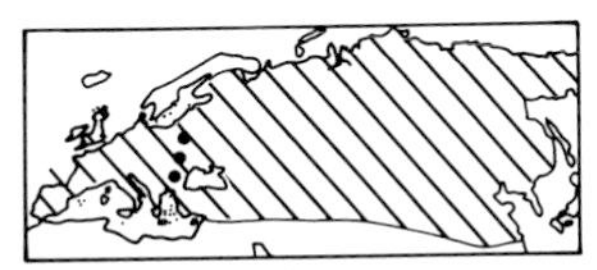

1083 1083 (ca. 0 AD)

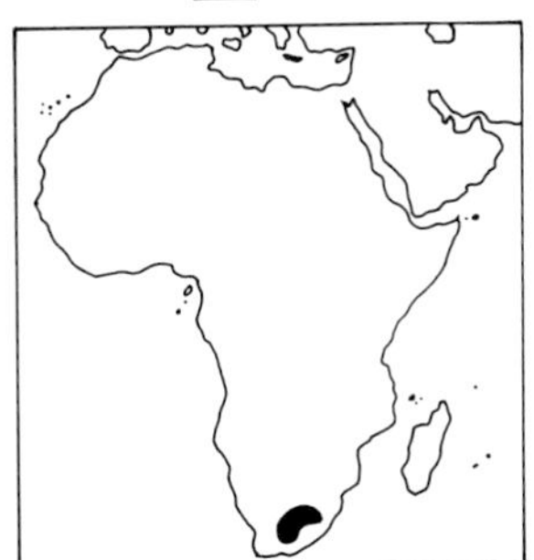

1084

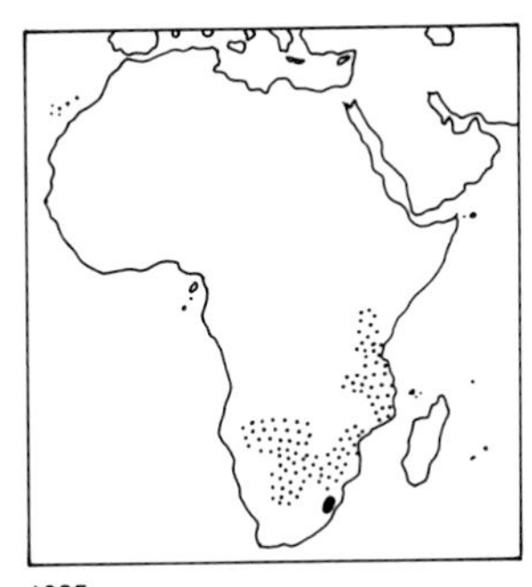

1085

*** BLACKBUCK

Antilope cervicapra (**1086**)

HB up to 47.5ins; T ca.7.1ins; SH ca.31.5ins; WT ca.88lb.

One of the few antelope with a noticeable sexual difference; the males are blackish-brown on upperparts, females (and young) sandy-brown; the males carry spiral, twisted horns up to nearly 28ins long (normally 20ins). Occur mainly in open plains, particularly with thorn and scattered dry forest, where they feed mostly on grasses. Much of their habitat has been lost to agriculture; they have also been extensively hunted. Formerly the most abundant hoofed animal, for most of its range in herds of thousands, it is now reduced to scattered populations of 5–50. Herds consist of females and young led by a single male, and also bachelor herds. Blackbuck are (or were) an important prey item for most of the larger carnivores of India. Often kept as pets by Indian royalty; have lived over 16 years in captivity where are common and breed freely. Introduced into North America in 1932; by 1979 numbered nearly 10,000; and some descendants have been reintroduced into the Cholistan Desert, Sind, Pakistan. Have also been established in Argentina since 1906, and Australia since 1912; a few survive in both countries. The American populations probably now exceed those surviving in India. There are two distinct populations: *A.c.cervicapra*, in the southern and eastern parts of its range; and *A.c.rajputanae*, the ancestors of the Texan animals, from the northwest of its range.

/* IMPALA

Aepyceros melampus (**1087**)

HB ca.43.5ins; T 8.7–15.8ins; SH 30.3–39.4ins; WT 66.1–143.3lb.

Only the male carries the lyre-shaped horns, up to 29ins long (record 32ins). Graceful antelopes, with shining reddish-brown upperparts, pure white below, and a distinctive rump pattern. They are gregarious, females and young living in small herds of up to 15–20, which may congregate, during the dry months, into groups of more than 100; males live in bachelor herds. During breeding season males defend territories. Impala graze, depending on locality and season, on a wide variety of grasses, herbs and shrubs. The single young is vulnerable to predation by hyaenas, Cheetahs and other large predators. Found mainly in wooded habitats, although have vanished over much of their range, particularly where close to human habitation. In Namibia and Angola the **Black-faced Impala** *A.m.petersi* is considered particularly threatened. Elsewhere they have been extensively introduced, and in S and E Africa are found on many ranches, where they are cropped for meat. Also common in zoos.

*** ORIBI

Ourebia ourebi (**1088**)

HB 36.2–43.3ins; T 2.4–5.9ins; SH 19.7–27.6ins; WT 30.9–46.3lb.

Female is slightly larger than male, but only the male carries horns, up to ca.7.5ins long. The hair is fine and silky and they are rather long-legged. Found in open grassy habitats up to 9855ft; formerly widespread and abundant throughout sub-Saharan Africa. Although still numerous in some areas, their range is now much fragmented, particularly in the southeastern and western parts, and they are extinct in many localities. Decline due mainly to overhunting, but disease and other factors may be involved. They occur in many protected areas within their range, and are occasionally kept in zoos and are bred in small numbers.

*** GRYSBOK

Raphicerus melanotis (**1089**)

HB 25.6–29.5ins; T 2–3.1ins; SH 17.7–21.7ins; WT 17.6–50.7lb.

Only the males carry horns, up to 5ins long. Found mainly in dense scrub cover, thick veld and rocky gorges, in south Cape Province, S Africa, where they are relatively little-known because of their nocturnal habits, and are thought to be declining in many areas. (They are often considered conspecific with **Sharpe's Grysbok** *R. sharpei* from SE Africa.) The * **Steenbok** *R. campestris* (**1090**) although widespread is apparently declining in areas close to human habitation. It occurs in two separate populations, in E and S Africa.

*** SUNI

Neotragus moschatus (**1091**)

HB 22.5–24.4ins; T 3.1–5.1ins; SH 13–15ins; WT 8.8–13.2lb.

A very small antelope; only the males carry horns, up to 5ins long. They inhabit dry bush country in eastern Africa from Mt Kenya and Tana River south to N Natal; also islands of Zanzibar, Chapani, Bawi, and Mafia. They use regular paths through undergrowth, and consequently are easily snared. The Zanzibar population *N.m. moschatus* is thought to be threatened. Little is known of the status of this or the closely related * **Royal Antelope** *N. pygmaeus* (**1092**) from W Africa, or * **Bates's Dwarf Antelope** *N. batesi* (**1093**) which occurs east of the Niger River to N Zaire, and also in W Uganda. They are both forest-dwelling species, and like the Suni are frequently hunted. They are also preyed on by many carnivores, pythons and birds of prey. Suni are kept in a few zoos, and bred in small numbers.

*** BEIRA ANTELOPE

Dorcatragus megalotis (**1094**)

HB 29.9–33.9ins; T 2–3ins; SH 19.7–29.9ins; WT 19.8–24.2lb.

Only the males carry horns, up to 3.9ins long; the ears are also large, up to 5.9ins long and 3ins wide. They inhabit arid, stony habitats, and can survive without drinking water. They live at low densities, and are confined to a relatively small range, in N Somalia, and possibly E Ethiopia.

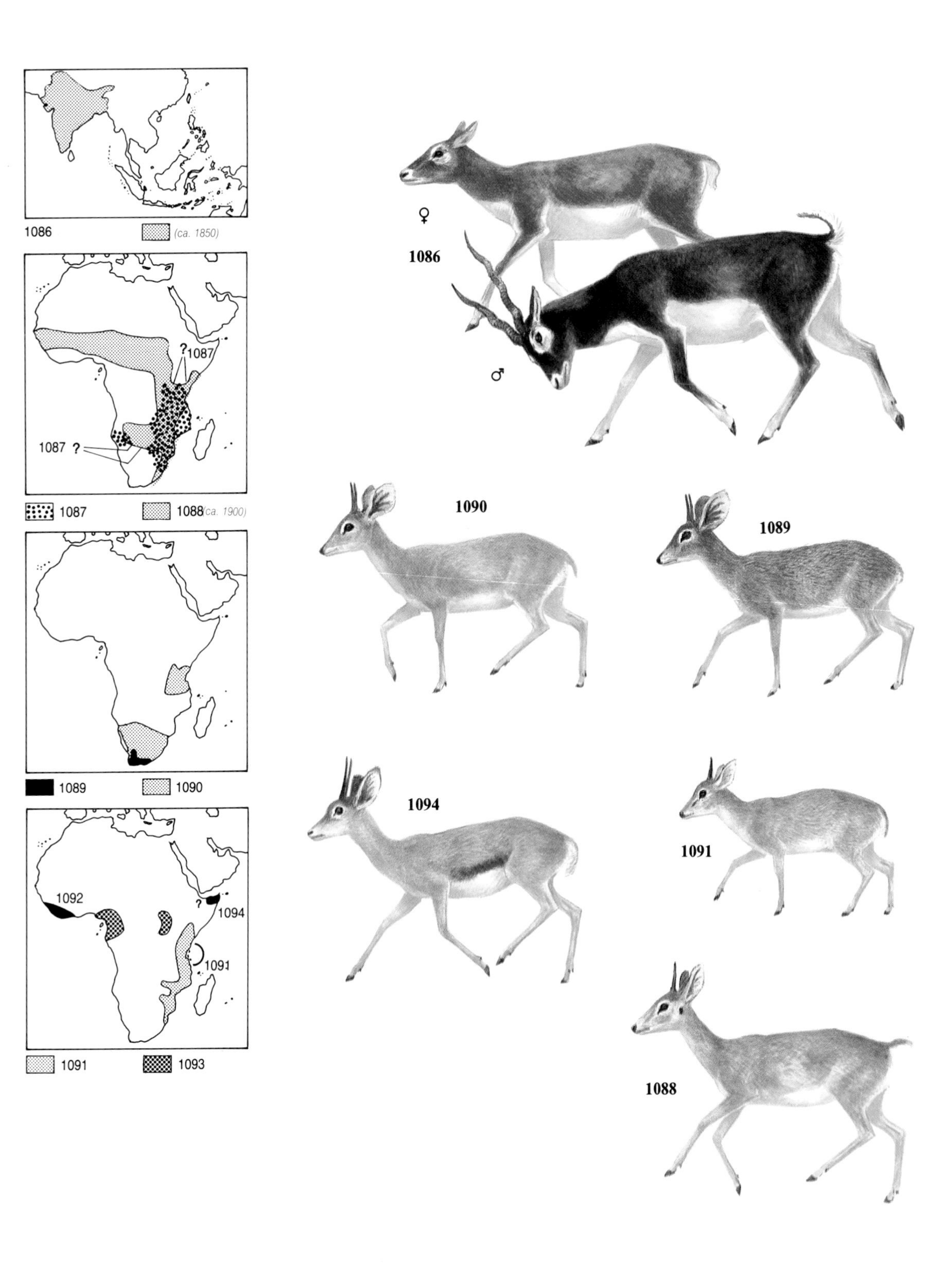
1086
(ca. 1850)
?1087
1087 ?
1087
1088(ca. 1900)
1089
1090
1092
?
1094
1091
1091
1093
♀
1086
♂
1090
1089
1094
1091
1088

*** DIBATAG or CLARKE'S GAZELLE**
Ammodorcas clarkei (**1095**)
HB 59–67ins; T 11.8–14.2ins; SH 31.5–34.7ins; WT 48.5–77.2lb.
Only the males carry horns, which grow to 13ins. A graceful long-legged, long-necked gazelle. Found in grassy plains with scattered trees and bush. They have a comparatively restricted range from Somalia and Ogaden, Ethiopia, east to the coast and south to near the Equator. They have been exterminated in many areas by poaching and through competition with domestic livestock for browse. Rarely seen in zoos.

***** DORCAS GAZELLE or CHINKARA**
Gazella dorcas (**1096**)
HB 35–43ins; T 5.9–25.6ins; SH 21.7–25.6ins; WT 33.1–50.7lb.
The males' horns are lyre-shaped, up to 16ins long; those of the female are smaller and straighter. Once extremely widespread and numerous in savannah, semi-desert and deserts, ranging over most of N Africa, in and around the Sahara, east through the Middle East, to Pakistan and India. Range now fragmented, scattered or extinct, particularly in N Africa.

***** ARABIAN GAZELLE or IDMI**
Gazella gazella (**1097**)
HB 35–41ins; T ca.3.9ins; SH 20–24ins; WT ca.44lb.
Both sexes carry horns and those of the males are up to 11ins from base to tip; the females' smaller. Similar in habits to other gazelles and like other large animals in the Arabian Peninsula, they have undergone a massive decline this century; following the introduction of modern firearms and motorised vehicles. There are several thriving herds in captivity. The **Chinkara** *G.d. bennetti* from India and Pakistan has often been classified as a subspecies of *G. gazella*, as has **Cuvier's Gazelle (1098)**.

***** EDMI or CUVIER'S GAZELLE**
Gazella cuvieri (**1098**)
HB 37–41ins; T 5.9–7.9ins; SH 23.6–31.5ins; WT 33.1–77.2lb.
The males' horns are only slightly larger than those of females, growing to a maximum of ca.15ins. Early this century were still quite widespread in N Africa, but now extinct over most of their former range, and extremely rare in the few isolated and scattered populations that survive. None known to be kept in zoos.

***** SAND GAZELLE or RHIM**
Gazella leptoceros (**1099**)
HB 39–43ins; T 5.9–7.9ins; SH 25.6–28.4ins; WT 44.1–66.1lb.
The males' horns grow to ca.16ins, the females' are smaller. One of the most desert-adapted of the gazelles, able to survive on dew and the moisture in vegetation, in the deserts of N Africa, from C Algeria to Egypt. Their populations are now highly fragmented and often isolated. Small numbers are kept in captivity, the majority in San Diego Wild Animal Park, USA.

***** SPEKE'S GAZELLE or DERO**
Gazella spekei (**1100**)
HB 37–41ins; T 5.9–7.9ins; SH 19.7–23.6ins; WT 33.1–55.1lb.
Both sexes carry horns, which are strongly S-shaped and slightly larger in the male, growing to ca.12ins. At the back of the nose there are inflatable protuberances. Little is known of its status or precise distribution. Found in arid habitats in E Ethiopia and Somalia.

***** GOITRED GAZELLE**
Gazella subgutturosa (**1101**)
HB ca.39ins; T 4.7–7.1ins; SH up to ca.24ins; WT 30.9–50.7lb.
Almost indistinguishable from *G. gazella*, but both sexes have a swelling around throat and only males carry horns, up to ca.16ins. Once widespread in arid regions from the Levant, to the Gobi Desert and N China, they have declined markedly, particularly in the west of their range, and in areas easily accessible to hunting parties.

***** SPRINGBOK**
Antidorcas marsupialis (**1102**)
HB 47–55ins; T 7.5–10.8ins; SH 28.8–34.3ins; WT 70.6–79.4lb.
Both sexes carry horns, which in males grow to a maximum of 20ins. Formerly extremely abundant in the savannah and veldt of southern Africa, but were hunted in enormous numbers by the European colonists. Herds of more than a million made migratory 'treks' which took several days to pass. Although still present in much of their former range, there are now very few herds of more than 1500. Exterminated in most of the Orange Free State, Transvaal, and Cape Province, but have been extensively reintroduced throughout S Africa. The largest population is in the Kalahari/Gemsbok NP, on the border of Botswana and S Africa, estimated at ca.20,000 in mid-1970s.

*** SAIGA ANTELOPE**
Saiga tatarica (**1103**)
HB 39–55ins; T 2.4–4.7ins; SH 23.6–31.5ins; WT 57.3–152.1lb.
Formerly occurred from the steppes of the Ukraine, east through C Asia, but by the early 20th century only ca.1000 survived. After total protection in 1923, their numbers rose, to ca. 2 million by 1958, and harvesting is now allowed. The Mongolian subspecies *S.t. mongolica* is still very much reduced, with ca.200 in the late 1970s. The ***** Zeren** *Procapra gutturosa* (**1104**) from the steppes of Siberia, China, and Mongolia is considered threatened in the USSR, and little is known of its status, or that of the other Central Asian gazelles, *Procapra* spp.

Many other *Gazella* spp. are declining in areas close to human habitation, including widespread and often abundant species such as Thomson's Gazelle *G.thomsoni*, and Soemmering's Gazelle *G.soemmeringi*. The ***Gerenuk** *Litocranius walleri* (**1105**), which is found from the Horn of Africa, south to Kenya and N Tanzania, was formerly more widespread, occurring in Egypt; it continues to decline around the southern edge of its range. The ***** Red-fronted Gazelle** *Gazella rufifrons* (**1106**) was formerly widely distributed across sub-Saharan Africa, but is increasingly rare, particularly in the west of its range, where many populations are now isolated and fragmented; *G.r. rufina*, often treated as a full species, from Algeria, is extinct. The ***** Dama Gazelle** *G.dama* (**1107**) once one of the most numerous and widespread gazelles from Senegal to the Sudan, is now extinct over most of its range, though perhaps still locally abundant.

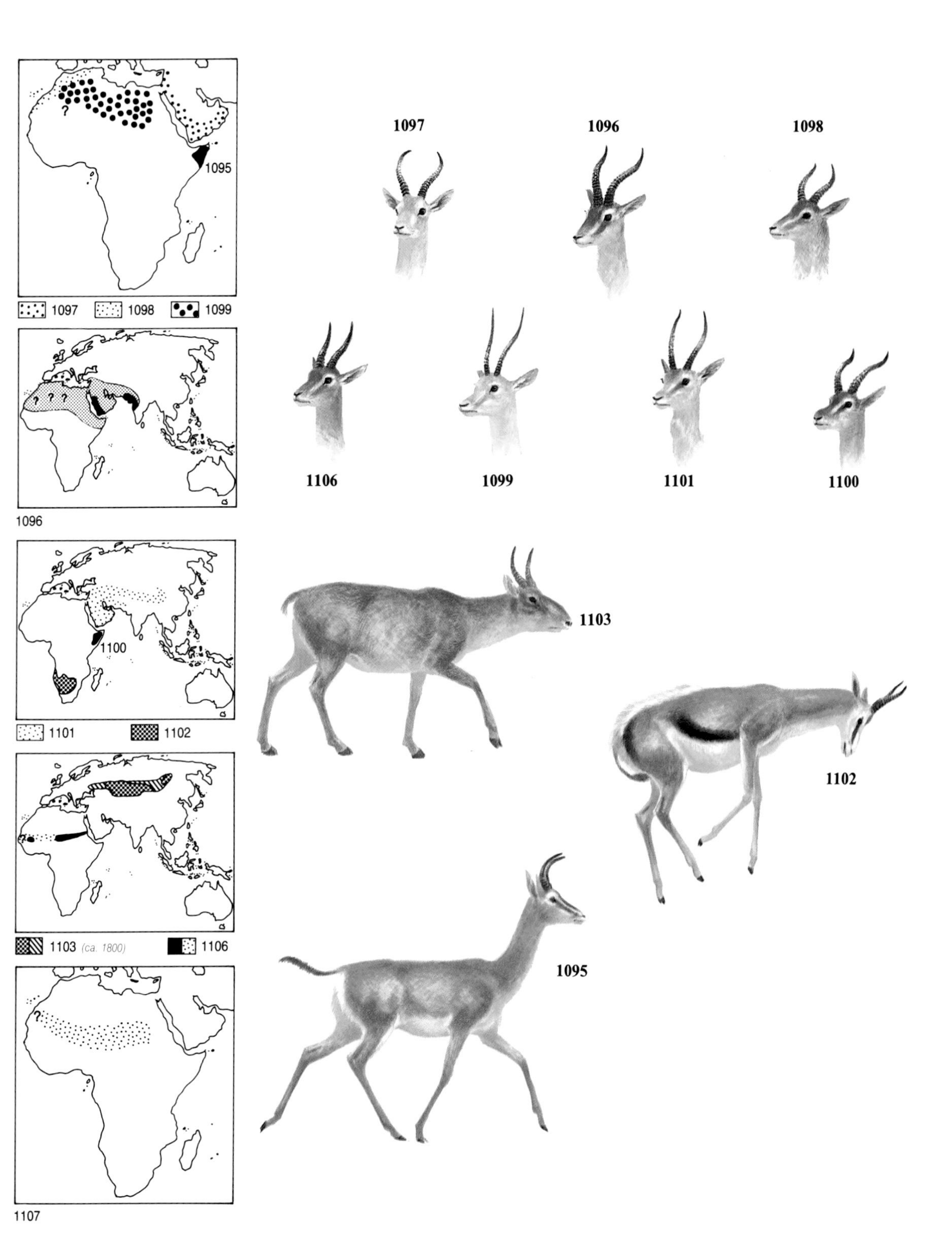
1095
1097
1098
1099
1096
1100
1101
1102
1103 (ca. 1800)
1106
1107
1097
1096
1098
1106
1099
1101
1100
1103
1102
1095

*** CHAMOIS**
Rupicapra rupicapra (**1108**)
HB 35.5–51ins; T 1.2–1.6ins; SH 30–32ins; WT up to 110lb.
Blackish-brown in winter, brownish in summer. Both sexes have horns, up to 8ins long. Females and young live in herds of up to 30, males are generally solitary, but join the females during the autumn rut. They live in mountains, migrating to the higher meadows during summer and descending into the valleys during winter. Formerly found over most of the Alpine regions of C and S Europe, Asia Minor and the Caucasus, but have been extensively hunted. Their skin is prized as 'Chamois' (*shammy*) leather, and tufts of hair are used as cockades in alpine hats. They are also hunted for meat and trophies. Many populations are considerably reduced or extinct. Several subspecies (often poorly defined) have been described and the Abruzzo NP (Italy) population of *R.r.ornata* numbers less than 400 and is listed on CITES App.I.

*** TAKIN**
Budorcas taxicolor (**1109**)
HB ca.47.5ins; T ca.3.9ins; SH 39.4ins; WT up to 606.5lb.
The shaggy fur varies from yellowish-white to dark brown, with a dark stripe along the back. Both sexes carry heavy horns, up to 24ins long. They live in dense bamboo and rhodedendron woodland, near the tree line, feeding on grasses and herbage in summer, and bamboo and tree shoots in winter. Stay in small herds, though old bulls are solitary. The single kid is able to follow its mother from about 3 days. Found in montane areas at altitudes of 7000–10,000ft, from Bhutan, SW China and Burma. They are hunted for their meat, and some populations have very restricted ranges. The Takin is rare in zoos, but appears to breed regularly.

**** ARABIAN TAHR**
Hemitragus jayakari (**1110**)
HB ca.34.5ins; T ca.3.3ins; SH ca.24.5ins; WT up to 66lb.
A rather small goat; the male has short horns, up to 10ins, and lacks a beard. The striped facial markings are distinctive. Confined to the mountains of Oman, in precipitous rocky areas with only very sparse vegetation. Nothing known of their former distribution; although scarce and confined to a very limited area, they occur within protected areas, and are under no immediate threat. Following their discovery in the 1890s they were unknown to western biologists until the 1940s. In the mid-1970s the population was estimated at less than 2000. A small, isolated population was discovered in the United Arab Emirates in 1949, and in 1983 was estimated at about 20.

***** NILGIRI TAHR**
Hemitragus hylocrius (**1111**)
SH up to ca.43ins; WT ca.220lb.
Larger and darker than the Arabian Tahr, with horns up to 18ins. Confined to hills in S India, from the Nilgiris and Anaimalais south in the Western Ghats, at altitudes of 4000–6000ft. They occur in Eravikulam Rajmallay NP, Kerala, and are strictly protected; their numbers are very depleted and although they are not thought to be in any immediate danger, the fragmented nature of their range gives cause for concern; in the mid 1970s the total population was estimated at 2230. The more widespread and abundant * **Himalayan Tahr** *H. jemlahicus* (**1112**) – found below the tree-line from Pir Panjal to Sikkim, is also threatened in India through over-hunting for meat and hides; has been introduced into New Zealand and S Africa where is flourishing.

***** COMMON GORAL**
Nemorhaedus goral (**1113**)
HB up to ca.51ins; T up to 7.9ins; SH 25.6–27.6ins; WT 55.1–66.1lb.
Both sexes carry short horns up to 5ins. Buff-grey above, paler below with a white throat patch and a dark stripe down the back. Groups up to 12 live in rugged, wooded mountainous areas at altitudes of up to 13,950ft. Their range extends from SE Siberia through the Himalayas to N Thailand and Burma. Despite their small horns, Goral are often hunted for sport, and in many parts of their range have declined; they occur within Khangchendzena NP, Sikkim, many of the protected areas of Himachal Pradesh and elsewhere in India, and are protected by law in India. Listed on CITES App.I.

*** RED GORAL**
Nemorhaedus cranbrooki (**1114**)
Very closely related to the Common Goral (1113); the general colouring is a bright rufous or brownish, with a dark stripe down the back. Although specimens had been collected in 1913, 1922 and 1931, only recognised as a distinct species in 1961. Confined to Tibet, Assam and N Burma, the Red Goral remains little-known, but is not thought to be in any immediate danger. The ** **Brown Goral** *N. baileyi* (**1115**) is only known from a single specimen described in 1914 from Tibet, and may be conspecific with Common Goral (1113).

***** MAINLAND SEROW**
Capricornis sumatrensis (**1116**)
HB up to 71ins; T up to 6.3ins; SH up to 39.5ins; WT up to 309lb.
Both sexes carry small horns up to 9.8ins long. Generally greyish or black above, whitish below. Several subspecies have been described. They inhabit thickly vegetated hills and mountains up to 10,000ft, but are not particularly agile. Generally solitary, feeding in the morning and evening on grasses, shoots and leaves. Their range extends from Kashmir to Arunachal Pradesh and Szechuan, and south through Burma, Thailand to Peninsular Malaysia and Sumatra. The Sumatran population, *C.s. sumatraensis* is very much reduced in numbers. Other populations are more abundant, but are generally declining. In India occur in several protected areas, notably Dachigam NP, Kashmir, and several reserves in Himachal Pradesh. Rare in zoos, but are held in captivity in small numbers and have bred. Listed on CITES App.I.

***** MOUNTAIN GOAT**
Oreamnos americanus (**1117**)
HB 47–63ins; T 3.9–7.9ins; SH 35.5–41.4ins; WT 101.5–308.7lb.
The fur is white or yellowish, with a thick woolly underfur. Live in alpine and tundra zone in uplands and mountains, where they feed on grasses, mosses, lichens and other plants. In winter congregate in flocks, but at other times of the year rarely in groups of more than 4. Their range extends from SE Alaska to Oregon and Montana. While not as depleted as many other species in N America, some populations have declined through over-hunting. There have been numerous introductions and translocations, including several Alaskan islands and Olympic NP. In the mid-1970s the total population was estimated at ca.100,000.

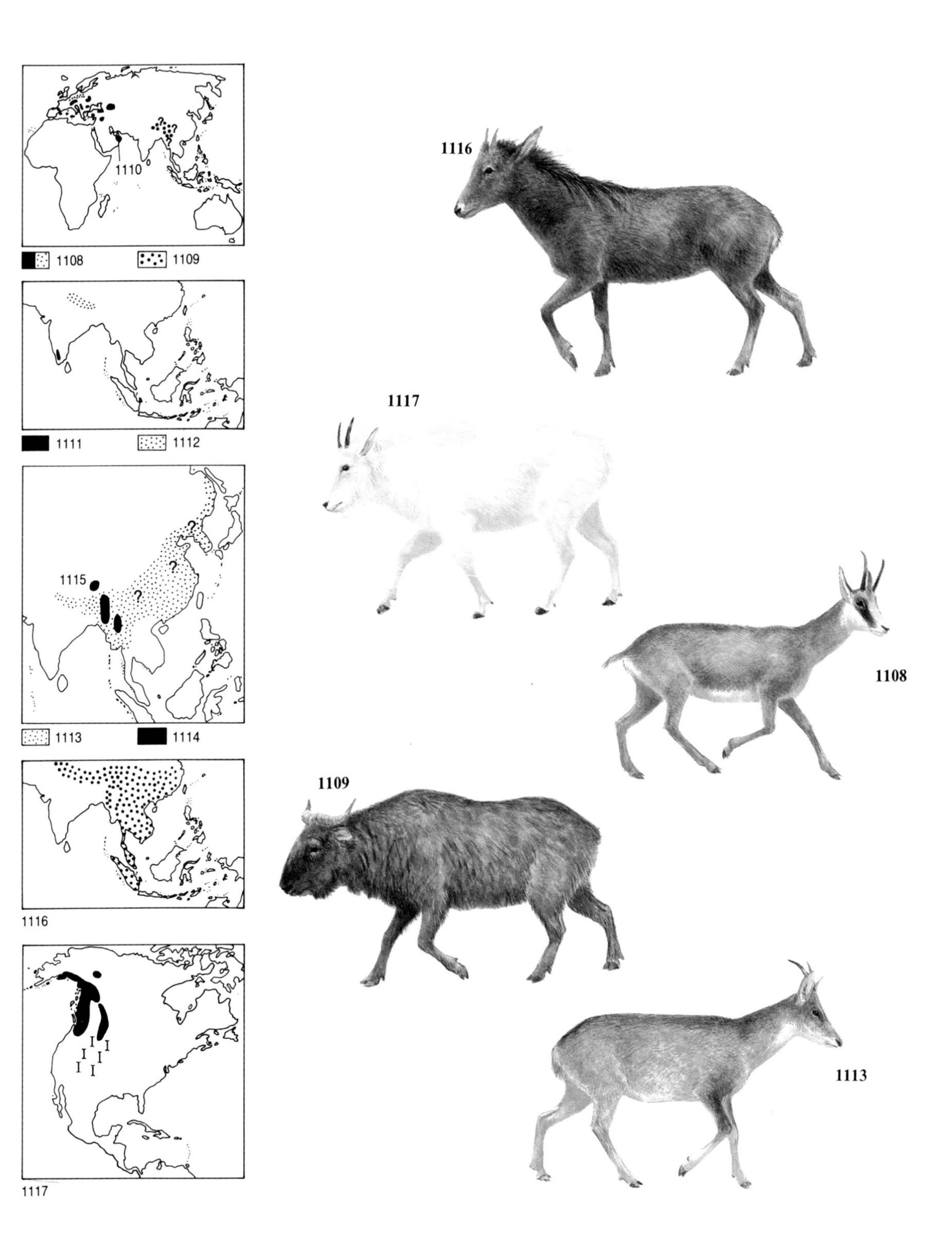
1110
1108
1109
1111
1112
1115
1113
1114
1116
1117
1116
1117
1108
1109
1113

GOATS

Capra spp.

The precise relationships between the various species and populations of goats are not fully understood. All ibexes are sometimes classified as a single species, and some populations such as the Cretan Goat may be relict, partially domesticated populations of *C. hircus*, the domestic Goat.

*** WILD GOAT

Capra aegagrus **(1118)**

HB up to 55ins; T up to 5.9ins; SH 21.7–37.4ins.; WT up to 265lb.

The horns, larger in males, grow to 51ins. Males have a beard, and both sexes have a dark stripe along back and across shoulders. Flocks of 20 or more are found in rather arid, remote mountainous areas. Their original range is unknown, and it is doubtful if any pure-bred wild goats survive, having been persecuted since antiquity, and frequently interbred with domesticated and feral goats. Comparatively pure populations survive on Crete and small islands nearby, and elsewhere in Greece, as well as Asia Minor, eastwards to Afghanistan and Pakistan and also Oman. They were kept in one of the first zoological gardens known, that of Tiglatpileser I (ca.1000 BC), but are uncommon in modern zoos. They occur in the Saimiri NP, Crete and other reserves, notably in the USSR.

*** IBEX

Capra ibex **(1119)**

HB 51–63ins; T 4.7–5.9ins; SH up to 39.5ins; WT up to 276lb.

Similar to Wild and domestic goats, but more uniformly coloured, with horns that are less curved, and broader at the front. Herds live at high altitudes (over 22,000ft in the Himalayas), above the tree-line where they graze on grasses, shrubs and other alpine vegetation. They are extremely agile and surefooted. There are normally 1 or 2 young, which are active and follow their mother within a few hours of birth. Their range extends from the Alps in Europe, the mountains of C Asia, Siberia, the Himalayas, Arabia, and N Africa, with some of these populations often being treated as separate species. In many parts of their range their numbers are depleted, largely through overhunting. The European population was almost exterminated, but survivors in the Gran Paradiso NP, Italy, have been used for translocations throughout the Alps and Yugoslavia, and their numbers are increasing. The **Nubian Ibex** *C.i. nubiana* from N Africa, the Near East and Arabia, declined rapidly with the spread of modern firearms. Ibex are commonly kept in zoos and there are thriving populations of the European Ibex *C.i. ibex* as well as some Nubian Ibex.

*** SPANISH IBEX

Capra pyrenaica **(1120)**

HB 39–59ins; T 4.7–5.9ins; SH 25.6–29.5ins.

Similar in general appearance to the Ibex, but with horns curved outwards and backwards. This character may be the result of interbreeding with domestic goats in the past, and there is considerable variation in the populations of Spanish Ibex. During the 19th century most populations were heavily depleted, and several became extinct; since then many have recovered and there have been extensive reintroductions and translocations. The subspecies *C.p. pyrenaica*, found in the Pyrenees, thought to number less than 20, is the rarest and most threatened.

***** WALIA IBEX

Capra walie **(1121)**

HB 59–67ins; T 7.9–9.8ins; SH 35–43ins; WT 176.5–275.5lb.

Like other ibex the horns of the males are much larger than those of the females, growing to 45ins. The Walia may be only a very well marked subspecies of *C. ibex*. In habits they are similar to other ibex, living at altitudes of 8200–14,750ft. Confined to the Simien Mountains in Ethiopia. In the mid-1970s out of a total population of 300, about 240 were in the Simien Mountains NP; however protection in the park is not assured.

*** MARKHOR

Capra falconeri **(1122)**

HB up to 73ins; SH up to ca.39.5ins; WT up to 240lb.

Females are much smaller than males but they have larger horns, growing to a maximum of 65ins along the curves. The spiralled horns are rather variable in shape and size, and about 7 subspecies are separated on the basis of horns, but are difficult to define. Markhor occur in arid upland habitats from 1950–11,800ft, but avoid the higher altitudes inhabited by Ibex. They are one of the most sought-after of all hunting trophies, and have been exterminated in many parts of their range, which formerly extended from the mountains of Uzbekistan and Tadzhikistan in USSR, to Afghanistan, Pakistan, Kashmir and Ladakh. Although they still occur in all these countries their range is very much reduced and many populations are endangered or extinct. In the early 1970s the world population was estimated at 20,000–25,000. They are popular and common in zoos, where they breed freely. Listed on CITES App.II, with 3 subspecies (probably impossible to identify) being listed on App.I.

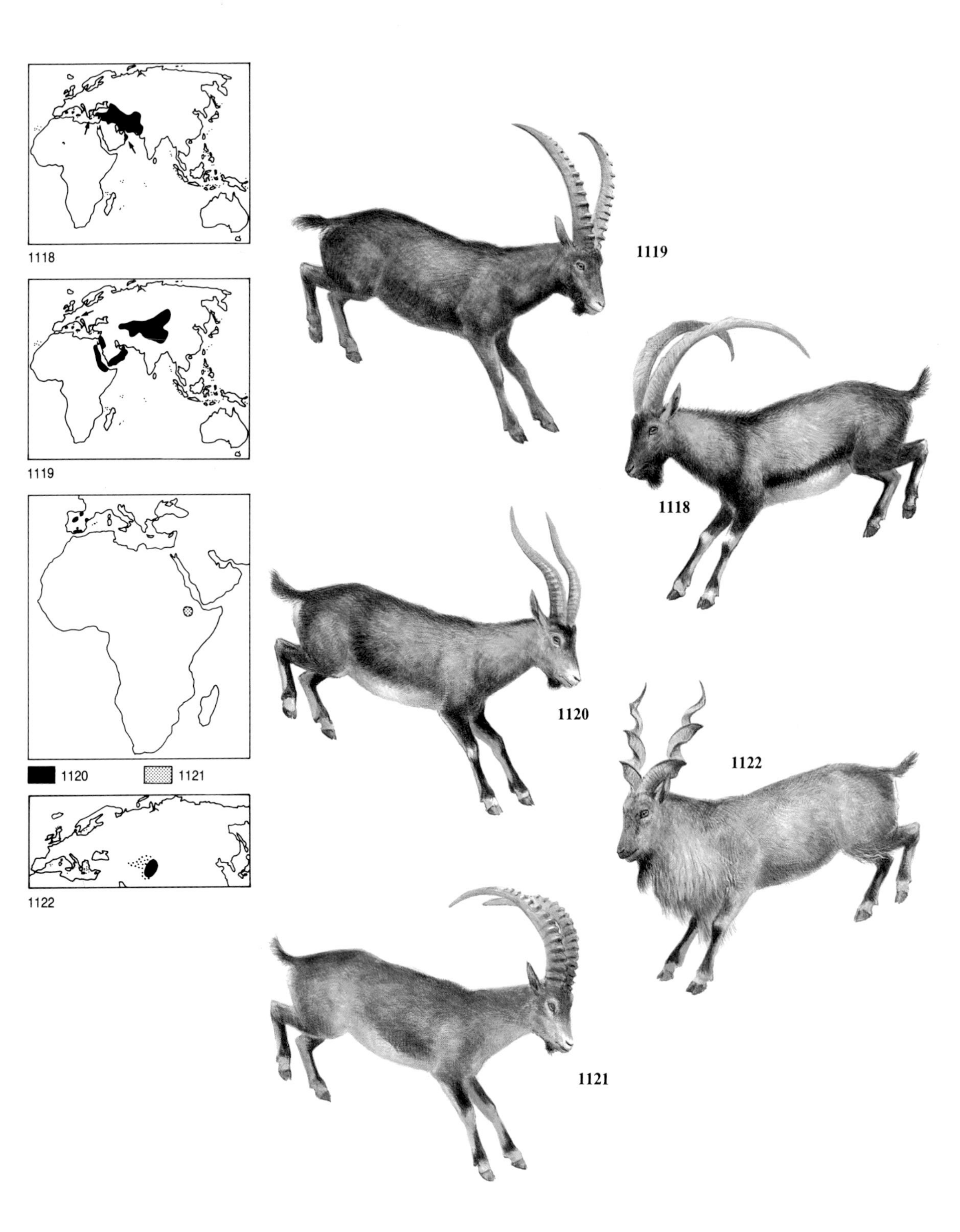
1118
1119
1120
1121
1122
1119
1118
1120
1122
1121

WILD SHEEP
Ovis spp.
There is no general agreement over the classification of Eurasian species of sheep, with some taxonomists preferring to regard the four species as populations of a single species, *O.ammon*. Since their ranges are considerably fragmented and isolated we have treated them separately. A factor confusing the taxonomy of Eurasian sheep (and goats) is the fact that they have been domesticated since prehistoric times, and interbred and transported by man.

*** **URIAL or SHAPU or ASIATIC MOUFLON**
Ovis vignei (**1123**)
SH up to 35.5ins; WT up to ca.132lb.
Males carry horns up to 39ins long around curves. Similar to the closely related Argali, but smaller, with reddish fur. They live in flocks, and outside the breeding season the mature males form flocks of up to 30–40. Once widespread and abundant in a wide variety of open habitats from S Soviet Central Asia, to Iran, the Himalayas, and Oman, but ousted from most areas by competition for forage with domesticated sheep. Disease and hunting for meat and trophies have also been instrumental in their decline. Their range is now extremely fragmented, and many populations and subspecies are threatened with extinction. They occur in many parks and reserves and breed freely in captivity, but there are relatively few of known ancestry kept in zoos. Listed on CITES App.I.

*** **BIGHORN SHEEP**
Ovis canadensis (**1124**)
HB ca.63ins; T ca.4.3ins; SH up to 39.5ins; WT up to 302lb.
Large, with massive curving horns in males, smaller horns in females. Originally found in mountains including foothills, but now restricted mostly to the more inaccessible higher altitudes, where they live in small flocks of up to 15 or more. Extinct over most of its former range, which once extended over a wide area in SW Canada, south to Baja California and N Mexico. Occurs in many parks and protected areas, notably Yellowstone NP, Glacier NP and Death Valley. They are kept in zoos, mostly in USA, where some of the rarer subspecies are being bred. The closely related * **Thinhorn** *O. dalli* (**1125**) occurs farther north to Alaska. Both species were formerly extremely numerous, with perhaps 1.5–2 million in the mid-19th century. During the early 20th century over-hunting, disease and competition with domestic livestock caused numerous local declines and extinctions. By the mid-1970s Bighorn populations were estimated at less than 42,000 and Thinhorn at less than 150,000. Active management and reintroduction programmes are in progress.The closely related *** **Snow Sheep** *O. nivicola* (**1126**) from Siberia is often regarded as conspecific with the Bighorn Sheep; several populations are considered threatened in the USSR.

*** **ARGALI or NAYAN**
Ovis ammon (**1127**)
SH up to 71ins; WT up to 254lb.
A large sheep, with massive horns in the male, up to 1.5m around the curves, otherwise very similar to the mouflons, with which it is often considered conspecific. They inhabit high mountain plateaux, up to ca.16,400ft, with seasonal movements to lower altitudes to avoid thick snow. A large number of subspecies have been described, but they probably all intergraded, at least until recent times when their range became fragmented. The best known subspecies is the **Marco Polo Sheep,** *O.a. polli* from Pakistan, Chinese Turkestan, Soviet Central Asia, and Afghanistan. Supposedly protected throughout most of their range they are still extensively hunted, under management in only some areas. They occur within many protected areas. Rarely kept in zoos, but would probably breed freely. *O.a. hodgsoni* from the Himalayas and Tibet is listed on CITES App.I, and all others are listed on App.II.

*** **BARBARY SHEEP or AOUDAD**
Ammotragus lervia (**1128**)
HB 51–63ins; T 5.9–9.8ins; SH 29–47ins; WT up to 320lb.
Male considerably larger than female, and carries horns up to 33ins long; females have smaller horns. Found in arid, rocky country with low vegetation where they hide from predators by remaining motionless. They feed on grass and other desert vegetation, and rely on dew for moisture. Barbary Sheep are often solitary, or live in small groups, and older males fight for control of groups of females. The 1–2 kids (occasionally 3) are fully active within hours of birth. Formerly widespread in and around the Sahara, they have been extensively hunted for meat, hides and almost all other parts, and are now extinct over much of their natural range. However, they have been introduced into the USA and Mexico for sport hunting, where their range is currently expanding. Frequently kept in zoos, where they breed freely, and are entirely self-sustaining. Have also been successfully introduced into several ranches in South Africa, where they are fenced. Listed on CITES App.II.

*** **ASIATIC MOUFLON**
Ovis orientalis (=aries) (**1129**)
Very similar in size and appearance to the Urial and Argali. Occurs in fragmented and isolated populations in Asia Minor and Iran, where there is a zone of hybridisation with the Urial.

*** **MOUFLON**
Ovis musimon (**1130**)
Very closely related to the Asiatic Mouflon, and probably descended from early domesticated stocks introduced by man; until the 19th century confined to Cyprus, Sardinia, and Corsica. During this century numbers have declined drastically within this range, but substantial numbers are kept in captivity, where they breed freely. They have also been extensively introduced into mainland Europe and N America.

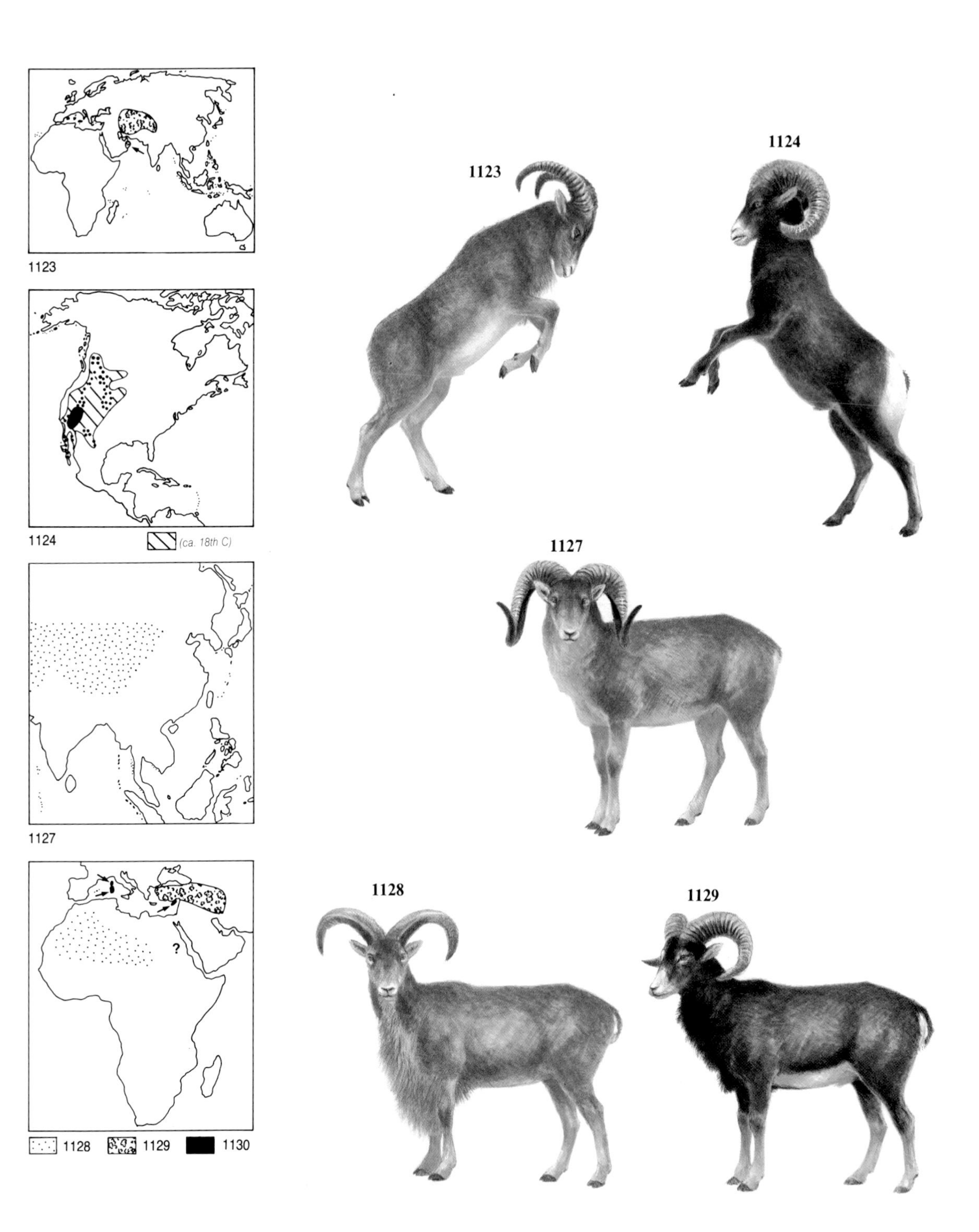
1123
1124
1127
1128
1129
1123
1124
(ca. 18th C)
1127
?
1128
1129
1130

CETACEANS

There is considerable public interest in the conservation status of Cetaceans, and also a very extensive literature. It is arguable that the majority of species are threatened and most may have declined. Those described are only those most threatened, better documented, or of general interest. When distribution refers to 'all seas', the inland seas, such as the Aral and Caspian Seas, are not included.

*** AMAZON DOLPHIN or BOUTU

Inia geoffrensis **(1131)**

TL 5 ft 7ins–9ft 10ins; WT up to 269lb.

A rather thick-set dolphin with a very mobile head – the neck is more pronounced than in any other species of dolphin. The snout is covered with short stiff bristles. Confined to freshwaters in the Amazon and Orinoco River systems, extending over 1740 mls from the sea to Peru and Bolivia. They generally live in pairs or alone, feeding on bottom-dwelling fish. Until recently they were very common in many parts of their range, and local superstitions attached to them protected them in some areas. In recent years, hunting has increased, and they are also threatened by motor boats, pollution and the construction of dams. Three subspecies have been described, and *I.g. boliviensis* from the Madeira River system of Bolivia, which is isolated from *I.g. geoffrensis* by 250 ml. stretch of rapids, has been extensively hunted for its meat and hides. *I.g. humboldtiana* is confined to the Orinoco River system. They have occasionally been exhibited in zoos and aquaria, where they have survived for over 16 years, and have given birth to young.

***** BAIJI or YANGTZE RIVER DOLPHIN

Lipotes vexillifer **(1132)**

TL 6ft 6ins–8ft 2ins; WT up to 353lb.

Similar to other river dolphins, but with slightly up-curved snout. Confined to the Yangtze (Chang Jiang) River system and its tributaries, and during floods travel upstream into lakes and small rivers. They feed on fish, and probably probe mud with their long snouts for crustaceans. Usually live in groups of 2–6, but may congregate in larger numbers when feeding. In the past have been revered as an incarnation of a drowned princess. Since 1975 they have had full legal protection. Some are killed accidentally by drowning in nets and on hooks; they are also killed in collisions with boats, and sedimentation of Lake Dongting, the type locality, has led to their disappearance. Protected in China and listed on CITES App.I.

*** FRANCISCANA or LA PLATA RIVER DOLPHIN

Pontoporia blainvillei **(1133)**

TL 4ft 3ins–5ft 9ins; WT up to 134.5lb.

Female slightly larger than male. The only member of the river dolphin family (Platanistidae) to occur in saltwater. Found along the coast of Argentina from the Valdez Peninsula north through Uruguay to S Brazil; they occur in the La Plata estuary, but do not normally venture into fresh water. It is estimated that 1500–2000 are killed by shark fishermen each year; the meat is used as pig-feed, and some of the blubber rendered for oil.

*** GANGES SUSU

Platanista gangetica **(1134)**

TL up to 9ft 10ins; WT up to 185lb.

Females are slightly larger than males. A small and rather angular dolphin, with a long (up to 18ins) slender snout. Habitually swims on its side; the eye is barely visible externally and lacks a lens; it is effectively blind, although it has been suggested that eyes are used for direction-finding. Found in the murky waters of the Ganges, Brahmaputra, Meghna and Karnaphuli Rivers and their tributaries, where the construction of dams and barrages has fragmented their populations into isolated stocks which can no longer interbreed. They have become extinct in many of the upper reaches of the rivers, such as Chittagong Hills, and are declining rapidly in the Padma River, Bangladesh. In the lower reaches of the rivers they are often remarkably widespread, even in apparently highly polluted stretches, such as those of the Hooghly River near Calcutta. In the late 1970s the total population was estimated at 4000–5000. They are only rarely hunted, but significant numbers are drowned in fishing nets. Found in some protected areas including the Chambal Sanctuary, a tributary of the Ganges, where they are threatened by falling river levels causing sand bars to isolate sections of the river. Protected throughout their range, and listed on CITES App.I.

***** INDUS DOLPHIN

Platanista indi (=minor) **(1135)**

Only readily distinguishable from the Gangetic species (1134) by differences in the skull; otherwise similar in appearance and biology. Confined to freshwaters in the Indus River system in Pakistan. The construction of dams and barrages has brought this species close to extinction. By the late 1970s the total population was estimated at possibly only 400, with the largest population isolated between the Kotri and Sukkur barrages, numbering some 70–80.

* IRRAWADDY DOLPHIN

Orcaella brevirostris **(1136)**

TL 6ft 7ins–8ft 3ins.

Uniformly slaty grey, with a low dorsal fin and a neck crease. Usually live singly or in schools of up to 10, keeping close to the coast and often ascending rivers. Widely distributed in warm tropical waters in the Indian and Pacific Oceans. They occur in the larger rivers, such as the Irrawaddy, Mekong, Ganges and Brahmaputra. Protected in most of the major river systems in which they occur, and although still locally abundant, are likely to be affected by increasing pollution, the construction of barrages such as dams, and build-ups of silt.

*** ESTUARINE DOLPHIN or TUCUXI

Sotalia fluviatilis **(1137)**

TL up to 6ft 3ins; WT up to 110lb.

One of the smallest living cetaceans, rather like a small Bottle-nosed Dolphin in appearance. Found in tropical coastal waters of South America and in the large river systems and their tributaries, as far as 1550 mls up the Amazon, to the base of the Andes. Although it is still locally common, the effects of numerous barrages and dams are not known in detail, but they are likely to isolate populations. They are also drowned in fishing nets, and hunted in some areas for their alleged medicinal value. Have occasionally been exhibited in zoos and aquaria.

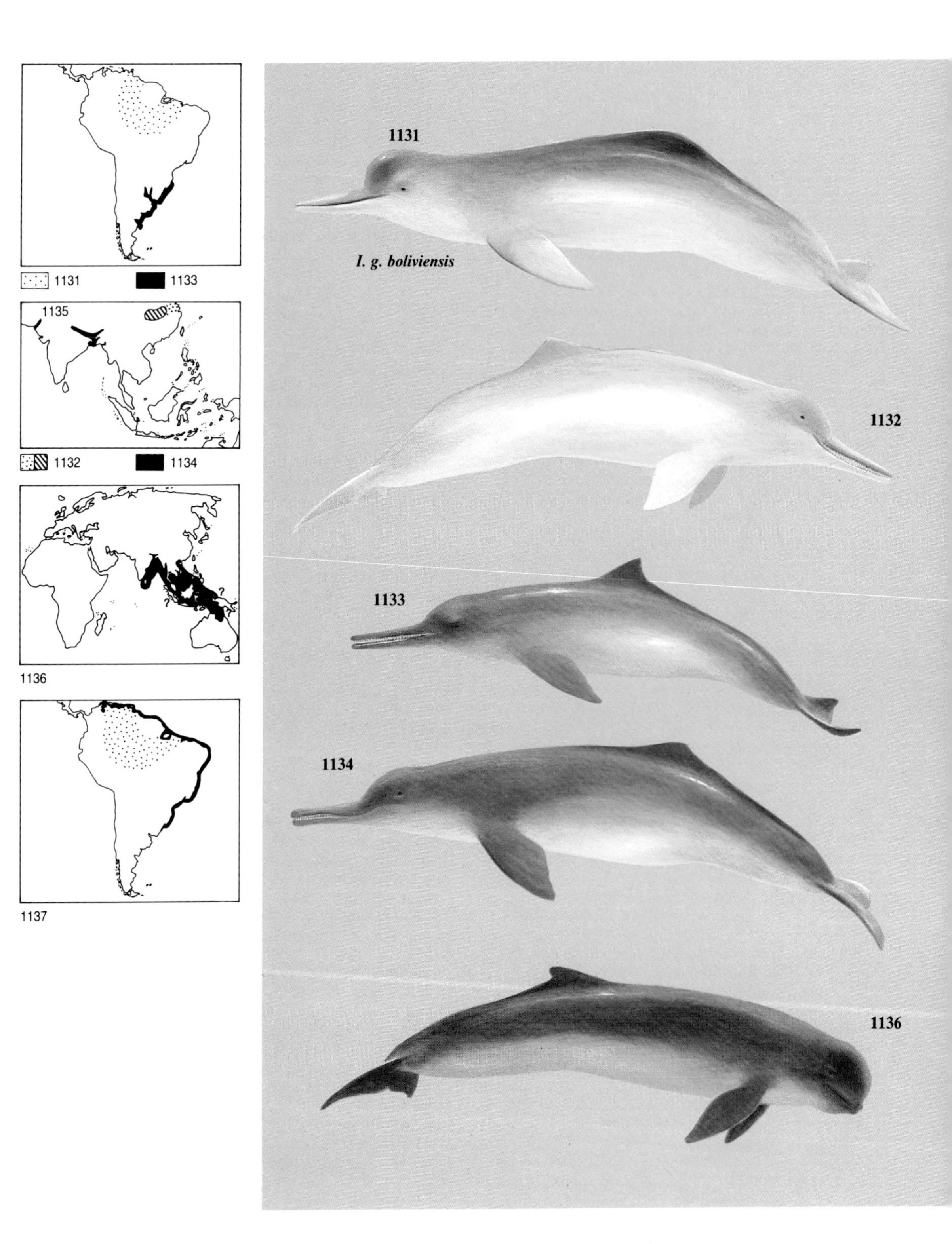

1131
1133
1135
1132
1134
1136
1137
1131
I. g. boliviensis
1132
1133
1134
1136

*** STRIPED, SPOTTED and SPINNER DOLPHINS**
Stenella spp.
TL up to 11ft 5ins; WT up to 364lb.
Approximately five species exist, but there is much confusion over delineating the species. The **Atlantic Spotted Dolphin** *S. plagiodon* (**1138**) and **Atlantic Spinner Dolphin** *S. clymene* (**1139**) occur in the warmer waters of the Atlantic; the **Striped Dolphin** *S. coeruleoalba* (**1140**), **Tropical Spinner Dolphin** *S. longirostris* (**1141**) and **Tropical Spotted Dolphin** *S. attenuata* (**1142**) occur in warm waters of the Atlantic, Pacific, and Indian Oceans and adjoining seas. There are or have been fisheries for *Stenella* dolphins in many parts of the world, the largest being in Japan where ca.20,000 a year were being killed in the 1970s, for human consumption. The populations off Japan are believed to have declined by at least 50%. *Stenella* dolphins have also been very seriously affected by incidental killing in tuna fishing operations; the dolphins associate with tuna, and fishermen set their purse seine nets around them. In the early 1970s over 250,000 were being drowned every year. Although techniques have been developed to reduce the kill, large numbers are still drowned, and some populations may be threatened.

***** COMMON DOLPHIN**
Delphinus delphis (**1143**)
TL up to 8ft 3ins; WT up to 165lb.
One of the most colourful of dolphins, with stripes of grey, yellow, and white on the sides, which may be absent in some populations. Mainly pelagic, occurring in all but the colder waters of the world. One of the most numerous of the cetaceans, and may occur in schools of up to 300,000. Also one of the most heavily exploited, and have been hunted since prehistoric times. The Black Sea population was estimated to have originally numbered ca.one million, but in the 1930s up to 200,000 were being killed a year by fishermen from USSR. This was banned in 1966, although Turkey continued, with 88,000 being killed in 1971; the hunt was banned in 1984. Also affected, together with *Stenella* dolphins, by incidental killing in tuna fishing operations. Often exhibited in oceanaria and aquaria, but rarely bred.

**** FRASER'S DOLPHIN**
Lagenodelphis hosei (**1144**)
TL up to 8ft 6ins; WT up to 463lb.
Distinguished by the three stripes running along its sides. One of the most recently discovered large mammals, Fraser's Dolphin was described in 1956 from a skeleton collected in the 19th century, in Borneo, which had remained unnoticed in the collections of the British Museum (Natural History). In 1971 strandings occurred on Cocos Island, E Pacific; Durban, S Africa; and New South Wales, Australia; since then they have been observed in schools of up to 500, usually in deep tropical waters. They may be more common than their recorded occurrence suggests.

**** HEAVISIDE'S or BENGUELA DOLPHIN**
Cephalorhynchus heavisidii (**1145**)
TL 4ft–4ft 7ins; WT up to 121lb.
A small dolphin, with no beak, confined to coastal waters of S Africa, in the Benguela Current system, from the Cape north to Namibia. First decribed in 1828, another found in 1856, then none observed until 1965, and 3 in 1969. One of the least-known dolphins, but around 100 are believed to be killed annually in purse seine fisheries, which could have a serious impact on such an apparently rare species.

**** CHILEAN or BLACK DOLPHIN**
Cephalorhynchus eutropia (**1146**)
TL up to 5ft 3ins; WT up to 132lb.
Similar to Heaviside's Dolphin, but with a smaller area of white on belly, and rounded dorsal fin. Only known from the cold coastal waters of S Chile, where they feed on cuttlefish and crustaceans. Often seen in the channels of Tierra del Fuego. They are killed in fishing nets and occasionally hunted for food or fishing bait; the effects of this on what must be regarded as a comparatively rare dolphin are not known.

***** COMMON or HARBOUR PORPOISE**
Phocoena phocoena (**1147**)
TL 5ft–6ft 6ins; WT 99–198.5lb.
Small and stocky with a small triangular dorsal fin. The colour is variable, and there is no regular pattern. They usually live in pairs or small groups, only rarely gathering in groups of 50 or more. They eat a wide variety of fish and cephalopods. Found in the cooler coastal waters of the northern hemisphere, but declining in many areas; pollution by pesticides in the Baltic has caused a considerable decline, and in the NE Atlantic many are drowned in fishing tackle. Also hunted, and taken in large numbers in the Black Sea until 1960s; the hunt did not end until the 1980s. Protected in most coastal waters, but receive little protection outside territorial waters.

***** COCHITO**
Phocoena sinus (**1148**)
TL up to 5ft; WT up to 121lb.
The smallest living cetacean, very similar in general appearance to the Common Porpoise. Confined to the waters of the upper quarter of the Gulf of California; substantial numbers have been drowned in fishing tackle, and pollution with pesticides may be having an adverse effect on them. The related **** Spectacled Porpoise** *P. dioptrica* (**1149**) is only known from a handful of observations around the coasts of Argentina, the Falkland Islands and South Georgia; there are also sightings elsewhere in Antarctic waters and NewZealand. Nothing in known of its status. Similarly, *** Burmeister's Porpoise** *P. spinipinnis* (**1150**) from the Atlantic and Pacific coasts of South America, is probably much more abundant than published information suggests, since around 2000 a year have been sold in Peruvian markets.

*** DALL'S PORPOISE**
Phocoenoides dalli (**1151**)
TL up to 7ft 3ins; WT up to 485lb.
Very stocky, powerfully built porpoise, with a bold black and white pattern and distinctive white marking on the dorsal fin. Confined to the cooler waters of the N Pacific, and while not known to be in any immediate danger, has been extensively exploited, and the effects on the various populations is not known. During the 1960s Japanese fishermen were killing ca.2500 a year, and up to 20,000 were being drowned in fishing tackle. Others are drowned in US waters.

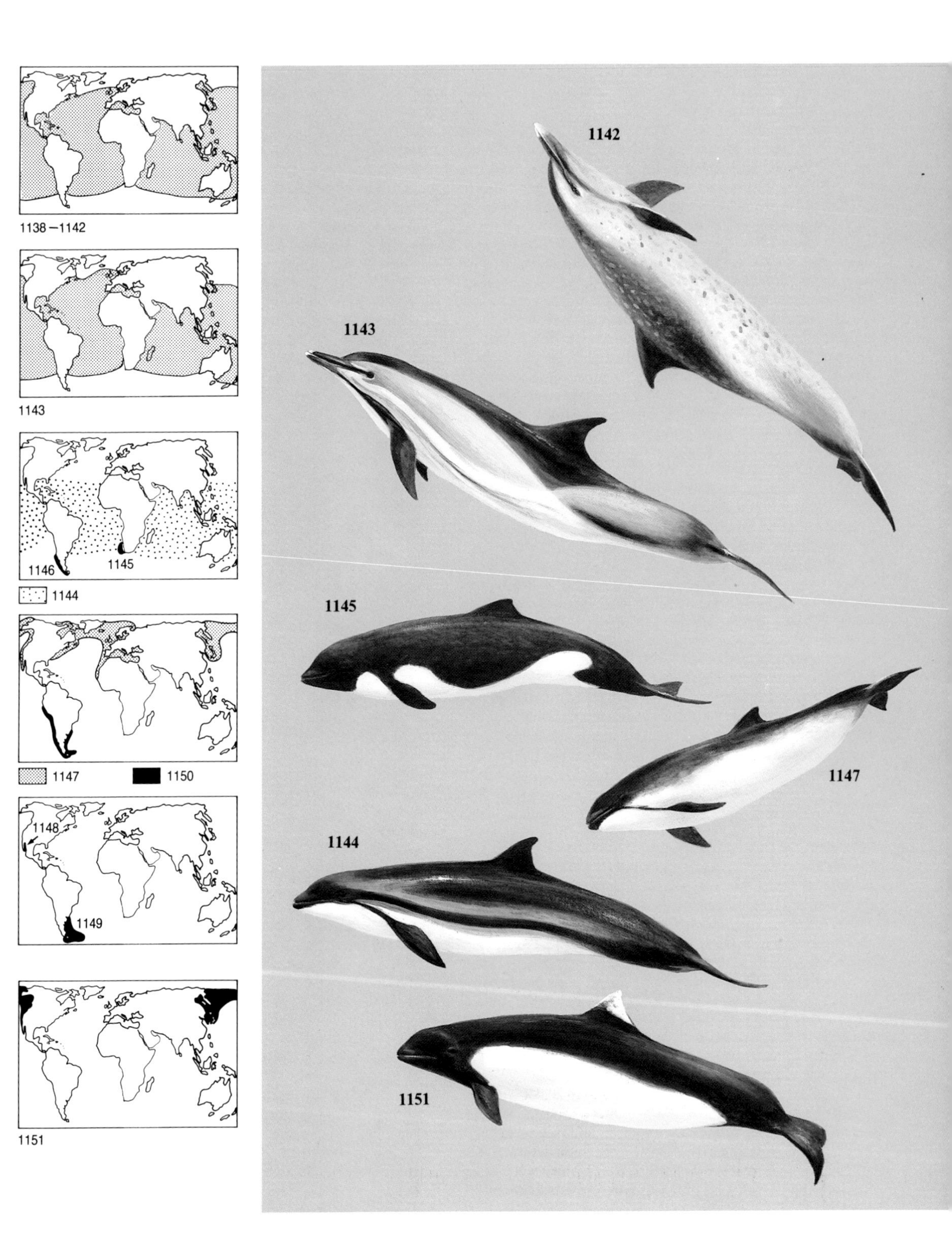
1138 – 1142
1143
1146
1145
1144
1147
1150
1148
1149
1151
1142
1143
1145
1147
1144
1151

*** KILLER WHALE or ORCA**
Orcinus orca (**1152**)
TL up to 31ft 3ins; WT up to 8.8 tons.
Male much larger than female. The most conspicuous features are the large dorsal fin and the black and white patch behind the eye. They are cosmopolitan, occurring in all seas, from the polar regions to the tropics, where they feed on a wide variety of fish, seabirds (including penguins), seals, and cetaceans. Publicity campaigns attacking the keeping of cetaceans in captivity have often focussed on Killer Whales, and consequently there is sometimes a supposition that they are in some way threatened in the wild. However, although they cannot be classed as endangered (there is some evidence that they have actually increased in the Antarctic), they are subject to local exploitation. Like most other cetaceans, very little is known of the structure of the various populations and stocks, and it is possible that localised populations may be being adversely affected by exploitation. Large numbers have been killed as part of so-called fisheries protection measures, and they have also been captured for display in zoos and aquaria.

***** LONG-FINNED PILOT WHALE**
Globiocephala melaena (**1153**)
TL up to 20ft 5ins; WT up to 3.3 tons.
Characterised by bulbous 'pothead' and long sickle-shaped flippers. Gregarious, sometimes forming schools of several hundreds, even more than a thousand, often mixing with other species of cetacean. They feed mainly on squid, and fish such as cod, when squid are not available. Widespread in the colder waters of the southen hemisphere and the N Atlantic. Have been hunted since prehistoric times, and in the Faeroes records of the hunt go back to 1584. In Newfoundland 10,000 were killed in 1956 alone, leading to the collapse of that population. Throughout the N Atlantic they are probably depleted, but southern hemisphere populations are relatively abundant. There is evidence from 10th-century manuscripts that they once occurred in the N Pacific and were hunted by the Japanese.

***** NARWHAL**
Monodon monoceros (**1154**)
TL up to ca.19ft 8ins, excluding tusk; WT up to 2 tons.
Male much larger than female, and carries a spiralled tusk, up to about 9ft 10ins long, and weighing up to 22lb. Occasionally the female has the tusk, and males may lack one. No dorsal fin. The skin colour changes with age, beginning a blotchy slaty grey, becoming completely greyish as juveniles, to almost pure white in old age. The tusk is used by males for fighting. They eat squid, crabs, shrimps and fish. Groups of 6–10 are led by an adult male, but during migration they may gather into much larger herds. Normally found only in the colder waters of the northern hemisphere, usually close to pack ice. They are extensively hunted by Inuits (Eskimos), and it is estimated that with modern rifles 3 or 4 are hit for every one landed. In 1981 the world population was estimated at less than 20,000 and still declining. Probably extinct in the waters around Novaya Zemlya and Franz Josef Land, USSR. In addition to subsistence hunting, there has been a commercial trade in tusks since at least the 10th century, and in medieval times were often traded as the horns of the unicorn. They are listed on CITES App.II.

***** WHITE WHALE or BELUGA**
Delphinapterus leucas (**1155**)
TL up to 15ft; WT up to 1.6 tons, possibly more.
The young are slaty grey, becoming paler with age until adult, when they are almost pure white. Some populations appear to have very restricted ranges, such as those in the Gulf of Alaska, while others undergo extensive migrations. Mostly confined to the Arctic and adjacent waters, occasionally ascending larger rivers such as the St Lawrence, Yenesi, Yuon, and Amur. They normally live in groups of about 10, but during migration as many as 10,000 have been recorded. They have been exploited since prehistoric times in the Arctic, and in the 20th century have been subjected to commercial fisheries, with the skin being marketed as 'porpoise leather'. Between the 1950s and 1970s the annual kill dropped from ca.1000 to 500 in Canada, and 4000 to 700 in the USSR, due to overfishing. In the mid-1970s the world population was estimated at less than 80,000, of which at least 30,000 were in North American waters.

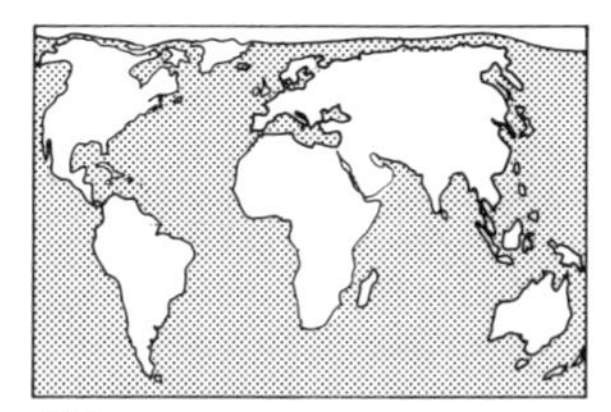

1152

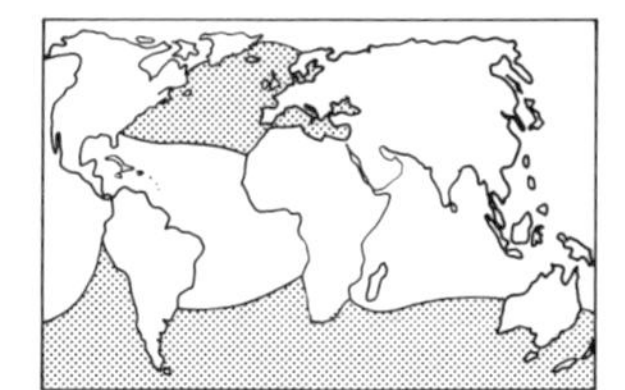

1153.

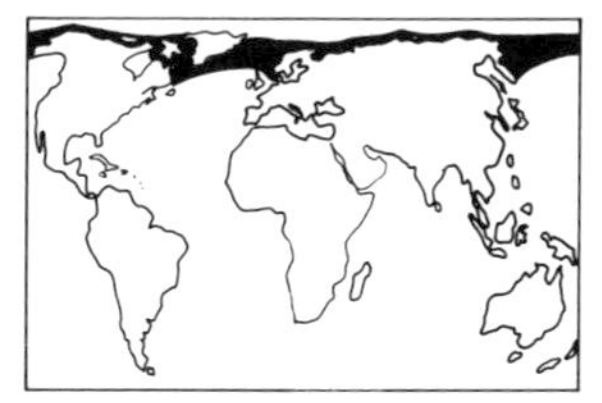

1154

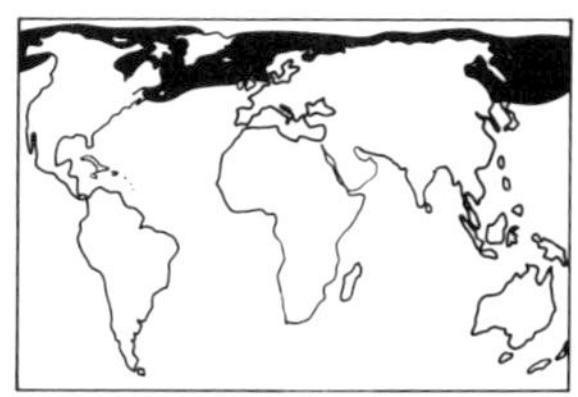

1155

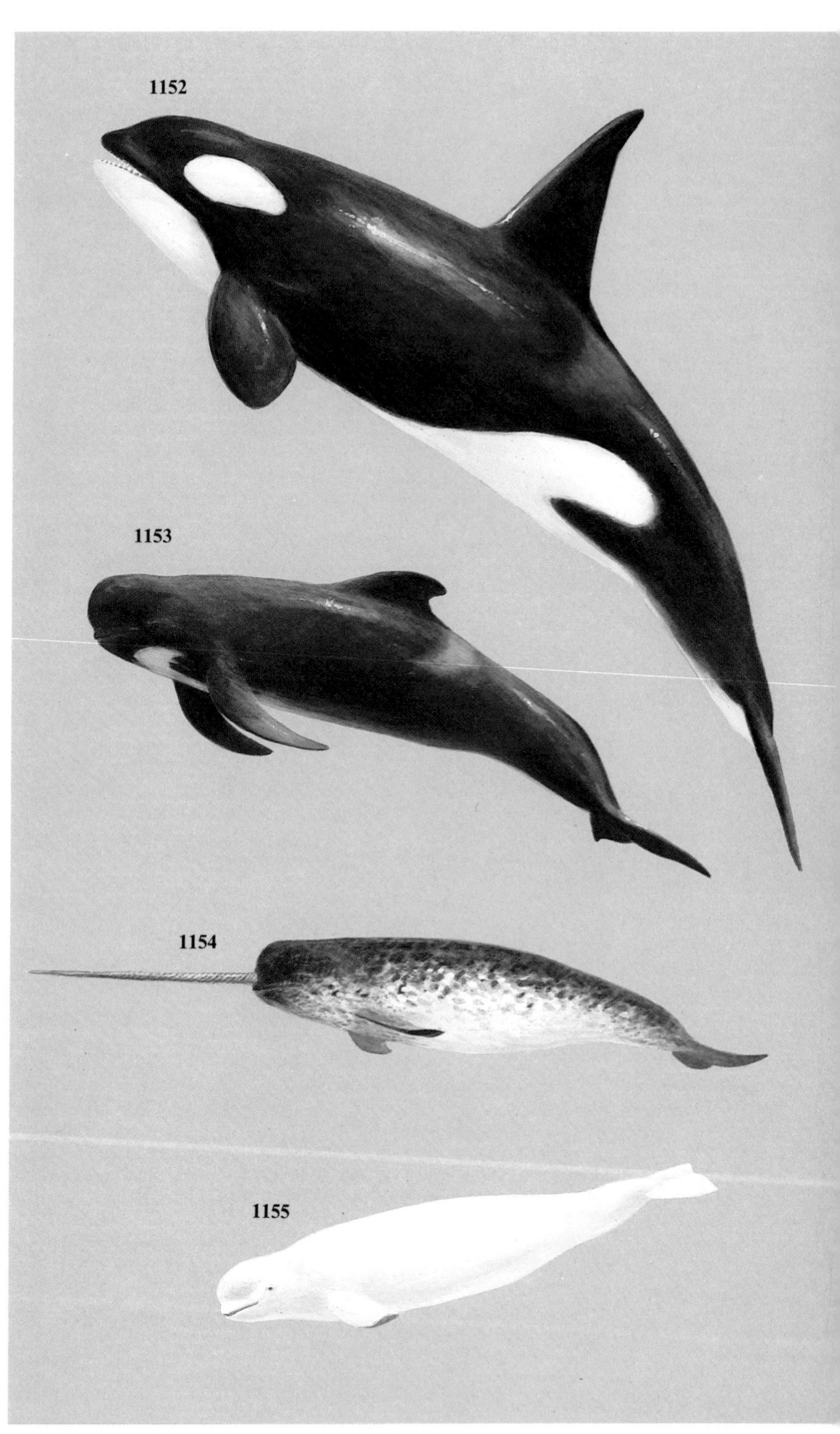

BEAKED WHALES
Mesoplodon spp.
A relatively large group of whales growing to 9ft 10ins and weighing up to 4.4 tons. They are probably not nearly as rare as they appear to be – are merely rare in museum collections. Do not get stranded very often, avoid ships and keep to deep waters. Those which are little-known include: ** **Hector's Beaked Whale** *M. hectori* (**1156**), ** **Hubbs's Beaked Whale** *M. carlhubbsi* (**1157**), ** **Stejneger's Beaked Whale** *M. stejnegeri* (**1158**), and ** **Andrews's Beaked Whale** *M. bowdoini* (**1159**).

**** GINKGO-TOOTHED WHALE**
Mesoplodon ginkgodens (**1160**)
TL up to 16ft 5ins; WT up to 1.6 tons.
One of the largest mammals to be discovered in the 20th century; it was described in 1958, relatively few specimens found, and nothing is known of its status. Occurs from the northern Indian Ocean, eastwards through the waters of SE Asia, to Japan, and also to the west coast of N America. Further study will almost certainly show them to be more abundant and widespread than available data suggest.

**** LONGMAN'S BEAKED WHALE**
Indopacetus pacificus (**1161**)
Only known from two beach-worn skulls, from Somalia and Australia.

**** DWARF SPERM WHALE**
Kogia simus (**1162**)
TL up to 8ft 10ins; WT up to 600lb.
A rather shark-like head, with 1–3 pairs of teeth on the upper jaw. Usually found in groups of less than 10, in warm and temperate waters, where they feed on squid, fish and crustaceans. They have been hunted, are slow-moving and apparently easily harpooned; it has been suggested that the rarity of the Dwarf Sperm Whale and the closely related * **Pygmy Sperm Whale** *K. breviceps* (**1163**) may be the result of earlier exploitation by whalers, which has gone unrecorded.

*** SPERM WHALE or CACHALOT**
Physeter catodon (=macrocephalus) (**1164**)
TL up to 65ft 8ins; WT up to 55.1 tons.
Male considerably larger than female. The largest of the toothed whales, with a massive squarish head, the most asymmetrical of any mammal, containing the largest mammalian brain, weighing nearly 22lb. They feed mainly on cephalopods including giant squid, diving to depths of more than 8200ft. Usually found in groups of 20–40 (pods), which may aggregate on migration, when up to 3000–4000 have been seen. They occur in all seas. Extensively hunted since 1712, mostly for their oil, and spermaceti wax from the head, but also for ambergris (a product of the gut used in perfumery), and occasionally for meat. The highest kill was nearly 30,000 in the 1963/64 season, and it remained at over 20,000 until 1976. By the 1980s they were considered so depleted that no quota was set by the International Whaling Commission (IWC) in 1981. Even so they are still the most numerous of the great whales. Listed on CITES App.I and regulated by the IWC.

****** BAIRD'S BEAKED WHALE**
Berardius bairdii (**1165**)
TL up to 38ft 9ins; WT up to 12.6 tons.
Large whales with four teeth, the lower pair visibly protruding. They are deep divers, feeding mostly on squid, but also on a wide range of crustaceans, octopuses, fish and even starfish. Confined to the northern Pacific, usually in water at least 3300ft deep. They live in groups of up to 30, often led by a large male. Have been hunted in Japan since at least 1612, but the annual kill was generally rather small, with a peak of 382 in 1952. The subsequent decline in catches, to less than 70 a year in the 1970s, was probably associated with a decline in their populations.

*** ARNOUX' BEAKED WHALE**
Bernardius arnuxii (**1166**)
Similar to Baird's Beaked Whale and often considered to be only a subspecies; it is smaller with proportionally longer flippers. Confined to the cooler waters of the southern hemisphere with a circumpolar distribution. Although only relatively few specimens are known, there is no particular evidence that this species is rare or declining.

**** SHEPHERD'S BEAKED WHALE**
Tasmacetus shepherdi (**1167**)
TL up to 23ft; WT up to 3 tons.
Distinguished from other beaked whales by its unique dentition; both jaws are lined with many (17–29) small teeth, and two larger teeth at the tips. Apparently confined to the temperate waters of the southern hemisphere. Since their discovery in 1933 only relatively few (less than 10) have been found, in New Zealand, the Stewart and Chatham Islands near New Zealand, Juan Fernandez Islands, off Chile, Tierra del Fuego and the Valdez Peninsula, Argentina. Nothing is known of its precise distribution, but it is presumed to be a rare species – although it could easily be commoner than data so far indicates.

****** NORTHERN BOTTLE-NOSED WHALE**
Hyperoodon ampullatus (**1168**)
TL up to 32ft 10ins; WT up to 5.9 tons.
Male larger than female. The very rounded forehead or 'melon' is characteristic. They are usually very curious about ships and approach them closely, particularly if engines are running, making them easy targets. Squid, herrings and starfish provide most of their food. They live in groups of 4–10, consisting of a male, along with a group of females and their young. Their precise population numbers are not known, but they are believed to be very heavily depleted, having been exploited since the 1870s. At the beginning of the 20th century, around 3000 were being killed each year, but by the late 1960s this had fallen to less than 100. The closely related * **Southern Bottle-nosed Whale** *H. planifrons* (**1169**) is very little studied; although it has been hunted occasionally it has not been reduced to the extent of *H. ampullatus*.

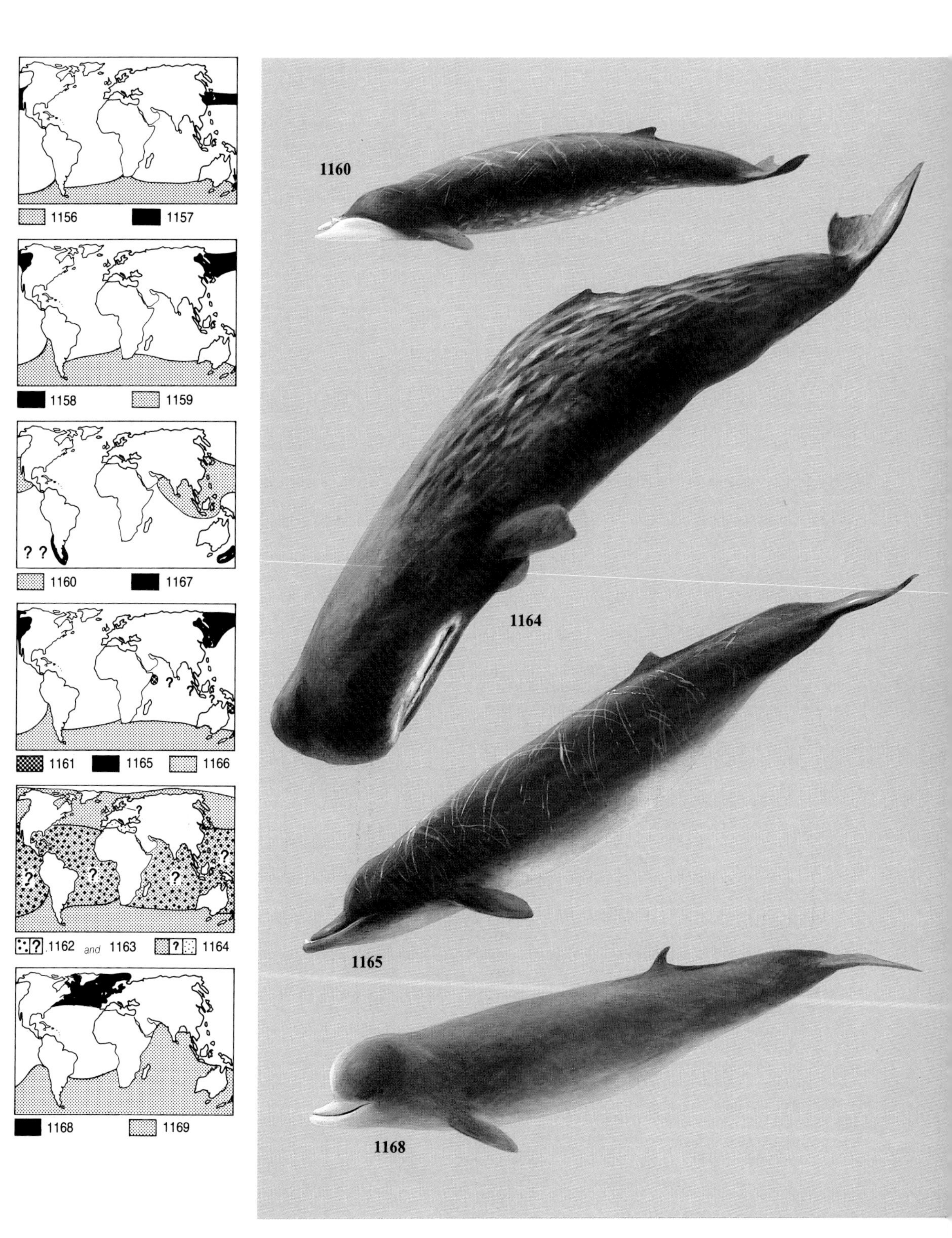
1156
1157
1158
1159
? ?
1160
1167
?
1161
1165
1166
?
?
?
?
?
?
1162 and 1163
?
1164
1168
1169
1160
1164
1165
1168

RORQUALS

Balaenoptera spp.

The name rorqual is derived from the Norwegian, and refers to the 10–100 longitudinal furrows on the throat allowing it to expand when feeding. They generally feed on whatever animal life happens to be abundant in the area in which they are feeding; shoaling fish such as herrings and capelin, and krill are important parts of their diet. They scoop in huge quantities of seawater containing their food, filtering it by forcing it out through the baleen plates. Rorquals were hardly exploited until the 1860s, when the invention of the explosive harpoon, powered boats, and techniques for inflating whales to prevent them sinking, allowed the large, fast-swimming rorquals to be taken. There was no effective regulation of hunting until the 1960s, by which time many populations were rapidly declining. In the 1980s a moratorium on commercial whaling came into force.

*** **MINKE or PIKED WHALE**

Balaenoptera acutorostrata (**1170**)

TL up to 35ft; WT up to 11 tons (usually slightly smaller).

The smallest of the rorquals, with a distinctive whitish band on the flipper of the northern hemisphere population; the southern hemisphere population is slightly smaller.They generally have 230–360 baleen plates. Found singly, in pairs, or trios, and also gather in larger groups around abundant food supplies during the polar summer. They are hunted by killer whales. Widely distributed in all oceans and seas, and believed to exist in at least three separate populations: N Pacific, N Atlantic, and southern hemisphere. All three have been exploited by numerous coastal and pelagic operations, and the two northern populations are generally considered depleted. The southern hemisphere population may have increased because of the reduction in competion from larger rorquals. In the late 1970s, the population in the south was estimated at 200,000 or more, and that of the two northern populations at ca.120,000. The only great whale that has been exhibited in captivity.

***** **BLUE WHALE**

Balaenoptera musculus (**1171**)

HB up to 100ft; WT up to 196 tons.

The largest living animal, although they no longer reach the size attained before exploitation; females are slightly larger than males. They are about 23ft long at birth. Usually occur singly or in pairs, although several may gather in a rich feeding area. Originally found in all oceans and seas. They migrate long distances along well-known routes. By the time protection was introduced in 1965, they were heavily depleted. In the late 1970s, it was estimated that only a few hundred remained in the N Atlantic, 1200–1700 in the N Pacific, and perhaps 5000 in the southern hemisphere, of which about half are a pygmy subspecies; before the advent of commercial whaling there were probably over 200,000 in the southern Ocean.

*** **BRYDE'S WHALE**

Balaenoptera edeni (**1172**)

TL up to 48ft; WT up to 22 tons.

Distinguished from the similar Sei Whale (1174) by white on throat, three ridges on head, and paler baleen plates with very stiff bristles. Found in groups of 2–10, occasionally forming feeding concentrations of up to 100. The only rorqual confined to warmer waters; those in the tropics appear to be sedentary, while those in temperate waters undergo migrations in response to changes in food supply. Earlier this century, coastal hunting took place off California, Japan and South Africa. Large scale commercial whaling did not start until the 1970s, after most other rorquals had been over-fished, with a maximum of 1800 killed in the 1973/74 season. The North Pacific population is thought to have declined from ca.20,000 before exploitation to ca.16,000 at the end of the 1970s; no figures are available for other areas, but similar declines are believed likely.

***** **FIN WHALE or COMMON RORQUAL**

Balaenoptera physalus (**1173**)

TL up to 88ft; WT up to 76.6 tons.

Only slightly smaller than the Blue Whale (1171); very slender and probably the fastest swimming of the great whales. More gregarious than other rorquals, and may form concentrations of over 100 on rich feeding grounds. In the early 1980s the world population of mature Fin Whales was estimated at 470,000 before exploitation commenced; in the early 1980s, after protection had allowed a slight recovery in numbers, the population was estimated at over 100,000.

*** **SEI WHALE**

Balaenoptera borealis (**1174**)

HB up to 66ft; WT up to 32 tons.

Smaller than the Fin Whale (1173), and not as slender. Very similar to Bryde's Whale (1172) but more uniformly dark, the shiny black baleen plates having a large number of fine, soft white bristles. They live in family groups of 4 or 5 sometimes congregating to feed. They are found in all oceans and seas except the colder, polar waters. Have been hunted since the 1860s, with less than 1000 a year being taken in the Antarctic until the 1950s when the kill rose to a maximum of over 25,000 in 1964/65 season. Increasing protection has gradually reduced quotas. In the early 1980s, the total population of mature Sei Whales was probably 80,000–100,000.

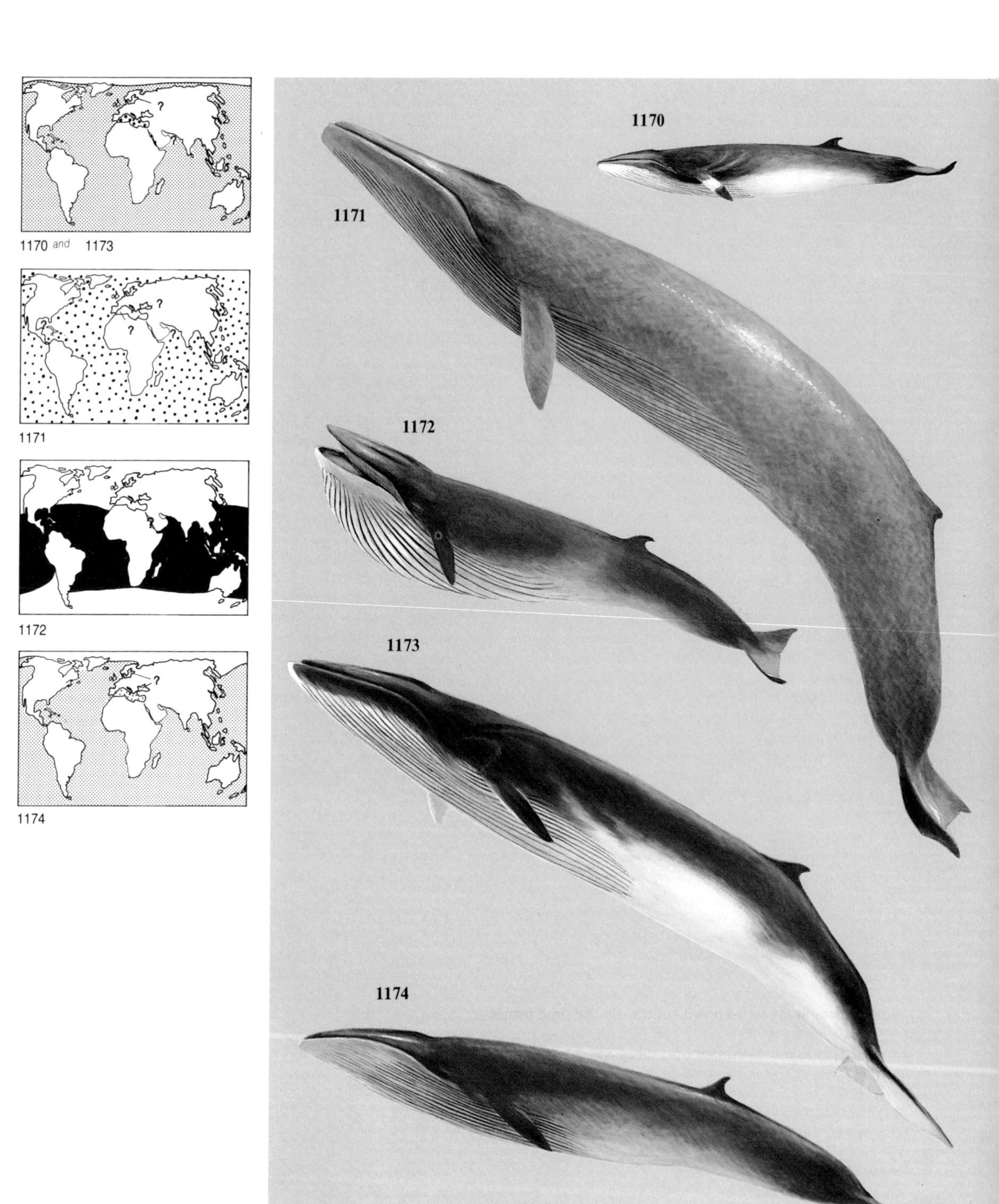
1170 and 1173
1171
1172
1174
1170
1171
1172
1173
1174

***** GREY WHALE**
Eschrichtius robustus (**1175**)
TL up to 49ft; WT up to 40.8 tons (but usually smaller).
Female larger than male. The dorsal fin is replaced by a series of bumps (up to 15) along the midline of the lower back. They are generally greyish, but extensively dappled due to patches of barnacles. Unlike any other baleen whales, they appear to be adapted for feeding on the seabed, using their short, stiff baleen plates to stir up sediment and then filtering out crustaceans, molluscs and bristle worms. They migrate, along well-established routes, from their summer feeding grounds in the Bering Straits area to the breeding lagoons in Baja California and Mexico. Three distinct populations are known: the Atlantic population, which was exterminated by about the late 18th century; the W Pacific population, which bred around Korea and was hunted until the mid-1960s, and is now extinct or close to extinction; and the E Pacific population, which was feared to be on the brink of extinction at the turn of the century, but has slowly increased under protection since 1946. By the early 1980s the population was estimated at ca.16,000. Disturbance from tourists visiting the breeding lagoons has become a significant threat. Still hunted in small numbers in Alaska and the USSR. A young one was kept for a few months in captivity, where it ate 1985lb of squid a day until it was released.

***** HUMPBACK WHALE**
Megaptera novaeangliae (**1176**)
TL up to 58ft; WT up to 52.9 tons (usually smaller).
Female larger than male. They have exceptionally long flippers (nearly one third of total body length). Generally blackish above and white below, but variable, and there are often patches of barnacles on them. There are 10–36 grooves on the throat, extending onto the belly. Extremely agile, often leaping completely clear of the water, plunging back headfirst. They are well-known for their extensive vocalisations. Originally they occurred in nearly all seas and oceans. During the summer months they feed in cold polar waters, following well-established migration routes to their relatively few breeding lagoons in the tropics. Breeding areas in the southern hemisphere include the Mozambique Channel, coasts of N Australia, Vanuatu, Fiji, Samoa, Ecuador, and NE Brazil. In the northern hemisphere they breed around the Marianas, Hawaii, Cape Verdes, the Lesser Antilles and Baja California. The northern and southern populations are believed to be discrete. Original populations are believed to have numbered ca.100,000 in the southern hemisphere, and ca.15,000 in the northern hemisphere. The world total in the early 1980s was estimated at under 7000 and is only slowly increasing. Still threatened by accidental entanglements in fishing tackle and competition from commercial fisheries.

******* RIGHT WHALE**
Eubalaena (=Balaena) glacialis (**1177**)
TL up to 60ft; WT up to 73.8 tons (usually much smaller: TL 43–53ft; WT ca.24.2 tons).
Mainly black, but sometimes have white patches on belly. Around the head there are protuberances infested with barnacles and parasitic crustaceans; the largest of these protuberances, known as the 'bonnet' is on the tip of the upper jaw. Individuals can be recognised by the callosities. Until the beginning of the 20th century, groups of up to 100 were commonly seen at the feeding grounds, but groups of 2–12 are now more common. Once abundant along all the major land masses in both hemispheres, but have been systematically over-exploited to the point of extinction or near extinction almost everywhere. The northern (*E.g. glacialis*) and southern (*E.g. australis*) populations are isolated from each other. The southern population possibly exceeded 100,000 at the beginning of the 19th century, but was rapidly depleted. During the past few decades there has been a slow recovery; a wintering population was discovered in 1969 around the Valdez Peninsula, Argentina. In the Atlantic and North Sea they were first hunted in the 9th century, and by 1700 were too rare to be of economic importance; they are on the brink of extinction in the Atlantic. The Pacific population was exploited in the 19th century, and only small numbers now survive. In the late 1970s the world population was estimated at ca.4000, with over 3000 in the south. Listed on CITES App.I. The documentation of the destruction of the Right Whales is extremely extensive, and one of the most detailed available for any endangered species.

******* BOWHEAD or GREENLAND RIGHT WHALE**
Balaena mysticetus (**1178**)
TL up to 66ft; WT up to 76 tons.
Similar to the Right Whale, but larger, with a proportionally larger head (a third of the total length). Usually completely black, except for a patch of white on the chin. They are slow-swimming, remaining submerged for up to 40 minutes. Once found in almost all Arctic waters, and commercial whaling commenced in the 16th century off Greenland. Because of the thickness of their blubber and the length of baleen, they were the most valuable of all whales and the 'right' ones to kill. They were too rare to be of any economic importance by the beginning of the 20th century. In the N Pacific, hunting continued, and even after total protection by the IWC in 1946, 10–15 were killed every year by Eskimos in Alaska. The kill rose in the 1970s, with increasing affluence, with 111 killed or injured in 1977 – a level which would probably prevent any recovery of the very low populations. In 1981 the population was estimated at 2264, and continuing to decline. Elsewhere Bowhead sightings are too sporadic for any estimate of populations to be made. Listed on CITES App.I.

**** PYGMY RIGHT WHALE**
Caperea marginata (**1179**)
TL up to 21ft; WT 5 tons.
The smallest baleen whale with a jaw curved like the Bowhead. Found in widely scattered localities in the cooler waters of the southern hemisphere. Relatively rarely recorded. This may be because they do not spend long at the surface of the water, and in the past have been confused with Minke Whales (1170); or it may reflect genuine rarity.

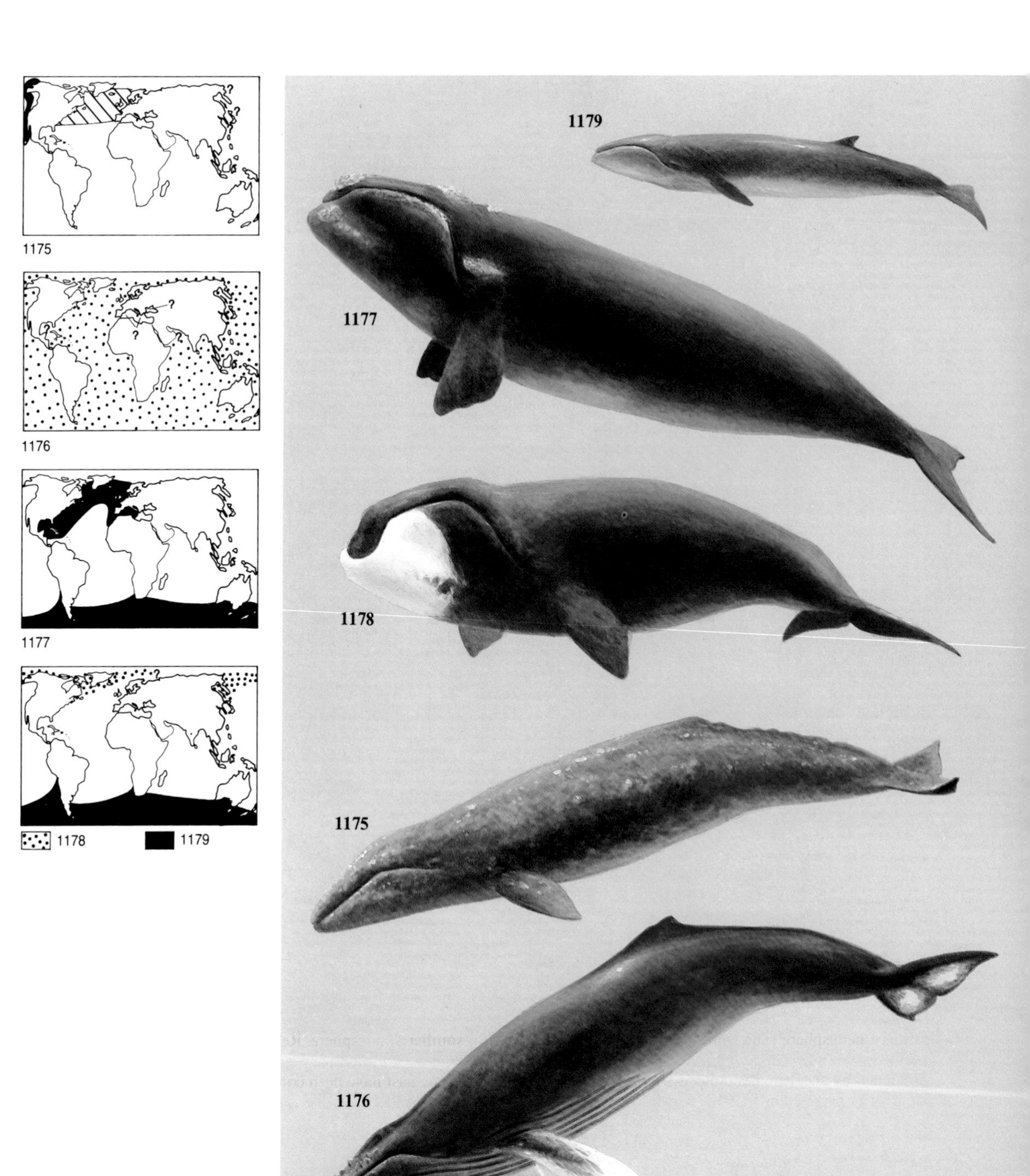
1175
1176
1177
1178
1179
1179
1177
1178
1175
1176

APPENDIX I
Checklist of Species

The checklist follows the systematic order used by Corbet & Hill in *A World List of Mammalian Species* (*see* bibliography).
The IUCN categories in column 5 are fully explained in the Introduction on p. 12, but can be interpreted as follows:

Ext.: extinct; E: endangered; V: vulnerable; R: rare; I: indeterminate; K: insufficiently known.

The categories in brackets indicate that only certain populations are under that particular threat.

STAR RATING	VERNACULAR NAME	SCIENTIFIC NAME	SPECIES NUMBER	IUCN CATEGORY	CITES APPENDIX
**	**Long-nosed Echidna**	*Zaglossus bruijni*	18	V	2
*	**Duck-billed Platypus**	*Ornithorhynchus anatinus*	19		
**	**Mouse Opossum**	*Marmosa andersoni*	44		
**	**Mouse Opossum**	*Marmosa cracens*	45		
**	**Mouse Opossum**	*Marmosa handleyi*	46		
**	**Mouse Opossum**	*Marmosa tatei*	47		
**	**Bolivian Short-tailed Opossum**	*Monodelphis kunsi*	49		
**	**Patagonium Opossum**	*Lestodelphys halli*	43		
**	**Black Four-eyed Opossum**	*Philander mcilhennyi*	42		
*	**Water Opossum** or **Yapok**	*Chironectes minimus*	41		
**	**Bushy-tailed Opossum**	*Glironia venusta*	48		
*	**Common Shrew-opossum**	*Caenolestes caniventer*	52		
*	**Common Shrew-opossum**	*Caenolestes tatei*	53		
**	**Peruvian Shrew-opossum**	*Lestoros inca*	50		
**	**Chilean Shrew-opossum**	*Rhyncholestes raphanurus*	51		
***	**Broad-striped Marsupial Mouse**	*Murexia rothschildi*	1		
*	**Narrow-striped Marsupial Mouse**	*Phascolosorex dorsalis*	2		
*	**Red-bellied Marsupial Mouse**	*Phascolosorex doriae*	3		
***	**Ingram's Planigale**	*Planigale ingrami*	9		
*	**Narrow-nosed Planigale**	*Planigale tenuirostris*	10		
*	**Paucident Planigale**	*Planigale gilesi*	11		
**	**Cinnamon Antechinus**	*Antechinus leo*	4		
*****	**Dibbler** or **Freckled Marsupial Mouse**	*Antechinus apicalis*	5	I	
*	**Godman's** or**Atherton Antechinus**	*Antechinus godmani*	6		0
*	**Swamp Antechinus**	*Antechinus minimus*	7		0
**	**Kultarr**	*Antechinus laniger*	8		
**	**Red-tailed Phascogale** or **Wambenger**	*Phascogale calura*	26	I	
*	**Brush-tailed Phascogale**	*Phascogale tapoatafa*	27		
***	**Eastern Quoll** or **Native Cat**	*Dasyurus viverrinus*	13		
*	**Tiger Cat**	*Dasyurus maculatus*	15		
***	**Mulgara**	*Dasycercus cristicauda*	16		
***	**Western Quoll, Chuditch** or **Native Cat**	*Dasyurus geoffroii*	17		
***	**Kowari**	*Dasyuroides byrnei*	14		
*	**Ningaui**	*Ningaui timealeyi*	28		
*	**Ningaui**	*Ningaui ridei*	29		
*	**Ningaui**	*Ningaui yvonneae*	30		
**	**Julia Creek Dunnart**	*Sminthopsis douglasi*	21	I	
*	**White-tailed Dunnart**	*Sminthopsis granulipes*	22		
*	**Hairy-footed Dunnart**	*Sminthopsis hirtipes*	23		
*****	**Long-haired Dunnart**	*Sminthopsis longicaudata*	24	K	1
*****	**Sandhill Dunnart**	*Sminthopsis psammophila*	25	K	1
***	**Numbat** or **Banded Anteater**	*Myrmecobius fasciatus*	20	E	
*****	**Tasmanian Wolf** or **Thylacine**	*Thylacinus cynocephalus*	12	Ext.	1
**	**New Guinea Mouse Bandicoot**	*Microperoryctes murina*	31		
****	**Western Barred Bandicoot**	*Perameles bougainville*	32	R	1
***	**Eastern Barred** or **Gunn's Bandicoot**	*Perameles gunnii*	33		
*	**White-lipped Bandicoot**	*Echymipera clara*	34	R	
*	**Rufous Bandicoot**	*Echymipera rufescens*	35		
**	**Seram Island Long-nosed Bandicoot**	*Rhynchomeles prattorum*	38		
*****	**Golden Bandicoot**	*Isoodon auratus*	36		
***	**Southern Brown Bandicoot**	*Isoodon obesulus*	37		
***	**Common** or **Greater Rabbit-bandicoot** or **Bilby**	*Macrotis lagotis*	39	E	1
*****	**Lesser Rabbit-bandicoot** or **Bilby**	*Macrotis leucura*	40	Ext.	1
**	**Scaly-tailed Possum**	*Wyulda squamicaudata*	64		
***	**Grey Cuscus**	*Phalanger orientalis*	72		2
***	**Woodlark Cuscus**	*Phalanger lullulae*	73	R	
*	**Black-spotted Cuscus**	*Phalanger rufoniger*	74	R	
*	**Stein's Cuscus**	*Phalanger interpositus*	75		
***	**Little Pygmy Possum**	*Cercartetus lepidus*	77		
*	**Long-tailed Pygmy Possum**	*Cercartetus caudatus*	78		

STAR RATING	VERNACULAR NAME	SCIENTIFIC NAME	SPECIES NUMBER	IUCN CATEGORY	CITES APPENDIX
***	**Burramys** or **Mountain Pygmy Possum**	*Burramys parvus*	76		2
*****	**Leadbeater's Possum**	*Gymnobelideus leadbeateri*	69	E	
*	**Yellow-bellied** or **Fluffy Glider**	*Petaurus australis*	70		
***	**Lemur-like Ringtail**	*Pseudocheirus lemuroides*	65		
*	**Arfak Ringtail**	*Pseudocheirus schlegeli*	66		
*	**Common Ringtail**	*Pseudocheirus peregrinus*	67		
*	**Toolah** or **Green Ringtail**	*Pseudocheirus archeri*	68		
*	**Musky Rat-kangaroo**	*Hypsiprymnodon moschatus*	86		
***	**Long-nosed Potoroo**	*Potorous tridactylus*	84		
**	**Long-footed Potoroo**	*Potorous longipes*	85	I	
*****	**Woylie, Bettong** or **Brush-tailed Rat-kangaroo**	*Bettongia penicillata*	80	E	1
***	**Eastern Bettong**	*Bettongia gaimardi*	81		1
***	**Boodie**	*Bettongia lesueur*	82	R	1
*****	**Northern Rat-kangaroo**	*Bettongia tropica*	83		1
*****	**Desert Rat-kangaroo**	*Caloprymnus campestris*	79	I	1
***	**Red-bellied** or **Tasmanian Pademelon**	*Thylogale billardierii*	98		
*	**Red-legged Pademelon**	*Thylogale stigmatica*	99		
*	**Red-necked Pademelon**	*Thylogale thetis*	100		
***	**Little Rock Wallaby** or **Nabarlek**	*Petrogale concinna*	97		
***	**Yellow-footed Rock Wallaby**	*Petrogale xanthopus*	106		
***	**Brush-tailed Rock Wallaby**	*Petrogale penicillata*	107		
*****	**Proserpine Rock Wallaby**	*Petrogale pesephone*	108	R	
*	**Godman's Rock Wallaby**	*Petrogale godmani*	109		
*	**Black-footed Rock Wallaby**	*Petrogale lateralis*	110		
***	**Rothschild's Rock Wallaby**	*Petrogale rothschildi*	111		
**	**Warabi**	*Petrogale burbidgei*	112		
***	**Western, Rufous Hare Wallaby** or **Mala**	*Lagorchestes hirsutus*	89	R	1
***	**Spectacled Hare Wallaby**	*Lagorchestes conspicillatus*	90		
*****	**Central Hare Wallaby**	*Lagorchestes asomatus*	91		
***	**Quokka**	*Setonix brachyurus*	92		
***	**Banded Hare Wallaby**	*Lagostrophus fasciatus*	93	R	1
*	**Eastern Grey Kangaroo**	*Macropus giganteus*	94		
*	**Western Grey Kangaroo**	*Macropus fuliginosus*	95		
*	**Red Kangaroo**	*Macropus rufa*	96		
***	**Tammar** or **Dama Wallaby**	*Macropus eugenii*	103		
***	**Black Wallaby**	*Macropus bernardus*	104		
*	**Parma** or **White-throated Wallaby**	*Macropus parma*	105		
*****	**Bridled Nail-tailed Wallaby, Flashjack** or **Merrin**	*Onychogalea fraenata*	87	E	1
*****	**Crescent Nail-tailed Wallaby** or **Mala**	*Onychogalea lunata*	88	Ext.	1
***	**Grizzled Tree Kangaroo**	*Dendrolagus inustus*	58		2
***	**Vogelkop Tree Kangaroo**	*Dendrolagus ursinus*	59		2
***	**Bennett's Tree Kangaroo**	*Dendrolagus bennettianus*	60		2
***	**Lumholtz's Tree Kangaroo**	*Dendrolagus lumholtzi*	61		2
***	**Unicoloured Tree Kangaroo**	*Dendrolagus dorianus*	62	(V)	
***	**Ornate Tree Kangaroo**	*Dendrolagus goodfellowi*	63	(V)	
***	**Matschie's Tree Kangaroo**	*Dendrolagus matschiei*			
*	**Black Forest-wallaby**	*Dorcopsis atrata*	101	R	
*	**Papuan Forest-wallaby**	*Dorcopsulus macleayi*	102	R	
*	**Koala** or **Native Bear**	*Phascolarctos cinereus*	57		
*	**Common Wombat**	*Vombatus ursinus*	56		
*****	**Northern Hairy-nosed Wombat**	*Lasiorhinus krefftii*	54	E	1
***	**Southern Hairy-nosed Wombat**	*Lasiorhinus latifrons*	55		
*	**Honey Possum**	*Tarsipes spenserae*	71		
***	**Giant Anteater**	*Myrmecophaga tridactyla*	490	V	2
***	**Pale-throated Sloth**	*Bradypus tridactylus*	496		
***	**Maned Sloth**	*Bradypus torquatus*	497	E	
*	**Two-toed Sloth**	*Choloepus hoffmanni*	498		
***	**Brown-throated Sloth**	*Bradypus variegatus*	495		2
***	**Giant Armadillo**	*Priodontes maximus*	491	V	1
**	**Brazilian Three-banded Armadillo**	*Tolypeutes tricinctus*	492	I	
**	**Pink Fairy Armadillo**	*Chlamyphorus truncatus*	493	K	
**	**Burmeister's Fairy Armadillo**	*Chlamyphorus retusus*	494	K	
***	**Hispaniolan Solenodon**	*Solenodon paradoxus*	115	E	
***	**Cuban Solenodon**	*Solenodon cubanus*	116	E	
*****	**Web-footed Tenrec**	*Dasogale fontoynonti*	128		
**	**Rice Tenrecs**	*Oryzorictes hova*	119		
**	**Rice Tenrecs**	*Oryzorictes talpoides*	120		
**	**Rice Tenrecs**	*Oryzorictes tetradactylus*	121		
**	**Shrew-like Tenrecs**	*Microgale crassipes*	122		
**	**Shrew-like Tenrecs**	*Microgale melanorrhachis*	123		
*****	**Shrew-like Tenrecs**	*Microgale pusilla*	124		
**	**Shrew-like Tenrecs**	*Microgale longicaudata*	125		
**	**Shrew-like Tenrecs**	*Microgale cowani*	126		
***	**Web-footed Tenrec** or **Voalavonddrano**	*Limnogale mergulus*	127		

STAR RATING	VERNACULAR NAME	SCIENTIFIC NAME	SPECIES NUMBER	IUCN CATEGORY	CITES APPENDIX
*	**Giant African Water Shrew** or **Giant Otter Shrew**	*Potamogale velox*	117		
*	**Dwarf Otter Shrew**	*Micropotamogale ruwenzorii*	118		
**	**Visagie's Cape Golden Mole**	*Chrysochloris visagei*	129	O	
*	**Golden Mole**	*Chrysochloris stuhlmanni*	130		
***	**Grant's Desert Golden Mole**	*Eremitalpa granti*	134		
*****	**De Winton's Golden Mole**	*Cryptochloris wintoni*	131		
***	**Van Zyl's Golden Mole**	*Cryptochloris zyli*	132		
**	**Gunning's Golden Mole**	*Amblysomus gunningi*	137		
**	**Juliana's Golden Mole**	*Amblysomus julianae*	138	R	
**	**Duthie's Golden Mole**	*Amblysomus duthiae*	135		
**	**Arend's Golden Mole**	*Amblysomus arendsi*	136		
**	**Somali Golden Mole**	*Amblysomus tytonis*	139		
***	**Giant Golden Mole**	*Chrysospalax trevelyani*	133	E	
*	**Mindanao Moon Rat**	*Podogymnura truei*	113	V	
**	**Hainan Moon Rat**	*Neohylomys hainanensis*	114		
***	**American Water Shrew**	*Sorex palustris*	142		
*	**Merriam's Shrew**	*Sorex merriami*	143		
*	**South-eastern** or **Bachman's Shrew**	*Sorex longirostris*	144		
*	**Preble's Shrew**	*Sorex preblei*	145		
**	**Ashland Shrew**	*Sorex trigonirostris*	146		
**	**Suisan Shrew**	*Sorex sinuosus*	147		
**	**Socorro Shrew**	*Sorex juncensis*	148		
**	**Mt Lyell Shrew**	*Sorex lyelli*	149		
**	**Pribilof Shrew**	*Sorex pribilofensis*	150		
**	**St Lawrence Shrew**	*Sorex jacksoni*	151		
*	**Rock Shrew**	*Sorex dispar*	152		
**	**Gaspar Shrew**	*Sorex gaspensis*	153		
**	**Shrew**	*Sorex stizodon*	154		
**	**Shrew**	*Sorex sclateri*	155		
**	**Shrew**	*Sorex milleri*	156		
***	**Hoy's Pygmy Shrew**	*Microsorex hoyi*	140		
*	**Southern Short-tailed Shrew**	*Blarina carolinensis*	157		
**	**Ender's Small-eared Shrew**	*Cryptotis endersi*	141		
**	**White-toothed Shrew**	*Crocidura beatus*	168		
**	**White-toothed Shrew**	*Crocidura bloyeti*	169		
**	**White-toothed Shrew**	*Crocidura bovei*	170		
**	**White-toothed Shrew**	*Crocidura caliginea*	171		
**	**White-toothed Shrew**	*Crocidura cinderella*	172		
**	**White-toothed Shrew**	*Crocidura crenata*	173		
**	**White-toothed Shrew**	*Crocidura edwardsiana*	174		
**	**White-toothed Shrew**	*Crocidura eisentrauti*	175		
**	**White-toothed Shrew**	*Crocidura elongata*	176		
**	**White-toothed Shrew**	*Crocidura grandis*	177		
**	**White-toothed Shrew**	*Crocidura lea*	178		
**	**White-toothed Shrew**	*Crocidura luluae*	179		
**	**White-toothed Shrew**	*Crocidura macowi*	180		
**	**White-toothed Shrew**	*Crocidura mindorus*	181		
**	**White-toothed Shrew**	*Crocidura palawanensis*	182	O	
**	**White-toothed Shrew**	*Crocidura parvacauda*	183		
**	**White-toothed Shrew**	*Crocidura phaeura*	184		
**	**White-toothed Shrew**	*Crocidura picea*	185		
**	**White-toothed Shrew**	*Crocidura susiana*	186		
**	**White-toothed Shrew**	*Crocidura tephra*	187		
**	**White-toothed Shrew**	*Crocidura vulcani*	188		
**	**White-toothed Shrew**	*Crocidura zimmeri*	189		
**	**White-toothed Shrew**	*Crocidura levicula*	190		
**	**Bornean Musk (Black) Rat**	*Suncus ater*	158		
**	**Kelaart's Long-clawed Shrew**	*Feroculus feroculus*	159		
*	**Pearson's Long-clawed Shrew**	*Solisorex pearsoni*	160		
**	**Mouse Shrew**	*Myosorex geata*	161		
**	**Mouse Shrew**	*Myosorex zinki*	162		
**	**Mouse Shrew**	*Myosorex norae*	163		
**	**Mouse Shrew**	*Myosorex polulus*	164		
**	**Mouse Shrew**	*Myosorex preussi*	165		
**	**Mouse Shrew**	*Myosorex polli*	166		
**	**Mouse Shrew**	*Myosorex longicaudatus*	167	O	
***	**Russian Desman**	*Desmana moschata*	192	V	
***	**Pyrenean Desman**	*Galemys pyrenaica*	194	V	
*	**Greater Japanese Mole**	*Talpa robusta*	193		
**	**Long-tailed Mole**	*Scaptonyx fusicaudus*	195		
**	**Kansu Mole**	*Scapanulus oweni*	191		
*	**Sao Thome Little Collared Fruit Bat**	*Myonycteris brachycephala*	197		
*	**Relict Little Collared Fruit Bat**	*Myonycteris relicta*	198		
**	**Bone's Fruit Bat**	*Boneia bidens*	199		

STAR RATING	VERNACULAR NAME	SCIENTIFIC NAME	SPECIES NUMBER	IUCN CATEGORY	CITES APPENDIX
***	Indian Ocean Island Flying Fox	*Pteropus rodricensis*	200		
***	Indian Ocean Island Flying Fox	*Pteropus niger*	201		
***	Indian Ocean Island Flying Fox	*Pteropus rufus*	202		
***	Indian Ocean Island Flying Fox	*Pteropus seychellensis*	203		
***	Indian Ocean Island Flying Fox	*Pteropus livingstonei*	204		
***	Indian Ocean Island Flying Fox	*Pteropus voeltzkowi*	205		
*	Pacific Ocean Island Flying Fox	*Pteropus mariannus*	206		
*	Pacific Ocean Island Flying Fox	*Pteropus tonganus*	207		
*	Pacific Ocean Island Flying Fox	*Pteropus tokudae*	208		
*	Pacific Ocean Island Flying Fox	*Pteropus ornatus*	209		
*	Pacific Ocean Island Flying Fox	*Pteropus vetulus*	210		
*	Pacific Ocean Island Flying Fox	*Pteropus auratus*	211		
*	Pacific Ocean Island Flying Fox	*Pteropus pelewenis*	212		
*	Pacific Ocean Island Flying Fox	*Pteropus pilosus*	213		
*	Pacific Ocean Island Flying Fox	*Pteropus yapensis*	214		
*	Pacific Ocean Island Flying Fox	*Pteropus ualanus*	215		
*	Pacific Ocean Island Flying Fox	*Pteropus phaeocephalus*	216		
*	Pacific Ocean Island Flying Fox	*Pteropus molossinus*	217		
*	Pacific Ocean Island Flying Fox	*Pteropus vanikorensis*	218		
*	Pacific Ocean Island Flying Fox	*Pteropus tuberculatus*	219		
*	Pacific Ocean Island Flying Fox	*Pteropus pselaphon*	220		
*	Pacific Ocean Island Flying Fox	*Pteropus nitendiensis*	221		
*	Pacific Ocean Island Flying Fox	*Pteropus fundatus*	222		
*	Pacific Ocean Island Flying Fox	*Pteropus samoensis*	223		
*	Pacific Ocean Island Flying Fox	*Pteropus howensis*	224		
*	SE Asian Flying Fox	*Pteropus speciosus*	225		
*	SE Asian Flying Fox	*Pteropus mearnsi*	226		
*	SE Asian Flying Fox	*Pteropus pumilus*	227		
*	SE Asian Flying Fox	*Pteropus balutus*	228		
*	SE Asian Flying Fox	*Pteropus tablasi*	229		
*	Fruit Bat	*Acerodon celebensis*	237		
*	Fruit Bat	*Acerodon humilis*	238		
*	Fruit Bat	*Acerodon lucifer*	239		
**	Bare-backed Fruit Bat	*Neopteryx frosti*	230		
*	Bare-backed Fruit Bat	*Dobsonia moluccense*	240		
**	Short-palate Fruit Bat	*Casinycteris argynnis*	232		
**	Polillo Dog-faced Fruit Bat	*Cynopterus archipelagus*	241		
*	Salim Ali's Fruit Bat	*Latidens salimalii*	242		
**	Short-nosed Fruit Bat	*Thoopterus nigrescens*	233		
***	Fossil Fruit Bat	*Aproteles bulmerae*	231	O	
*****	Halcon Fruit Bat	*Haplonycteris fischeri*	234		
**	Flying Fox	*Alionycteris paucidentata*	236		
**	Long-haired Fruit Bat	*Otopteropus cartilagonodus*	235		
*	Tube-nosed Fruit Bats	*Nyctimene* spp.	246		
**	Sulawesi Dawn Bat	*Eonycteris rosenbergi*	244		
**	Northern Blossom Bat	*Macroglossus lagochilus*	243		
**	Pacific Sheath-tailed Bat	*Emballonura sulcata*	247		
***	Seychelles Sheath-tailed Bat	*Coleura seychellensis*	245		
*	Papuan Sheath-tailed Bat	*Taphozous mixtus*	248		
*	Naked-rumped Sheath-tailed Bat	*Taphozous nudicluniatus*	249		
*	Little Sheath-tailed Bat	*Taphozous australis*	250		
**	White-striped Sheath-tailed Bat	*Taphozous kapalgensis*	251		
***	Kitti's Hog-nosed (Bumblebee) Bat	*Craseonycteris thonglongyai*	252	K	
***	Ghost Bat (Australian False Vampire)	*Macroderma gigas*	253	V	
***	Greater Horseshoe Bat	*Rhinolophus ferrumequinum*	254		
***	Lesser Horseshoe Bat	*Rhinolophus hipposideros*	255		
*	Meheley's Horseshoe Bat	*Rhinolophus mehelyi*	256		
*	Blasius's Horseshoe Bat	*Rhinolophus blasii*	257		
*	Mediterranean Horseshoe Bat	*Rhinolophus euryale*	258		
**	Iriomote Horseshoe Bat	*Rhinolophus imaizumii*	259		
*	Philippine Horseshoe Bat	*Rhinolophus philippinensis*	260		
**	Horseshoe Bat	*Rhinolophus marshalli*	261		
**	Horseshoe Bat	*Rhinolophus paradoxolophus*	262		
**	Cox's Leaf-nosed Bat	*Hipposideros coxi*	263		
*	Ridley's Leaf-nosed Bat	*Hipposideros ridleyi*	264	E	
**	Boonsong's Leaf-nosed Bat	*Hipposideros lekaguli*	265		
*	Cantor's Leaf-nosed Bat	*Hipposideros galeritus*	266		
*	Leaf-nosed Bat	*Hipposideros turpis*	267		
*	Lesser Wart-nosed Horseshoe Bat	*Hipposideros stenotis*	268		
*	Greater Wart-nosed Bat	*Hipposideros semoni*	269		
**	Flower-faced Bat	*Anthops ornatus*	280		
*	African Trident-nosed Bat	*Cloeotis percivali*	281		
***	Golden (Orange) Horseshoe Bat	*Rhinonicteris aurantius*	282		
***	Triple Nose-leaf Bat	*Triaenops persicus*	283		

STAR RATING	VERNACULAR NAME	SCIENTIFIC NAME	SPECIES NUMBER	IUCN CATEGORY	CITES APPENDIX
**	Funnel-eared Bat	*Paracoelops megalotis*	284		
**	Sword-nosed Bat	*Lonchorhina marinkellei*	270		
*	Sword-nosed Bat	*Lonchorhina aurita*	271		
**	Nose-leaf Bat	*Platalina genovensium*	272		
*	Jamaican Fig-eating Bat	*Ariteus flavescens*	273		
**	Red Fruit Bat	*Stenoderma rufum*	274		
*	Brown Flower Bat	*Erophylla sezekorni*	279		
*****	Flower Bat	*Phyllonycteris major*	275		
*	Flower Bat	*Phyllonycteris poeyi*	276		
***	Flower Bat	*Phyllonycteris aphylla*	277		
***	Flower Bat	*Phyllonycteris obtusa*	278		
***	Indiana Bat	*Myotis sodalis*	290	V	
***	Grey Bat	*Myotis grisescens*	291	E	
**	Miller's Bat	*Myotis milleri*	292		
**	Sakhalin Bat	*Myotis abei*	293		
***	Large Mouse-eared Bat	*Myotis myotis*	294		
*	Nathalina Bat	*Myotis nathalinae*	295		
***	Bechstein's Bat	*Myotis bechsteini*	296		
*	Island Bat	*Myotis insularum*	297		
***	Lesser Mouse-eared Bat	*Myotis blythi*	298		
***	Long-fingered Bat	*Myotis capaccini*	299		
**	Wing-gland or Hairy Bat	*Cistugo seabrai*	304		
**	Wing-gland or Hairy Bat	*Cistugo lesueuri*	305		
**	Disc-footed Bat	*Eudiscopus denticulatus*	306		
*	Pipistrelle	*Pipistrellus pipistrellus*	307		
*	Pipistrelle	*Pipistrellus nathusii*	308		
*	Pipistrelle	*Pipistrellus kuhli*	309		
*	Pipistrelle	*Pipistrellus savii*	310		
**	Pipistrelle	*Pipistrellus societas*	311		
**	Pipistrelle	*Pipistrellus bodenheimeri*	312		
**	Pipistrelle	*Pipistrellus aero*	313		
**	Pipistrelle	*Pipistrellus arabicus*	314		
*	Pipistrelle	*Pipistrellus anthonyi*	315		
*	Pipistrelle	*Pipistrellus joffrei*	316		
*	Pipistrelle	*Pipistrellus permixtus*	317		
*	Pipistrelle	*Pipistrellus maderensis*	318		
*	Pipistrelle	*Pipistrellus tenuis*	319		
**	Pipistrelle	*Pipistrellus kitcheneri*	320		
**	Pipistrelle	*Pipistrellus cuprosus*	321		
**	Pipistrelle	*Pipistrellus vordermanni*	322		
***	Greater Noctule	*Nyctalus lasiopterus*	323		
*	Leisler's Bat	*Nyctalus leisleri*	324		
*	Noctule	*Nyctalus nyctalus*	325		
**	Euphrates Serotine	*Eptesicus walli*	326		
**	Serotine	*Eptesicus loveni*	327		
**	Yellow-lipped Serotine	*Eptesicus douglasi*	328		
**	Serotine	*Eptesicus platyops*	329		
*	Serotine	*Eptesicus guadaloupensis*	330		
**	Serotine	*Eptesicus demissus*	331		
*	Northern Serotine	*Eptesicus nilssonii*	332		
*	Serotine	*Eptesicus serotinus*	333		
**	Namib Long-eared Bat	*Laephotis namibensis*	285		
**	Botswana Long-eared Bat	*Laephotis botswanae*	286		
**	de Winton's Long-eared Bat	*Laephotis wintoni*	287		
*	Angolan Long-eared Bat	*Laephotis angolensis*	288		
**	Big-eared Brown Bat	*Histiotus* sp.nov.	289		
**	False Serotine Bat	*Hesperoptenus doriae*	334		
**	Pied Butterfly Bat	*Glauconycteris superba*	335		
**	Bibundi Bat	*Glauconycteris egeria*	336		
*	New Zealand Lobe-lipped Bat	*Chalinolobus tuberculatus*	301		
**	Little Yellow Bat	*Rhogeesa mira*	343		
**	Little Yellow Bat	*Rhogeesa gracilis*	344		
**	Indian Harlequin Bat	*Scotomanes emarginatus*	345		
***	Hoary Bat	*Lasiurus cinereus*	302	(I)	
***	Townsend's Big-eared Bat	*Plecotus townsendii*	346	(I/E)	
*	Brown Long-eared Bat	*Plecotus auritus*	347		
*	Grey Long-eared Bat	*Plecotus austriacus*	348		
*	Mexican Long-eared Bat	*Plecotus mexicanus*	349		
*	Spotted (Pinto) Bat	*Euderma maculatum*	350		
*	Schreiber's Long-fingered Bat	*Miniopterus schreibersi*	351		
***	Little Long-fingered Bat	*Miniopterus australis*	352		
*	Long-fingered Bat	*Miniopterus robustior*	353		
*	Long-fingered Bat	*Miniopterus tristis*	354		
**	Tube-nosed Insectivorous Bat	*Murina florium*	337		

STAR RATING	VERNACULAR NAME	SCIENTIFIC NAME	SPECIES NUMBER	IUCN CATEGORY	CITES APPENDIX
**	Tube-nosed Insectivorous Bat	*Murina tenebrosa*	338		
**	Tube-nosed Insectivorous Bat	*Murina canascens*	339		
*	Tube-nosed Insectivorous Bat	*Murina puta*	340		
*****	Tanzanian Woolly Bat	*Kerivoula africana*	300		
**	Lamington Free-eared or Long-eared Bat	*Lamingtona lophorhina*	303		
**	Pygmy or Little Territory Long-eared Bat	*Nyctophilus walkeri*	341		
**	Big-eared Bat	*Pharotis imogene*	342		
*	New Zealand Short-tailed Bat	*Mystacina tuberculata*	355		
**	Large-eared Free-tailed Bat	*Tadarida lobata*	356		
**	African Free-tailed Bat	*Tadarida africana*	357		
**	Natal Free-tailed Bat	*Tadarida acetabulosa*	358		
***	Brazilian Free-tailed Bat	*Tadarida brasiliensis*	359		
**	Bini Free-tailed Bat	*Myopterus whitleyi*	360		
**	Surinam Free-tailed Bat	*Molossops neglectus*	361		
*	Panamanian Domed-palate Mastiff Bat	*Promops panama*	362		
***	Lesser Mouse-lemur	*Microcebus murinus*	369		1
*****	Coquerel's Mouse-lemur	*Microcebus coquereli*	370	K	1
***	Greater Dwarf Lemur	*Cheirogaleus major*	373		1
***	Fat-tailed Lemur	*Cheirogaleus medius*	374		1
*****	Hairy-eared Dwarf Lemur	*Allocebus trichotis*	372	E	1
***	Fork-marked Lemur	*Phaner furcifer*	375	K	1
***	Black Lemur	*Lemur macaco*	371	(E/V)	1
*	Ring-tailed Lemur	*Lemur catta*	376	K	1
***	Brown Lemur	*Lemur fulvus*	377		1
***	Mongoose Lemur	*Lemur mongoz*	378	V	1
***	Crowned Lemur	*Lemur coronatus*	379	K	1
*****	Red-bellied Lemur	*Lemur rubriventer*	380	I	1
*	Grey Gentle Lemur	*Hapalemur griseus*	394	K	1
**	Broad-nosed Gentle Lemur	*Hapalemur simus*	395	E	1
***	Ruffed or Variegated Lemur	*Varecia variegata*	386	I	1
***	Greater Weasel Lemur	*Lepilemur mustelinus*	387	K	1
***	Island Weasel Lemur	*Lepilemur dorsalis*	388	K	1
***	Light-necked Weasel Lemur	*Lepilemur microdon*	389	K	1
***	Dry-bush Weasel Lemur	*Lepilemur leucopus*	390	K	1
***	Lesser Weasel Lemur	*Lepilemur ruficaudatus*	391	K	1
***	Milne-Edwards' Weasel Lemur	*Lepilemur edwardsi*	392	K	1
***	Northern Weasel Lemur	*Lepilemur septentrionalis*	393	K	1
***	Avahi or Woolly Lemur	*Avahi laniger*	381	K	1
***	Verreaux's Sifaka	*Propithecus verreauxi*	384	K	1
***	Diademed Sifaka	*Propithecus diadema*	385	V	1
***	Indris	*Indri indri*	383	E	1
*****	Aye-aye	*Daubentonia madagascariensis*	382	E	1
*	Slender Loris	*Loris tardigradus*	366		2
***	Slow Loris	*Nycticebus coucang*	367		2
*	Lesser Slow Loris	*Nycticebus pygmaeus*	368		2
***	Western or Horsfield's Tarsier	*Tarsius bancanus*	363	(I)	2
*	Eastern or Sulawesi Tarsier	*Tarsius spectrum*	364		2
***	Philippine Tarsier	*Tarsius syrichta*	365	E	2
***	Silvery Marmoset	*Callithrix argentatus*	400	(V)	2
***	Common or Tufted Marmoset	*Callithrix jacchus*	401	(E)	1/2
***	Santarem Marmoset	*Callithrix humeralifer*	402	V	2
*	Pygmy Marmoset	*Cebuella pygmaea*	403		2
***	Cotton-top or Crested Tamarin	*Saguinus oedipus*	396	(E)	1
***	Emperor Tamarin	*Saguinus imperator*	397	I	2
*	Bare-faced Tamarin	*Saguinus bicolor*	398	I	1
***	White-footed Tamarin	*Saguinus leucopus*	399	V	1
*****	Golden Lion Tamarin	*Leontopithecus rosalia*	406	E	1
*****	Golden-headed Lion Tamarin	*Leontopithecus chrysomelas*	407	E	1
*****	Golden-rumped or Black Lion Tamarin	*Leontopithecus chrysopygus*	408	E	1
***	Goeldi's Marmoset	*Callimico goeldii*	404	R	1
***	Masked Titi	*Callicebus personatus*	409	V	2
***	Red-backed Squirrel Monkey	*Saimiri oerstedii*	405	E	1
***	Red and White Uakaris	*Cacajao calvus*	415	V	1
***	Black-headed Uakari	*Cacajao melanocephalus*	416	V	1
***	White-nosed Saki	*Chiropotes albinasus*	413	V	1
***	Black-bearded Saki	*Chiropotes satanas*	414	(E)	2
***	Mantled Howler	*Alouatta palliata*	410		1
***	Guatemalan Howler	*Alouatta pigra*	411	I	2
***	Brown Howler	*Alouatta fusca*	412	I	2
***	Black-handed or Geoffroy's Spider Monkey	*Ateles geoffroyi*	420	V	1/2
***	Long-haired Spider Monkey	*Ateles belzebuth*	421	V	2
***	Brown-headed Spider Monkey	*Ateles fusciceps*	422	I	2
***	Black Spider Monkey	*Ateles paniscus*	423	V	2
***	Woolly Spider Monkey	*Brachyteles arachnoides*	419	E	1

APPENDIX I

STAR RATING	VERNACULAR NAME	SCIENTIFIC NAME	SPECIES NUMBER	IUCN CATEGORY	CITES APPENDIX
***	**Common** or **Humboldt's Woolly Monkey**	*Lagothrix lagothrica*	417	V	2
***	**Yellow-tailed** or **Hendee's Woolly Monkey**	*Lagothrix flavicauda*	418	E	1
*****	**Lion-tailed Macaque** or **Wanderoo**	*Macaca silenus*	438	E	1
*	**Stump-tailed Macaque**	*Macaca arctoides*	439		2
*	**Assam Macaque**	*Macaca assamensis*	440		2
*	**Pere David's Macaque**	*Macaca thibetana*	441		2
***	**Pig-tailed Macaque**	*Macaca nemestrina*	442	(I)	2
***	**Barbary Ape**	*Macaca sylvanus*	443	V	2
*	**Crab-eating Macaque**	*Macaca fascicularis*	444		2
*	**Sulawesi Black Ape**	*Macaca nigra*	445		2
*	**Japanese Macaque**	*Macaca fuscata*	446		2
*	**Rhesus Macaque**	*Macaca mulatta*	447		2
*	**Toque Monkey**	*Macaca sinica*	448		2
*	**Formosan Rock Macaque**	*Macaca cyclopis*	449		2
**	**Agile** or **Crested Mangabey**	*Cercocebus galeritus*	436	(E)	1/2
*	**White-collared Mangabey**	*Cercocebus torquatus*	437	V	2
***	**Hamadryas** or **Sacred Baboon**	*Papio hamadryas*	451		2
***	**Mandrill**	*Papio sphinx*	452	V	1
***	**Drill**	*Papio leucophaeus*	453	E	1
*	**Chacma Baboon**	*Papio ursinus*	454		2
*	**Guinea Baboon**	*Papio papio*	455		2
***	**Gelada Baboon**	*Theropithecus gelada*	450		2
***	**Diana Monkey**	*Cercopithecus diana*	424	V	1
**	**Red-eared Monkey**	*Cercopithecus erythrotis*	425	V	2
***	**l'Hoest's Monkey**	*Cercopithecus lhoesti*	426	V	2
*****	**Red-bellied Monkey**	*Cercopithecus erythrogaster*	427	V	2
***	**Blue Monkey**	*Cercopithecus mitis*	428		2
*	**de Brazza's Monkey**	*Cercopithecus neglectus*	429		2
*	**White-nosed Monkey**	*Cercopithecus nictitans*	430		2
*	**Spot-nosed Monkey**	*Cercopithecus petaurista*	431		2
*	**Dryas Guenon**	*Cercopithecus dryas*	432	K	2
*	**Owl-faced Monkey**	*Cercopithecus hamlyni*	433	K	2
**	**Salongo Monkey**	*Cercopithecus salongo*	434	K	2
*	**Allen's Monkey**	*Allenopithecus nigroviridis*	435	K	2
***	**Western Colobus** or **Western Black and White Monkey**	*Colobus polykomos*	473		2
***	**Black and White Colobus**	*Colobus angolensis*	474		2
***	**Abyssinian Black and White Colobus**	*Colobus guereza*	475		2
***	**Black Colobus**	*Colobus satanas*	476	V	2
***	**Red Colobus**	*Colobus badius*	477	(E/V)	2
***	**Kirk's** or **Zanzibar Red Colobus**	*Colobus kirkii*	478	E	1
***	**Olive Colobus**	*Procolobus verus*	479		2
***	**Douc Langur**	*Pygathrix nemaeus*	468	E	1
***	**Golden Snub-nosed Monkey**	*Pygathrix roxellanae*	469	K	1
**	**Guizhou Snub-nosed Monkey**	*Pygathrix brelichi*	470		1
*	**Tonkin Snub-nosed Monkey**	*Pygathrix avunculus*	471		1
**	**Pig-tailed Langur** or **Simakobu Monkey**	*Nasalis concolor*	467	E	1
***	**Proboscis Monkey**	*Nasalis larvatus*	472	V	1
***	**Dusky** or **Spectacled Langur**	*Presbytis obscura*	456		2
***	**Phayre's Leaf Monkey**	*Presbytis phayrei*	457		2
***	**Silvered Langur**	*Presbytis cristata*	458		2
**	**Golden Langur**	*Presbytis geei*	459	R	1
***	**Common Langur** or **Hanuman**	*Presbytis entellus*	460		1
***	**Nilgiri Langur**	*Presbytis johnii*	461	V	2
*	**Banded Leaf Monkey**	*Presbytis aygula/melalophus* group	462		2
**	**Francois's Leaf Monkey**	*Presbytis francoisi*	463		2
*	**Mentawi Island Langur**	*Presbytis potenziani*	464	I	1
*	**Capped Langur**	*Presbytis pileata*	465		1
***	**Purple-faced Langur**	*Presbytis senex*	466		2
***	**Kloss's** or **Dwarf Gibbon**	*Hylobates klossi*	480	V	1
***	**Common** or **Lar Gibbon**	*Hylobates lar*	481	(E)	1
***	**Pileated Gibbon**	*Hylobates pileatus*	482	E	1
***	**Siameng**	*Hylobates syndactylus*	483		1
*	**Crested** or **Black Gibbon**	*Hylobates concolor*	484	I	1
***	**Hoolock Gibbon**	*Hylobates hoolock*	485		1
****	**Orang Utan**	*Pongo pygmaeus*	486	E	1
***	**Chimpanzee**	*Pan troglodytes*	488	V	1
***	**Pygmy Chimpanzee**	*Pan paniscus*	489	V	1
***	**Gorilla**	*Gorilla gorilla*	487	V(E)	1
*	**Aardvark**	*Orycteropus afer*	505		
*	**Temminck's Ground Pangolin**	*Manis temmincki*	499		1
*	**Giant Pangolin**	*Manis gigantea*	500		
*	**Tree Pangolin**	*Manis tricuspis*	501		
*	**Malayan Pangolin**	*Manis javanica*	502		2
*	**Chinese Pangolin**	*Manis pentadactyla*	503		2

STAR RATING	VERNACULAR NAME	SCIENTIFIC NAME	SPECIES NUMBER	IUCN CATEGORY	CITES APPENDIX
*	**Indian Pangolin**	*Manis crassicaudata*	504		2
*	**Tassel-eared Squirrel**	*Sciurus aberti*	531		
*	**Peter's Squirrel**	*Sciurus oculatus*	532		
*	**Allen's Squirrel**	*Sciurus alleni*	533		
*	**Arizona Gray Squirrel**	*Sciurus arizonensis*	535		
**	**Tree Squirrel**	*Sciurus richmondi*	536		
**	**Tree Squirrel**	*Sciurus sanborni*	537		
**	**Tree Squirrel**	*Sciurus stramineus*	538		
**	**Tree Squirrel**	*Sciurus colliaei*	539		
**	**Tree Squirrel**	*Sciurus deppei*	540		
***	**Fox Squirrel**	*Sciurus niger*	541	(E)	
*	**Eurasian Red Squirrel**	*Sciurus vulgaris*	542		
*	**Nayarit Squirrel**	*Sciurus nayaritensis*	534		
**	**Panama Mountain Squirrel**	*Syntheosciurus brochus*	563		
**	**Poas Mountain Squirrel**	*Syntheosciurus poasensis*	564		
*	**Giant Squirrel**	*Ratufa macroura*	534		2
*	**Giant Squirrel**	*Ratufa bicolor*	545		2
*	**Giant Squirrel**	*Ratufa affinis*	546		2
*	**Giant Squirrel**	*Ratufa indica*	544		2
*	**African Palm Squirrel**	*Epixerus ebii*	530		
*	**Black and Red Bush Squirrel**	*Paraxerus lucifer*	566		
**	**Swynnerton's Bush Squirrel**	*Paraxerus vexillarius*	567		
*	**Striped Bush Squirrel**	*Paraxerus flavivittis*	568		
*	**Vincent's Squirrel**	*Paraxerus vincenti*	569		
*	**Beautiful Squirrel**	*Callosciurus finlaysoni*	565		
**	**Sunda Tree Squirrel**	*Sundasciurus davensis*	554		
**	**Sunda Tree Squirrel**	*Sundasciurus hoogstraali*	555		
**	**Sunda Tree Squirrel**	*Sundasciurus mollendorffi*	556		
*	**Four-striped Ground Squirrel**	*Lariscus hosei*	550		2
**	**Grey-bellied Sculptor Squirrel**	*Glyphotes canalvus*	529		
*****	**Vancouver Marmot**	*Marmota vancouverensis*	518	E	
**	**Olympic Marmot**	*Marmota olympia*	519		
***	**Bobak Marmot**	*Marmota bobak*	520		
***	**Menzbier's Marmot**	*Marmota menzbieri*	521	V	
*	**Black-tailed Prairie Dog**	*Cynomys ludovicianus*	523		
***	**Utah Prairie Dog**	*Cynomys parvidens*	524	V	
*	**Mexican Prairie Dog**	*Cynomys mexicanus*	525		1
*	**European Souslik**	*Spermophilus citellus*	522		
***	**Giant Flying Squirrel**	*Petaurista petaurista*	551		
**	**Woolly Flying Squirrel**	*Eupetaurus cinereus*	526		
*	**Siberian Flying Squirrel**	*Pteromys volans*	553		
*	**Northern Flying Squirrel**	*Glaucomys sabrinus*	527		
*****	**Mindanao Arrow-tailed Flying Squirrel**	*Hylopetes mindanensis*	557		
*	**Arrow-tailed Flying Squirrel**	*Hylopetes alboniger*	558		
*	**Dwarf Flying Squirrel**	*Petinomys bartelsi*	559		
*	**Dwarf Flying Squirrel**	*Petinomys sagitta*	560		
*	**Dwarf Flying Squirrel**	*Petinomys crinitus*	561		
*	**Dwarf Flying Squirrel**	*Petinomys hageni*	562		
**	**Hairy-footed Flying Squirrel**	*Belomys pearsoni*	552		
**	**Pygmy Flying Squirrel**	*Petaurillus kinlochii*	547		
**	**Pygmy Flying Squirrel**	*Petaurillus hosei*	548		
**	**Pygmy Flying Squirrel**	*Petaurillus emiliae*	549		
***	**South-eastern Pocket Gopher**	*Geomys pinetis*	576		
**	**Tropical Pocket Gopher**	*Geomys tropicalis*	577		
*	**Pocket Gopher** or **Tuza**	*Pappogeomys tylorhinus*	572		
*	**Pocket Gopher**	*Pappogeomys neglectus*	573		
*	**Pocket Gopher**	*Pappogeomys merriami*	574		
**	**Big Pocket Gopher** or **Taltuza**	*Orthogeomys lanius*	570		
**	**Big Pocket Gopher** or **Taltuza**	*Orthogeomys cuniculus*	571		
*	**Michoacan Pocket Gopher**	*Zygogeomys trichopus*	575		
**	**Anthony's Pocket Mouse**	*Perognathus anthonyi*	591		
*	**Chisel-toothed Kangaroo Rat**	*Dipodomys microps*	578		
**	**Heermann's Kangaroo Rat**	*Dipodomys heermanni*	579	(E)	
**	**Stephens' Kangaroo Rat**	*Dipodomys stephensi*	581		
**	**Texas Kangaroo Rat**	*Dipodomys elator*	580	R	
**	**Big-eared Kangaroo Rat**	*Dipodomys elephantinus*	582		
**	**Giant Kangaroo Rat**	*Dipodomys ingens*	583		
**	**Fresno Kangaroo Rat**	*Dipodomys nitratoides exilis*	584		
*	**Phillip's Kangaroo Rate**	*Dipodomys phillipsii*	585		2
*	**Ord's Kangaroo Rat**	*Dipodomys ordii*	586		
**	**Island Kangaroo Rat**	*Dipodomys insularis*	587		
**	**Santa Margarita Kangaroo Rat**	*Dipodomys margaritae*	588		
**	**Forest Spiny Pocket Mouse**	*Heteromys nelsoni*	589		
**	**Forest Spiny Pocket Mouse**	*Heteromys longicaudatus*	590		

STAR RATING	VERNACULAR NAME	SCIENTIFIC NAME	SPECIES NUMBER	IUCN CATEGORY	CITES APPENDIX
***	Eurasian Beaver	*Castor fiber*	593		
***	American Beaver	*Castor canadensis*	594		
*	Zenker's Flying Squirrel	*Idiurus zenkeri*	592		
*****	Silver or Key Rice Rat	*Oryzomys argentatus*	622	I	
**	Rice Rat	*Oryzomys bauri*	623		
**	Rice Rat	*Oryzomys narboroughi*	624		
**	Rice Rat	*Oryzomys fernandinae*	625		
**	Rice Rat	*Oryzomys cozumelae*	626		
**	Rice Rat	*Oryzomys nelsoni*	627		
**	Rice Rat	*Oryzomys peninsulae*	628		
**	Rice Rat	*Oryzomys aphrastus*	629		
**	Brazilian Spiny Rice Rate	*Abrawayaomys ruschii*	686		
**	Ecuadorian Spiny Mouse	*Scolomys melanops*	687		
**	Guianan Water Rat	*Nectomys parvipes*	630		
**	Mr Pirri Climbing Mouse	*Rhipidomys scandens*	688		
**	Rio Rice Rat	*Phaenomys ferrugineus*	689		
**	Chiapas Climbing Rat	*Tylomys bullaris*	597		
**	Darien Climbing Rat	*Tylomys fulviventer*	598		
**	Tumbala Climbing Rat	*Tylomys tumbalensis*	599		
***	Saltmarsh Harvest Mouse	*Reithrodontomys raviventris*	600	E	
**	Sumichrast's Harvest Mouse	*Reithrodontomys sumichrasti*	601		
**	Small-toothed Harvest Mouse	*Reithrodontomys microdon*	602		
**	Rodriguez's Harvest Mouse	*Reithrodontomys rodriguezi*	603		
**	Harvest Mouse	*Reithrodontomys spectabilis*	604		
***	White-footed Mouse	*Peromyscus gossypinus*	605		
***	White-footed Mouse	*Peromyscus floridanus*	606		
***	Beach Mouse	*Peromyscus polionotus*	607		
**	Mexican White-footed Mouse	*Peromyscus aztecus*	608		
**	Jico Deer Mouse	*Peromyscus simulatus*	609		
**	White-footed Mouse	*Peromyscus chinanteco*	610		
**	Yellow Deer Mouse	*Peromyscus flavidus*	611		
**	Perote Deer Mouse	*Peromyscus bullatus*	612		
**	Burt's Deer Mouse	*Peromyscus caniceps*	613		
**	Dickey's Deer Mouse	*Peromyscus dickeyi*	614		
**	White-footed Mouse	*Peromyscus grandis*	615		
**	White-footed Mouse	*Peromyscus mayensis*	616		
**	White-footed Mouse	*Peromyscus pembertoni*	617		
**	White-footed Mouse	*Peromyscus pseudocrinitus*	618		
**	White-footed Mouse	*Peromyscus slevini*	619		
**	White-footed Mouse	*Peromyscus stephani*	620		
**	White-footed Mouse	*Peromyscus winkelmanni*	621		
**	South American Field Mouse	*Akodon llanoi*	631		
**	South American Field Mouse	*Akodon kempi*	632		
**	Roraima Mouse	*Podoxymys roraimae*	644		
**	Brasilia Burrowing Mouse	*Juscelinomys candango*	645		
**	Leaf-eared Mouse	*Galenomys garleppi*	633		
*	Chinchilla Mouse	*Chinchillula sahamae*	634		
**	Turner Island Wood Rat	*Neotoma varia*	635		
**	Bryant's Wood Rat	*Neotoma bryanti*	636		
**	Anthony's Wood Rat	*Neotoma anthonyi*	637		
**	San Martin Wood Rat	*Neotoma martinensis*	638		
**	Bunker's Wood Rat	*Neotoma bunkeri*	639		
**	Nelson's Wood Rat	*Neotoma nelsoni*	640		
*	Eastern Wood Rat	*Neotoma floridana*	641		
**	Magdalena Rat	*Xenomys nelsoni*	642		
**	Venezuelan Fish-eating Rat	*Icthyomys pittieri*	646		
**	Ecuador Fish-eating Rat	*Anotomys leander*	643		
**	Venezualan Fishing Rat	*Daptomys venezuelae*	648		
**	Peruvian Fishing Rat	*Daptomys peruviensis*	649		
**	Guyanan Fishing Rat	*Daptomys oyapocki*	650		
**	Goldman's Water Mouse	*Rheomys raptor*	647		
**	Mouse-like Hamster	*Calomyscus bailwardi*	651		
**	Mouse-like Hamster	*Camomyscus hotsoni*	652		
***	Common Hamster	*Cricetus cricetus*	653		
*	Golden Hamster	*Mesocricetus auratus*	654		
*	Romanian Hamster	*Mesocricetus newtoni*	655		
*	Greater Mole-rat	*Spalax microphthalmus*	665		
*	Giant Mole-rat	*Spalax giganteus*	666		
***	Malagasy Giant Rat	*Hypogeomys antimena*	664		
**	Varying Lemming	*Dicrostonyx unalescensis*	679		
**	Varying Lemming	*Dicrostonyx vinogradovi*	680		
*	Southern Bog Lemming	*Synaptomys cooperi*	684		
***	White-footed Vole	*Phenacomys albipes*	663		
**	Beach Vole	*Microtus breweri*	670	R	

STAR RATING	VERNACULAR NAME	SCIENTIFIC NAME	SPECIES NUMBER	IUCN CATEGORY	CITES APPENDIX
**	Field Vole	*Microtus abbreviatus*	671		
**	Field Vole	*Microtus canicaudus*	672		
**	Field Vole	*Microtus mujanensis*	673		
**	Field Vole	*Microtus oaxacensis*	674		
**	Field Vole	*Microtus umbrosus*	675		
***	California Vole	*Microtus californicus*	676		
***	Field Vole	*Microtus ochrogaster*	677		
*	Root Vole	*Microtus oeconomus*	678		
*	Steppe Lemming	*Lagurus luteus*	682		
*	Steppe Lemming	*Lagurus lagurus*	683		
**	Pygmy Gerbil	*Gerbillus mauritaniae*	656		
**	Pygmy Gerbil	*Gerbillus hoogstraali*	657		
**	Pygmy Gerbil	*Gerbillus occiduus*	658		
**	Dune Hairy-footed Gerbil	*Gerbillurus tytonis*	662		
**	Greater Short-tailed Gerbil	*Dipodillus maghrebi*	659		
**	Tunisian Short-tailed Gerbil	*Dipodillus zakariai*	660		
**	Somali Pygmy Gerbil	*Microdillus peeli*	661		
**	African Climbing Mouse	*Dendromus kahuziensis*	685		
**	Congo Tree Mouse	*Dendroprionomys rousseloti*	669		
*****	Groove-toothed Forest Mouse	*Leimacomys buettneri*	668		
*****	Ghana Fat Mouse	*Steatomys jacksoni*	667		
**	Dollman's Tree Mouse	*Prionomys batesi*	681		
**	Krk Wood Mouse	*Apodemus sylvaticus*	699		
***	Ryukyu Spiny Mouse	*Tokudaia osimensis*	700	(I)	
**	Luzon Rat	*Carpomys melanurus*	697		
**	Luzon Rat	*Carpomys phaeurus*	698		
**	Philippine Forest Rat	*Batomys granti*	692	I	
**	Philippine Forest Rat	*Batomys dentatus*	693		
**	Philippine Forest Rat	*Batomys salomonseni*	694		
*****	Central Rock-rat or MacDonnel Rat	*Zyzomys pedunculatus*	716	I	1
*	Woodward's Rock-rat	*Zyzomys woodwardi*	717		
***	Golden-backed Tree Rat	*Mesembriomys macrurus*	713		
***	Black-footed Tree Rat	*Mesembriomys gouldii*	714		
*	Lake Victoria Groove-toothed Rat	*Pelomys isseli*	726		
**	Groove-toothed Rat	*Pelomys rex*	727		
**	Rosevear's Striped Grass Mouse	*Lemniscomys roseveari*	722		
**	Striped Grass Mouse	*Lemniscomys mittendorfi*	723		
**	Sulawesi Soft-furred Rat	*Eropeplus canus*	691		
**	Mindoro Rat	*Anonymomys mindorensis*	751		
**	Ceylonese Rat	*Srilankamys ohiensis*	752		
*	Javan Grey Tree Rat	*Kadarsanomys sodyi*	753		
**	Margareta's Rat	*Margaretamys parvus*	754		
*	Limestone Rat	*Niviventer hinpoon*	755		
**	Long-tailed Giant Rat	*Leopoldamys neilli*	756		
*	Black Rat	*Rattus rattus*	757		
*	Canefield Rat	*Rattus sordidus*	758		
*	Pale Field Rat	*Rattus tunneyi*	759		
*	Rat	*Rattus burrus*	760		
*	Rat	*Rattus palmarum*	761		
*	Rat	*Rattus pulliventer*	762		
*	Rat	*Rattus culionensis*	763		
*	Rat	*Rattus elephinus*	764		
*	Rat	*Rattus enganus*	765		
*	Rat	*Rattus latidens*	766		
*	Rat	*Rattus nativitatis*	767		
*	Rat	*Rattus omichlodes*	768		
*	Rat	*Rattus owiensis*	769		
*	Rat	*Rattus ranjiniae*	770		
*	Rat	*Rattus rogersi*	771		
*	Rat	*Rattus stoicus*	772		
*	Rat	*Rattus tyrannus*	773		
*	Rat	*Rattus remotus*	774		
*	Rat	*Rattus cremoriventer*	775		
**	Philippine Rat	*Apomys sacobianus*	777		
**	Philippine Rat	*Apomys microdon*	778		
**	Philippine Rat	*Apomys littoralis*	779		
**	Chirinda Rock Rat	*Aethomys selindensis*	724		
**	Soft Furred Rat	*Praomys hartwigi*	725		
**	Mindanao Rat	*Limnomys sibuanus*	793		
**	Mount Apo Rat	*Tarsomys apoensis*	738		
**	Mearns' Luzon Rat	*Tryphomys adustus*	780		
*****	Stick-nest Rat	*Leporillus conditor*	715	R	1
**	Lakeland Downs Mouse	*Leggadina lakedownensis*	729		
*****	Shark Bay Mouse	*Pseudomys praeconis*	701	R	1

STAR RATING	VERNACULAR NAME	SCIENTIFIC NAME	SPECIES NUMBER	IUCN CATEGORY	CITES APPENDIX
*****	Hastings River Mouse	*Pseudomys oralis*	702	I	
**	Pilliag Mouse	*Pseudomys pilligaensis*	703		
***	Heath Rat	*Pseudomys shortridgei*	704		2
***	Ash-grey Mouse	*Pseudomys albocinereus*	705		
***	Western Mouse	*Pseudomys occidentalis*	706		
***	Desert Mouse	*Pseudomys desertor*	707		
*****	Gould's Mouse	*Pseudomys gouldii*	708		
*****	Alice Springs Mouse	*Pseudomys fieldi*	709		
***	Smokey Mouse	*Pseudomys fumeus*	710		1
*	Pebble-mound Mouse	*Pseudomys chapmani*	711		
**	Banana or Mouse-tailed Rat	*Melomys albidens*	731		
**	Banana or Mouse-tailed Rat	*Melomys arcium*	732		
*	Banana or Mouse-tailed Rat	*Melomys levipes*	733		
*	Fawn-footed Melomys	*Melomys cervinipes*	734		
**	Lowland Brush Mouse	*Pogonomelomys bruijni*	735		
**	White-tailed New Guinea Rat	*Xenuromys barbatus*	796		
**	African Swamp Rat	*Malacomys verschureni*	728		
**	Manipur or Crump's Mouse	*Diomys crumpi*	745		
**	Stripe-backed Mouse	*Muriculus imberbis*	792		
**	House Mouse	*Mus baoulei*	791		
**	Seram Rat	*Nesoromys ceramicus*	776		
**	Philippine Swamp Rat	*Crunomys fallax*	797		
**	Philippine Swamp Rat	*Crunomys melanius*	798		
**	Northern Hopping Mouse	*Notomys aquilo*	718	K	2
*	Dusky Hopping Mouse	*Notomys fuscus*	719		2
*	Fawn Hopping Mouse	*Notomys cervinus*	720		2
***	Mitchell's Hopping Mouse	*Notomys mitchelli*	721		2
*	Broad-toothed Rat	*Mastacomys fuscus*	730		
**	Sulawesi Lesser Shrew-rat	*Melasmothrix naso*	794		
**	Tate's Rat	*Tateomys rhinogradoides*	795		
**	Prehensile-tailed Rat	*Pogonomys mollipilosus*	696		
*	Rintja Mouse	*Komodomys rintjanus*	690		
*	Flores Giant Rat	*Papagomys armandvillei*	695		
**	Slender-tailed Cloud Rat	*Phloeomys elegans*	712		
*	Dinaget Island Cloud Rat	*Crateromys australis*	595		
*	Ilin Island Cloud Rat	*Crateromys paulus*	596		
**	Luzon Shrew-rat	*Celaenomys silaceus*	800		
***	False Water Rat	*Xeromys myoides*	739	R	1
**	Red-sided Hydromine	*Paraleptomys rufilatus*	799		
**	Eastern False Water Rat	*Pseudohydromys murinus*	740		
**	Western False Water Rat	*Pseudohydromys occidentalis*	741		
**	Groove-toothed Shrew Rat or Moss Mouse	*Microhydromys richardsoni*	742		
**	Shrew-like Rat	*Rhynchomys isarogensis*	736		
**	Shrew-like Rat	*Rhynchomys soricoides*	737		
*	Hazel Dormouse	*Muscardinus avellanarius*	781		
*	Garden Dormouse	*Eliomys quercinus*	782		
**	Woolly Forest Dormouse	*Dryomys laniger*	784		
**	Sechuan Dormouse	*Chaetocauda sechuanensis*	783		
**	Persian Dormouse	*Myominus setzeri*	785		
***	Desert Dormouse	*Selvinia betpakdalensis*	786		
***	Southern Birch Mouse	*Sicista subtilis*	743		
*	Birch Mouse	*Sicista betulina*	744		
***	Four-toed Jerboa	*Allactaga tetradactyla*	787		
**	Four-toed Jerboa	*Allactaga firouzi*	788		
**	Four-toed Jerboa	*Allactaga hotsoni*	789		
***	Five-toed Pygmy Jerboa	*Cardiocranius paradoxus*	790		
**	Dwarf Jerboa	*Salpingotus heptneri*	746		
**	Dwarf Jerboa	*Salpingotus thomasi*	747		
**	Dwarf Jerboa	*Salpingotus michaelis*	748		
**	Dwarf Jerboa	*Salpingotus crassicauda*	749		
**	Dwarf Jerboa	*Salpingotus kozlovi*	750		
*	Crested Porcupine	*Hystrix cristata*	801		
*	Hodgson's Porcupine	*Hystrix hodgsoni*	802		
***	North American Porcupine	*Erithizon dorsatum*	816		
*	Upper Amazon Porcupine	*Echinoprocta rufescens*	815		
***	Thin-spined Porcupine	*Chaetomys subspinosus*	803	I	
*	Patagonian Mara	*Dolichotis patagonum*	817		
*	Capybara	*Hydrochaeris hydrochaeris*	825		
***	Pacarana	*Dinomys branickii*	818		
***	Paca	*Agouti taczanowskii*	819		
***	Paca	*Agouti paca*	820		
*	Agouti	*Dasyprocta prymnolopha*	821		
*	Agouti	*Dasyprocta leporina*	822		
**	Agouti	*Dasyprocta ruatanica*	823		

STAR RATING	VERNACULAR NAME	SCIENTIFIC NAME	SPECIES NUMBER	IUCN CATEGORY	CITES APPENDIX
**	Agouti	*Dasyprocta coibae*	824		
***	Plains Viscacha	*Lagostomus maximus*	826		
*	Mountain Viscacha	*Lagidium peruanum*	827		
*	Mountain Viscacha	*Lagidium viscacia*	828		
*	Mountain Viscacha	*Lagidium wolffsohni*	829		
***	Short-tailed Chincilla	*Chinchilla brevicaudata*	830	I	1
***	Long-tailed Chincilla	*Chinchilla lanigera*	831	I	1
*****	Cabrera's Hutia	*Capromys angelcabrerai*	804	E	
*****	Large-eared Hutia	*Capromys auritus*	805	E	
**	Garrido's Hutia	*Capromys garridoi*	806	I	
***	Bushy-tailed Hutia	*Capromys melanurus*	807	I	
*****	Dwarf Hutia	*Capromys nanus*	808	E	
*****	Little Ground Hutia	*Capromys sanfelipensis*	809	E	
***	Cuban Hutia	*Capromys pilorides*	810		
***	Jamaican Hutia	*Capromys brownii*	811	I	
***	Bahamian Hutia	*Capromys ingrahami*	812	R	
***	Hispaniolan Hutia	*Plagiodontia aedium*	813	(I)	
***	Hispaniolan Hutia	*Plagiodontia hylaeum*	814		
**	Flightless Scaly-tailed Squirrel	*Zenkerella insignis*	528		
***	Amami Rabbit	*Pentalagus furnessi*	506	E	
***	Volcano Rabbit	*Romerolagus diazi*	511	E	1
***	Hispid Hare or Bristly Rabbit	*Caprolagus hispidus*	517	E	1
****	Bushman Rabbit	*Bunolagus monticularis*	512	E	
*	White-sided Jack Rabbit	*Lepus callotis*	513		
*	Island Jack Rabbit	*Lepus insularis*	514		
*	Snowshoe Hare	*Lepus americanus*	515		
*	Oriental Hare	*Lepus sinensis*	516		
*	New England Cottontail	*Silvilagus transitionalis*	508		
**	Omilteme Rabbit	*Silvilagus insonus*	509		
*	San José Rabbit	*Silvilagus mansuetus*	510		
*****	Sumatran Rabbit	*Nesolagus netscheri*	507	R	2
*	Yellow-rumped Elephant Shrew	*Rhynchocyon chrysopygus*	196		
***	Grey Wolf	*Canis lupus*	840	V	1/2
*****	Red Wolf	*Canis rufus*	841	E	
*****	Simien Jackal	*Canis simensis*	845	E	
*	Arctic Fox	*Alopex lagopus*	838		
***	Corsac Fox	*Vulpes corsac*	832		
***	Blanford's Fox	*Vulpes cana*	833		2
***	Swift Fox	*Vulpes velox*	834		
***	Kit Fox	*Vulpes macrotis*	835		
***	Fennec Fox	*Fennecus zerda*	836		2
**	Island Gray Fox	*Urocyon littoralis*	837		
***	Small-eared Dog or Zorro	*Atelocynus microtis*	842	K	
***	Small Grey Fox	*Pseudalopex griseus*	843		2
*	South American Red Fox	*Pseudalopex culpaeus*	844		2
***	Maned Wolf	*Chrysocyon brachyurus*	839	V	2
***	Bush Dog	*Speothos venaticus*	848	V	1
***	Dhole or Red Dog	*Cuon alpinus*	847	V	
*****	Hunting (Painted) Dog	*Lycaon pictus*	846	V	
***	Spectacled Bear	*Tremarctos ornatus*	852	V	1
***	Asiatic Black Bear	*Selenarctos thibetanus*	853	(E)	1
***	Brown or Grizzly Bear	*Ursus arctos*	851		1/2
*	American Black Bear	*Ursus americanus*	855		
*	Polar Bear	*Thalarctos maritimus*	856	V	2
***	Malayan Sun Bear	*Ursus malayanus*	854		1
*	Sloth Bear	*Melursus ursinus*	857	I	
*****	Barbados Raccoon	*Procyon gloveralleni*	858		
**	New Providence Island Raccoon	*Procyon maynardi*	859		
**	Guadaloupe Raccoon	*Procyon minor*	860		
**	Cozumel Raccoon	*Procyon pygmaeus*	861		
**	Tres Marias Raccoon	*Procyon insularis*	862		
**	Cozumel Coati	*Nasua nelsoni*	865		
**	Chiriqui Olingo	*Bassaricyon pauli*	863		
**	Harris's Olingo	*Bassaricyon lasius*	864		
***	Lesser or Red Panda or Cat-bear	*Ailurus fulgens*	850		2
*****	Giant Panda	*Ailuropoda melanoleuca*	849	R	1
*****	Black-footed Ferret	*Mustela nigripes*	866	E	1
*	European Polecat	*Mustela putorius*	867		
**	Water Weasel	*Mustela felipei*	868		
**	Black-striped weasel	*Mustela strigidorsa*	869		
***	European Mink	*Mustela lutreola*	871		
*	Siberian Weasel	*Mustela sibirica*	872		
*	Malaysian Weasel	*Mustela nudipes*	873		
*	Marbled Polecat	*Vormela peregusna*	870		

APPENDIX I

STAR RATING	VERNACULAR NAME	SCIENTIFIC NAME	SPECIES NUMBER	IUCN CATEGORY	CITES APPENDIX
*	**Japanese Marten**	*Martes melampus*	878		
*	**Yellow-throated Marten**	*Martes flavigula*	879		
*	**Nilgiri Marten**	*Martes gwatkinsi*	880		
*	**Sable**	*Martes zibellina*	881		
*	**Pine Marten**	*Martes martes*	882		
*	**Patagonian Weasel**	*Lyncodon patagonicus*	876		
***	**Wolverine** or **Glutton**	*Gulo gulo*	874		
*	**Ratel** or **Honey Badger**	*Mellivora capensis*	883		
*	**Eurasian Badger**	*Meles meles*	877		
**	**Ferret Badger**	*Melogale everetti*	884		
*	**Ferret Badger**	*Melogale personata*	885		
*	**Ferret Badger**	*Melogale moschata*	886		
***	**Patagonian Hog-nosed Skunk**	*Conepatus humboldti*	875		2
***	**Eurasian Otter**	*Lutra lutra*	887	(V)	1
*	**Hairy-nosed Otter**	*Lutra sumatrana*	888		2
*	**Spot-necked Otter**	*Lutra maculiocollis*	889		2
***	**North American Otter**	*Lutra canadensis*	890		2
***	**Smooth-coated Otter**	*Lutra perspicillata*	891		2
***	**South American River Otter**	*Lutra longicaudis*	892	(V)	1
***	**Southern River Otter**	*Lutra provocax*	893	I	1
***	**Marine Otter** or **Chingungo**	*Lutra felina*	894	V	1
***	**Giant Otter**	*Pteronura brasiliensis*	897	V	1
***	**Zaire Clawless Otter**	*Aonyx congica*	895		1/2
*	**African Clawless** or **Swamp Otter**	*Aonyx capensis*	896		2
***	**Sea Otter**	*Enhydra lutris*	898		1/2
*	**African Linsang**	*Poiana richardsoni*	905		
*	**Small Spotted Genet**	*Genetta genetta*	907		
**	**Aquatic Genet** or **Civet**	*Osbornictis piscivora*	904		
***	**Oriental** or **Large Spotted Civet**	*Viverra megaspila*	899	(E)	
*	**Large Indian Civet**	*Viverra zibetha*	900		
*	**African Civet**	*Civettictis civetta*	901		
***	**Spotted Linsang**	*Prionodon pardicolor*	921		1
*	**Three-striped Palm Civet**	*Arctogalidea trivirgata*	923		
*	**Palm Civet**	*Paradoxurus hermaphroditus*	924		
*	**Sulawesi Palm Civet**	*Macrogalidia musschenbroeki*	922		
*	**Malagasy Civet**	*Fossa fossa*	911	K	2
*	**Banded Palm Civet**	*Hemigalus derbyanus*	909		2
**	**Hose's Civet**	*Hemigalus hosei*	910		
**	**Owston's Palm Civet**	*Chrotogale owstoni*	908		
*	**Otter Civet**	*Cynogale bennettii*	902		2
**	**Vietnam Civet**	*Cynogale lowei*	903		
***	**Falanouc**	*Eupleres goudotii*	912	K	2
**	**Broad-striped Mongoose**	*Galidictis fasciata*	913	K	
**	**Brown-tailed Mongoose**	*Salanoia concolor*	914	K	
**	**Hose's Mongoose**	*Herpestes hosei*	915		
**	**Snouted Mongoose**	*Herpestes naso*	916		
*	**Mongoose**	*Herpestes smithi*	917		
*	**Mongoose**	*Herpestes vitticollis*	918		
*	**Mongoose**	*Herpestes brachyurus*	919		
**	**Liberian Mongoose**	*Liberiictis kuhni*	920		
***	**Fossa**	*Cryptoprocta ferox*	906	V	2
*	**Aardwolf**	*Proteles cristatus*	925		
*	**Spotted Hyaena**	*Crocuta crocuta*	926		
***	**Brown Hyaena**	*Hyaena brunnea*	927	V	1
***	**Striped Hyaena**	*Hyaena hyaena*	928	(E)	
***	**Margay** or **Tree Ocelot**	*Felis wiedii*	934	V	1/2
***	**Jaguarundi**	*Felis yagouaroundi*	935	I	1/2
***	**Geoffroy's Cat**	*Felis geoffroyi*	936		2
***	**Mountain Cat**	*Felis jacobita*	937	R	1
***	**Ocelot**	*Felis pardalis*	938	V	1/2
***	**Cougar, Puma** or **Mountain Lion**	*Felis concolor*	939	(E)	1/2
***	**Serval**	*Felis serval*	941		2
***	**Lynx**	*Felis lynx*	942	(E)	2
*	**Sand Cat**	*Felis margarita*	943	(E)	2
***	**Caracal**	*Felis caracal*	944	(R)	1/2
*	**Bobcat**	*Felis rufus*	945		1/2
***	**Pallas's Cat**	*Felis manul*	946		2
***	**Bay (Bornean Red) Cat**	*Felis badia*	948	R	2
*	**Flat-headed Cat**	*Felis planiceps*	949	I	1
*****	**Iriomote Cat**	*Felis iriomotensis*	950	E	2
*	**Asiatic Golden Cat**	*Felis temmincki*	951	I	1
*	**African Golden Cat**	*Felis aurata*	952		2
***	**Marbled Cat**	*Felis marmorata*	953	I	1
***	**Leopard Cat**	*Felis bengalensis*	954		1/2

STAR RATING	VERNACULAR NAME	SCIENTIFIC NAME	SPECIES NUMBER	IUCN CATEGORY	CITES APPENDIX
***	Fishing Cat	*Felis viverrinus*	955		2
*	Little-spotted Cat, Tiger Cat or Oncilla	*Felis tigrinus*	956	V	1/2
***	Lion	*Panthera leo*	929	(E)	1/2
***	Leopard	*Panthera pardus*	930	V	1
***	Tiger	*Panthera tigris*	931	E	1
***	Snow Leopard or Ounce	*Panthera uncia*	932	E	1
***	Jaguar	*Panthera onca*	933	V	1
***	Clouded Leopard	*Neofelis nebulosa*	947	V	1
***	Cheetah	*Acinonyx jubatus*	940	V(E)	1
***	Juan Fernandez Fur Seal	*Arctocephalus philippii*	970	V	2
*	Guadalupe Fur Seal	*Arctocephalus townsendi*	971	V	1
*	New Zealand Fur Seal	*Arctocephalus forsteri*	972		2
*	Galapagos Fur Seal	*Arctocephalus galapagoensis*	973		2
*	South American Fur Seal	*Arctocephalus australis*	974		
***	Californian Sea Lion	*Zalophus californianus*	959	(E)	
***	Northern or Steller's Sea Lion	*Eumetopias jubatus*	960		
***	South American Sea Lion	*Otaria flavescens*	961		
**	Australian Sea Lion	*Neophoca cinerea*	957		
**	New Zealand Sea Lion	*Neophoca hookeri*	958		
*	Walrus	*Odobenus rosmarus*	967	(K)	
***	Common or Harbour Seal	*Phoca vitulina*	962		
*	Pacific Common Seal	*Phoca richardi*	963		
*	Largha Seal	*Phoca largha*	964		
*	Ringed Seal	*Phoca hispida*	965	(R)	
***	Baikal Seal	*Phoca sibirica*	975		
*	Harp Seal	*Pagophilus groenlandicus*	977		
*	Grey Seal	*Halichoerus grypus*	966		
*****	Mediterranean Monk Seal	*Monachus monachus*	968	E	1
*****	Hawaiian Monk Seal	*Monachus schauinslandi*	969	E	1
***	Hooded Seal	*Cystomphora cristata*	976		
***	Franciscana or La Plata River Dolphin	*Pontoporia blainvillei*	1133		2
***	Amazon Dolphin or Bouta	*Inia geoffrensis*	1131		2
*****	Baiji or Yangtze River Dolphin	*Lipotes vexillifer*	1132	E	1
***	Ganges Susu	*Platanista gangetica*	1134		1
*****	Indus Dolphin	*Platanista indi*	1135	E	1
***	Estuarine Dolphin or Tucuxi	*Sotalia fluviatilis*	1137		1
*	Atlantic Spotted Dolphin	*Stenella plagiodon*	1139		2
*	Atlantic Spinner Dolphin	*Stenella clymene*	1139		2
*	Striped Dolphin	*Stenella coeruleoalba*	1140		2
*	Tropical Spinner Dolphin	*Stenella longirostris*	1141		2
*	Tropical Spotted Dolphin	*Stenella attenuata*	1142		2
***	Common Dolphin	*Delphinus delphis*	1143		2
**	Fraser's Dolphin	*Lagenodelphis hosei*	1144		2
**	Heaviside's or Benguela Dolphin	*Cephalorhynchus heavisidii*	1145		2
**	Chilean or Black Dolphin	*Cephalorhynchus eutropia*	1146		2
*	Irrawaddy Dolphin	*Orcaella brevirostris*	1136		2
*	Killer Whale or Orca	*Orcinus orca*	1152		2
***	Long-finned Pilot Whale	*Globiocephala melaena*	1153		2
***	Common or Harbour Porpoise	*Phocoena phocoena*	1147		2
***	Cochito	*Phocoena sinus*	1148	V	1
**	Spectacled Porpoise	*Phocoena dioptrica*	1149		2
***	Burmeister's Porpoise	*Phocoena spinipinnis*	1150		2
*	Dall's Porpoise	*Phocoenoides dalli*	1151		2
***	White Whale or Beluga	*Delphinapterus leucas*	1155		2
***	Narwhal	*Monodon monoceros*	1154		2
**	Dwarf Sperm Whale	*Kogia simus*	1162		2
*	Pygmy Sperm Whale	*Kogia breviceps*	1163		2
*	Sperm Whale or Cachalot	*Physeter catodon*	1164		1
**	Shepherd's Beaked Whale	*Tasmacetus shepherdi*	1167		2
***	Baird's Beaked Whale	*Berardius bairdii*	1165		1
*	Arnoux' Beaked Whale	*Berardius arnuxii*	1166		1
**	Hector's Beaked Whale	*Mesoplodon hectori*	1156		2
**	Hubb's Beaked Whale	*Mesoplodon carlhubbsi*	1157		2
**	Stejneger's Beaked Whale	*Mesoplodon stejnegeri*	1158		2
**	Andrews's Beaked Whale	*Mesoplodon bowdoini*	1159		2
**	Gingko-toothed Whale	*Mesoplodon ginkgodens*	1160		2
**	Longman's Beaked Whale	*Indopacetus pacificus*	1161		2
***	Northern Bottle-nosed Whaled	*Hyperoodon ampullatus*	1168	V	1
*	Southern Bottle-nosed Whale	*Hyperoodon planifrons*	1169		1
***	Grey Whale	*Eschrichtius robustus*	1175		1
***	Minke or Piked Whale	*Balaenoptera acutorostrata*	1170		1/2
*****	Blue Whale	*Balaenoptera musculus*	1171	E	1
***	Bryde's Whale	*Balaenoptera edeni*	1172		1
*****	Fin Whale or Common Roqual	*Balaenoptera physalus*	1173	V	1

STAR RATING	VERNACULAR NAME	SCIENTIFIC NAME	SPECIES NUMBER	IUCN CATEGORY	CITES APPENDIX
***	Sei Whale	*Balaenoptera borealis*	1174		1
***	Humpback Whale	*Megaptera novaeangliae*	1176	E	1
**	Pygmy Right Whale	*Caperea marginata*	1179		1
*****	Right Whale	*Eubalaena glacialis*	1177	E	1
*****	Bowhead or Greenland Right Whale	*Eubalaena mysticetus*	1178	E	1
***	Dugong	*Dugong dugong*	989	V	1/2
***	Amazon Manatee	*Trichechus inunguis*	990	V	1
***	American Manatee	*Trichechus manatus*	991	V	1
***	African Manatee	*Trichechus senegalensis*	992	V	2
***	African Elephant	*Loxodonta africana*	979	V	2
***	Asiatic Elephant	*Elephas maximas*	978	E	1
***	Kiang	*Equus kiang*	993		
***	Asiatic Wild Ass	*Equus hemionus*	994	V(E)	1/2
****	African Wild Ass or Donkey	*Equus asinus*	995	E	1
****	Wild or Przewalski's Horse	*Equus caballus*	996	Ext.	1
***	Grevy's Zebra	*Equus grevyi*	997	E	1
***	Mountain Zebra	*Equus zebra*	998	V/E	1/2
***	Common or Burchell's Zebra	*Equus burchelli*	999		
***	Mountain Tapir	*Tapirus pinchaque*	985	V	1
***	Baird's Tapir	*Tapirus bairdii*	986	V	1
*	Brazilian Tapir	*Tapirus terrestris*	987		1
***	Asian (Malayan) Tapir	*Tapirus indicus*	988	E	1
***	Great Indian Rhinoceros	*Rhinoceros unicornis*	983	E	1
*****	Javan or Lesser One-horned Rhinoceros	*Rhinoceros sondaicus*	984	E	1
*****	Sumatran or Asiatic Two-horned or Hairy Rhinoceros	*Dicerorhinus sumatrensis*	982	E	1
****	White Rhinoceros	*Ceratotherium simum*	981	(E)	1
****	Black Rhinoceros	*Diceros bicornis*	980	E	1
*****	Pygmy Hog	*Sus salvanius*	1001	E	1
***	Bearded Pig	*Sus barbatus*	1006		
*	Javan Warty Pig	*Sus verrucosus*	1007		
*	Sulawesi Warty Pig	*Sus celebensis*	1008		
*	Wild Boar	*Sus scrofa*	1009		
***	Wart Hog	*Phacochoerus aethiopicus*	1000		
***	Giant Forest Hog	*Hylochoerus meinertzhageni*	1005		
***	Babirusa	*Babyrousa babyrussa*	1002	V	1
***	White-lipped Peccary	*Tayassu pecari*	1004		
***	Chacoan Peccary	*Catagonus wagneri*	1003	V	
***	Hippopotamus	*Hippopotamus amphibius*	1010		
***	Pygmy Hippopotamus	*Choeropsis liberiensis*	1011	V	2
***	Guanaco	*Lama guanicoe*	1014		2
***	Vicuna	*Vicugna vicugna*	1015	V	1
****	Bactrian Camel	*Camelus bactrianus*	1012	V	
****	One-humped Camel	*Camelus dromedarius*	1013		
***	Musk Deer	*Moschus moschiferus*	1016		1/2
***	Musk Deer	*Moschus sifanicus*	1017		1/2
***	Musk Deer	*Moschus chrysogaster*	1018	(V)	1/2
***	Fea's Munjac	*Muntiacus feae*	1019	E	
***	Black or Hairy Fronted Muntjac	*Muntiacus crinifrons*	1020	I	1
*	Bornean yellow Muntjac	*Muntiacus atheroides*	1021		
***	Fallow Deer	*Cervus dama*	1022	(E)	(1)
**	Prince Alfred's or Spotted Deer	*Cervus alfredi*	1023		
***	Mariana Sambar	*Cervus mariannus*	1024		
*	Rusa	*Cervus timorensis*	1025		
***	Swamp Deer or Barasingha	*Cervus duvauceli*	1026	E	1
***	Thorold's Deer	*Cervus albirostris*	1027	I	
*	Hog Deer	*Cervus porcinus*	1028		(1)
**	Bawean Deer	*Cervus kuhli*	1029	R	1
****	Calamian Deer	*Cervus calamianensis*	1030	V	1
***	Brow-antlered Deer or Thamin	*Cervus eldi*	1031	(E)	1
***	Sika Deer	*Cervus nippon*	1032	(E)	
***	Red Deer or Wapiti	*Cervus elaphus*	1033	(E/V)	(1/2)
****	Pere David's Deer	*Elaphurus davidianus*	1034		
***	Reindeer or Caribou	*Rangifer tarandus*	1043		
*	Mule Deer	*Odocoileus hemionus*	1036	(R)	
***	White-tailed Deer	*Odocoileus virginianus*	1038	(R)	
***	Marsh Deer or Guasu Pucu	*Blastocerus dichotomus*	1035	V	1
***	Pampas Deer	*Ozotoceros bezoarticus*	1037	(E)	1
***	Peruvian Huemul	*Hippocamelus antisensis*	1041	V	1
*****	Chilean Huemul	*Hippocamelus bisulcus*	1042	E	1
***	Northern Pudu	*Pudu mephistophiles*	1039	I	2
*	Southern Pudu	*Pudu pudu*	1040		1
*	Okapi	*Okapi johnstoni*	1051		
***	Giraffe	*Giraffa camelopardalis*	1050		
***	Pronghorn	*Antilocapra americana*	1049	(E)	(1/2)

STAR RATING	VERNACULAR NAME	SCIENTIFIC NAME	SPECIES NUMBER	IUCN CATEGORY	CITES APPENDIX
***	**Common Eland**	*Taurotragus oryx*	1044		
****	**Giant Eland**	*Taurotragus derbianus*	1045	(E)	
*	**Mountain Nyala**	*Tragelaphus buxtonis*	1046		
*	**Sitatunga**	*Tragelaphus spekei*	1047		
***	**Bongo**	*Tragelaphus euryceros*	1048		
***	**Water Buffalo**	*Bubalus bubalis*	1053	E	
*****	**Tamaraw**	*Bubalus mindorensis*	1054	E	1
***	**Lowland Anoa**	*Bubalus depressicornis*	1055	E	1
***	**Mountain Anoa**	*Bubalus quarlesi*	1056	E	1
***	**Banteng**	*Bos javanicus*	1057	V	
***	**Gaur**	*Bos gaurus*	1058	V	1
*****	**Kouprey**	*Bos sauveli*	1059	E	1
***	**Wild Yak**	*Bos mutus*	1060	E	1
***	**African Buffalo**	*Syncerus caffer*	1052		
****	**American Bison**	*Bison bison*	1082		(1)
****	**European Bison**	*Bison bonasus*	1083		
***	**Jentink's Duiker**	*Cephalophus jentinki*	1061	E	2
***	**Blue Duiker**	*Cephalophus monticola*	1062		2
***	**Zebra** or **Banded Duiker**	*Cephalophus zebra*	1063	I	2
***	**Yellow-backed Duiker**	*Cephalophus sylvicultor*	1064		2
***	**Ader's Duiker**	*Cephalophus adersi*	1065	V	
*	**Abbott's Duiker**	*Cephalophus spadix*	1066	K	
***	**Lechwe**	*Kobus leche*	1070	V	2
***	**Common Reedbuck**	*Redunca arundinum*	1067		
*	**Mountain Reedbuck**	*Redunca fulvorufula*	1068		
*	**Bohor Reedbuck**	*Redunca redunca*	1069		
***	**Roan Antelope**	*Hippotragus equinus*	1071		2
***	**Sable Antelope**	*Hippotragus niger*	1072	(E)	(1)
****	**Arabian** or **White Oryx**	*Oryx leucoryx*	1073	E	1
****	**Scimitar Oryx**	*Oryx dammah*	1075	E	1
*	**Fringe-eared Oryx** or **Gemsbok**	*Oryx gazella*	1076		
****	**Addax**	*Addax nasomaculatus*	1074	E	1
****	**Black Wildebeest** or **White-tailed Gnu**	*Connochaetes gnou*	1084		
***	**Blue Wildebeest** or **Brindled Gnu**	*Connochaetes taurinus*	1085		
***	**Hartebeest** or **Kongoni**	*Alcelaphus buselaphus*	1080	(E)	
***	**Lichtenstein's Hartebeest**	*Alcelaphus lichtensteini*	1081		
****	**Bontebok/Blesbok**	*Damaliscus dorcas*	1077	(V)	(2)
***	**Hunter's Hartebeest**	*Damaliscus hunteri*	1078	R	
***	**Topi**	*Damaliscus lunatus*	1079		
***	**Beira Antelope**	*Dorcatragus megalotis*	1094	K	
***	**Oribi**	*Ourebia ourebi*	1088		
***	**Grysbok**	*Raphicerus melanotis*	1089		
*	**Steenbok**	*Raphicerus campestris*	1090		
***	**Suni**	*Neotragus moschatus*	1091	(E)	
*	**Royal Antelope**	*Neotragus pygmaeus*	1092		
*	**Bates's Dwarf Antelope**	*Neotragus batesi*	1093		
/*	**Impala**	*Aepyceros melampus*	1087	(E)	
***	**Blackbuck**	*Antilope cervicapra*	1086		
***	**Springbok**	*Antidorcas marsupialis*	1102		
*	**Gerenuk**	*Litocranius walleri*	1105		
*	**Dibatag** or **Clarke's Gazelle**	*Ammodorcas clarkei*	1095	V	
***	**Dorcas Gazelle** or **Chinkara**	*Gazella dorcas*	1096	(K)	
***	**Arabian Gazelle** or **Idmi**	*Gazella gazella*	1097	V	
***	**Edmi or Cuvier's Gazelle**	*Gazella cuvieri*	1098	E	
***	**Sand Gazelle** or **Rhim**	*Gazella leptoceros*	1099	V	
***	**Speke's Gazelle** or **Dero**	*Gazella spekei*	1100	K	
***	**Goitred Gazelle**	*Gazella subgutturosa*	1101	(E)	
***	**Red-fronted Gazelle**	*Gazella rufifrons*	1106	(K)	
***	**Dama gazelle**	*Gazella dama*	1107	V	1
***	**Zeren**	*Procapra gutturosa*	1104		
*	**Saiga Antelope**	*Saiga tatarica*	1103		
***	**Common Goral**	*Nemorhaedus goral*	1113		
*	**Red Goral**	*Nemorhaedus cranbrooki*	1114		
**	**Brown Goral**	*Nemorhaedus baileyi*	1115		
***	**Mainland Serow**	*Capricornis sumatrensis*	1116	(E)	1
***	**Mountain Goat**	*Oreamnos americanus*	1117		
*	**Chamois**	*Rupicapra rupicapra*	1108	(K/V)	(1)
*	**Takin**	*Budorcas taxicolor*	1109	(R/I)	2
**	**Arabian Tahr**	*Hemitragus jayakari*	1110	E	
***	**Nilgiri Tahr**	*Hemitragus hylocrius*	1111	V	
*	**Himalayan Tahr**	*Hemitragus jemlahicus*	1112		
***	**Wild Goat**	*Capra aegagrus*	1118		
***	**Ibex**	*Capra ibex*	1119		
***	**Spanish Ibex**	*Capra pyrenaica*	1120	(E)	

STAR RATING	VERNACULAR NAME	SCIENTIFIC NAME	SPECIES NUMBER	IUCN CATEGORY	CITES APPENDIX
*****	**Walia Ibex**	*Capra walie*	1121	E	
***	**Markhor**	*Capra falconeri*	1122	V/E	1/2
***	**Barbary Sheep** or **Aoudad**	*Ammotragus lervia*	1128	V	2
***	**Urial** or **Shapu** or **Asiatic Mouflon**	*Ovis vignei*	1123		1
***	**Bighorn Sheep**	*Ovis canadensis*	1124		2
*	**Thinhorn**	*Ovis dalli*	1125		
***	**Snow Sheep**	*Ovis nivicola*	1126		
***	**Argali** or **Nayan**	*Ovis ammon*	1127		1/2
***	**Asiatic Mouflon**	*Ovis orientalis*	1129		1
***	**Mouflon**	*Ovis musimon*	1130	E	

APPENDIX II

List of Additional Species

The following species were identified as threatened too recently for inclusion in the book, but probably deserve to be included in the ** or *** category.

Cheiromeles torquatus
Chiroderma improvisum
Chironax melanocephalus
Coelops frithi
Dyacopterus spadiceus
Ectophylla alba
Emballonura furax
Glischropus javanus
Harpiocephalus harpia
Hipposideros crumeniferus
Hipposideros wollastoni
Hipposideros papua
Kerivoula harwickei
Kerivoula papillosa
Kerivoula picta
Macaca pagensis (=nemestrina)
Megaerops kusnotoi
Miniopterus medius
Murina suilla
Myotis bartelsii
Myotis hasseltii
Myotis hermani
Neotoma phenax
Otomops formosus
Phalanger atrimaculatus (=maculatus)
Philetor brachypterus
Phoniscus atrox
Phoniscus jagorii
Pipistrellus circumdatus
Pipistrellus minahassae
Pipistrellus mordax
Rhinopoma microphyllum
Rhinolophus sedulus
Sorex nanus
Sorex tenellus
Soriculus salenskii
Sylvilagus graysoni
Tachyoryctes macrocephalus
Tadarida labiatus

BIBLIOGRAPHY

The bibliography below gives a selection of the works most frequently used when compiling the present volume. It is not intended to be an exhaustive list, but should provide a starting point for anyone wishing to further their research on endangered mammals. Among the most useful is Graham Hickman's in *Mammal Review*, which covers a large number of the regional guides to mammals. In addition to the books mentioned, many journals have published relevant articles and papers, among them *Oryx*, the journal of the Fauna and Flora Preservation Society. The IUCN and WWF (both international and national branches of WWF) publish a wide range of reports and yearbooks. The *Wildlife Review* published by the US Fish and Wildlife Service is a most useful and up-to-date general bibliographic source, while the *Zoological Record* is an essential tool for comprehensive research. The publications of the Mammal Society (UK), the American Society of Mammalogists (ASM), and similar bodies provide a wealth of information; the *Mammalian Species* series of the ASM is particularly useful.

The references marked with † proved particularly helpful.

† **Allen**, G. M., *Extinct and Vanishing Mammals of the Western Hemisphere*, New York, 1942 (rp.1972)

Allen, Thomas B., *Vanishing Wildlife of North America*, Washington D.C., 1974

Anon., *1986 IUCN Red List of Threatened Animals*, Cambridge, 1986

Anon., *Decade of Progress for South American National Parks 1974–1984*, Washington D.C., 1985

Anon., *List of Threatened and Endangered Plants and Wildlife of the U.S.*, Washington D.C., 1985

Anon., *Our Wildlife in Peril*, Wellington, N.Z., 1983

† **Archer**, Michael, *The Kangaroo*, New South Wales, 1985

Archer, Michael (ed.), *Carnivorous Marsupials* (2 vols.), Mosman, New South Wales, 1982

Barbour, R. W. & Davis, W. H., *Bats of America*, Lexington, 1969

Boonsong, Lekagul & McNeely, Jeffrey, *Mammals of Thailand*, Bangkok, 1977

† **Burt**, W. H., *A Field Guide to the Mammals*, Boston, 1964

† **Burton**, J. A., 'Bibliography of Red Data Books (Part 1: Animal Species)', *Oryx* 18:1, 1985

Burton, Robert, *Carnivores of Europe*, London, 1979

Cabrera, Angel & Yepes, José, *Mamíferos Sud-Americanos*, Buenos Aires, 1940

Chapman, Joseph A. & Feldhamer, George A., *Wild Mammals of North America*, Baltimore, 1982

Corbet, G. B., *The Terrestrial Mammals of Western Europe*, London, 1966

Corbet, G. B., *The Mammals of the Palaearctic Region: A Taxonomic Review*, London & Ithaca, 1978

† **Corbet**, G. B., *The Mammals of Britain and Europe*, London, 1980

Corbet, G. B. & Hill, J. E., *A World List of Mammalian Species*, London, 1980

Davey, Keith, *Australian Marsupials*, Melbourne, 1970

† **Day**, David, *The Doomsday Book of Animals*, New York, 1981

† **Fisher**, J., Simon, N. & Vincent, J., *The Red Book Wildlife in Danger*, London, 1969

Fitter, R. S. R., *Vanishing Wild Animals of the World*, London, 1968

Frith, H. J. & Calaby, J. H., *Kangaroos*, London & New York, 1969

† **Goodwin**, Harry A. *et al.*, *Red Data Book, Volume 1: Mammalia*, Morges, Switzerland, 1978

Gressitt, J. L. (ed.), *Biogeography and Ecology of New Guinea*, The Hague, 1982
Grimwood, I. R., *Notes on the Distribution and Status of Some Peruvian Mammals 1968*, New York, 1969
Hall, Leslie S. & Richards, G. C., *Bats of Eastern Australia*, Queensland Museum Booklet, 1979
† **Haltenorth**, Theodor & Diller, Helmut, *A Field Guide to the Mammals of Africa including Madagascar*, London, 1980
Hanks, John, *Mammals of Zambia*, Zambia National Tourist Board, 1972
Harris, C.J., *Otters*, London, 1978
Harrison, David L., *The Mammals of Arabia* (2 vols.), London, 1964
Harrison, David L., *Mammals of the Arabian Gulf*, London, 1981
Hayman, R. W. & Misonne, X., *The Bats of the Congo and of Rwanda and Burundi*, Tervuren, Belgium, 1966
Heaney, L. R. & Patterson, B. D. (eds.) *Symposium on Island Biogeography of Mammals*, London, 1986
Herschkovitz, P., *Living New World Monkeys*, Volume 1, Chicago, 1977
† **Hickman**, Graham C., 'National Mammal Guides: A Review of References to Recent Faunas', *Mammal Review* 11:2, 1981
Hill, John E. & Smith, James D., *Bats: A Natural History*, London, 1984
Hinton, H. E. & Dunn, A. M. S., *Mongooses*, London, 1967
† **Honacki**, James H. *et al.*, *Mammal Species of the World*, Kansas, 1982
Hornaday, William T., *Our Vanishing Wildlife*, New York, 1913
HRH Prince Philip Duke of Edinburgh & Fisher, James, *Wildlife Crisis*, London, 1970
Joslin, P. & Maryanka, D., *Endangered Mammals of the World*, Morges, Switzerland, 1968
Khan, Mohammad Ali Reza, *Mammals of Bangladesh*, Dhaka, Bangladesh, 1985
† **Kingdon**, Jonathan, *East African Mammals: An Atlas of Evolution in Africa*, London, 1971–1979
Lever, Christopher, *Naturalized Mammals of the World*, London & New York, 1985
Luard, Nicholas, *The Wildlife Parks of Africa*, London, 1985
† **Lyster**, Simon, *International Wildlife Law*, Cambridge, 1985
Mack, David & Mittermeier, R. A., *The International Primate Trade*, Volume 1, Washington D.C., 1984
Mallon, D. P., *The Mammals of the Mongolian Peoples Republic*, Oxford, 1985
Marlow, Basil J., *Marsupials of Australia*, Brisbane, 1962
Mason, C. E. & Macdonald S. M., *Otters, Ecology and Conservation*, Cambridge, 1986
Medway, Lord, *The Wild Mammals of Malaya*, Oxford, 1969
Menzies, J. I. & Dennis, Elizabeth, *Handbook of New Guinea Rodents*, Wau, PNG, 1979
Myers, K. & McInnes, C. D. (eds.), *Proceedings of the World Lagomorph Conference, August 12–16 1979*, Guelph, Canada, 1982
Neal, Ernest, *The Natural History of Badgers*, London, 1985
† **Nowak**, Ronald M. & Paradiso, John L., *Walker's Mammals of the World* (4th ed.), Baltimore & London, 1983
Oberle, Philippe (ed.), *Madagascar*, Paris, 1981
Osborn, J. Dale & Helmy, Ibrahim, *The Contemporary Land Mammals of Egypt (including Sinai)*, Chicago, 1980
† **Ovington**, Derrick, *Australian Endangered Species*, Melbourne, 1978
Payne, Junaidi; Francis, Charles M. & Phillipps, Karen, *A Field Guide to the Mammals of Borneo*, Kuala Lumpur, 1985
Prater, S. H., *The Book of Indian Animals*, Bombay, 1971
Ride, W. D. L., *A Guide to the Native Mammals of Australia*, Melbourne, 1970
Roberts, T. J., *The Mammals of Pakistan*, London, 1977
Smith, Andrew & Hume, Ian (eds.), *Possums and Gliders*, New South Wales, 1984
† **Smithers**, Reay H. N., *The Mammals of the Southern African Subregion*, Pretoria, 1983
† **Strahan**, Ronald (ed.), *Complete Book of Australian Mammals*, Sydney, 1983
† **Tattersall**, Ian, *The Primates of Madagascar*, New York, 1982
† **Thornback**, Jane & Jenkins, Martin, *The IUCN Mammal Red Data Book*, Part 1, Gland & Cambridge, 1982
Tyler, Michael J., *The Status of Endangered Australian Wildlife*, Adelaide, 1979
† **Whitaker**, John O. Jr., *The Audubon Society Field Guide to North American Mammals*, New York, 1980
Wolfheim, Jaclyn H., *The Primates of the World*, New York, 1983

INDEX

References in **bold** are to the page numbers of the illustrations; others are to the page numbers of the main entries in the text.

CONVERSION TABLE

From the Metric System to the Imperial System

To convert	Multiply by
millimetres (mm) to inches	0.039
centimetres (cm) to inches	0.394
metres (m) to feet	3.281
metres (m) to yards	1.094
kilometres (km) to miles	0.621
square kilometres (km^2) to square miles	0.386
hectares (ha.) to square miles	0.004
grammes (g) to ounces	0.035
kilogrammes (kg) to pounds	2.205
tonnes to tons	1.102
tonnes to long tons	0.984